T0136966

Lecture Notes in Networks and Systems

Volume 320

The series “Lecture Notes in Networks and Systems” publishes the latest developments in Networks and Systems—quickly, informally and with high quality. Original research reported in proceedings and post-proceedings represents the core of LNNS.

Volumes published in LNNS embrace all aspects and subfields of, as well as new challenges in, Networks and Systems.

The series contains proceedings and edited volumes in systems and networks, spanning the areas of Cyber-Physical Systems, Autonomous Systems, Sensor Networks, Control Systems, Energy Systems, Automotive Systems, Biological Systems, Vehicular Networking and Connected Vehicles, Aerospace Systems, Automation, Manufacturing, Smart Grids, Nonlinear Systems, Power Systems, Robotics, Social Systems, Economic Systems and other. Of particular value to both the contributors and the readership are the short publication timeframe and the world-wide distribution and exposure which enable both a wide and rapid dissemination of research output.

The series covers the theory, applications, and perspectives on the state of the art and future developments relevant to systems and networks, decision making, control, complex processes and related areas, as embedded in the fields of interdisciplinary and applied sciences, engineering, computer science, physics, economics, social, and life sciences, as well as the paradigms and methodologies behind them.

Indexed by SCOPUS, INSPEC, WTI Frankfurt eG, zbMATH, SCImago.

All books published in the series are submitted for consideration in Web of Science.

More information about this series at http://www.springer.com/series/15179

Javier Prieto · Alberto Partida ·
Paulo Leitão · António Pinto
Editors

Blockchain and Applications

3rd International Congress

Editors
Javier Prieto
Building Edificio I+D+i, Room 24.1
University of Salamanca
Salamanca, Salamanca, Spain

Paulo Leitão
Research Centre in Digitalization
Instituto Politécnico de Bragança
Bragança, Portugal

Alberto Partida
Universidad Rey Juan Carlos
Móstoles, Madrid, Spain

António Pinto
ESTG, Instituto Politécnico do Porto
Felgueiras, Portugal

ISSN 2367-3370 ISSN 2367-3389 (electronic)
Lecture Notes in Networks and Systems
ISBN 978-3-030-86161-2 ISBN 978-3-030-86162-9 (eBook)
https://doi.org/10.1007/978-3-030-86162-9

This Springer imprint is published by the registered company Springer Nature Switzerland AG
The registered company address is: Gewerbestrasse 11, 6330 Cham, Switzerland

Preface

The 3rd International Congress on Blockchain and Applications 2021 will be held in Salamanca from 6 to 8 of October. This annual congress will reunite blockchain and artificial intelligence (AI) researchers, who will share ideas, projects, lectures, and advances associated with those technologies and their application domains.

Among the scientific community, blockchain and AI are seen as a promising combination that will transform the production and manufacturing industry, media, finance, insurance, e-government, etc. Nevertheless, there is no consensus with schemes or best practices that would specify how blockchain and AI should be used together. Combining blockchain mechanisms and artificial intelligence is still a particularly challenging task.

The BLOCKCHAIN'21 congress is devoted to promoting the investigation of cutting-edge blockchain technology, to exploring the latest ideas, innovations, guidelines, theories, models, technologies, applications and tools of blockchain and AI for the industry, and to identifying critical issues and challenges those researchers and practitioner must deal with in the future research. We want to offer researchers and practitioners the opportunity to work on promising lines of research and to publish their developments in this area.

The technical program has been diverse and of high quality, and it focused on contributions to both, well-established and evolving areas of research. More than 44 papers have been submitted to 38 from over 20 different countries (Canada, France, Germany, India, Ireland, Italy, Jordan, Luxembourg, Malaysia, Malta, Morocco, Netherlands, Oman, Portugal, Slovenia, Spain, Sweden, United Arab Emirates, and USA).

We would like to thank all the contributing authors, the members of the Program Committee, the sponsors (IBM, Indra, EurAI, AEPIA, AFIA, APPIA, and AIR Institute), and the Organizing Committee for their hard and highly valuable work. We are especially grateful for the funding supporting by project "XAI - XAI - Sistemas Inteligentes Auto Explicativos creados con Módulos de Mezcla de Expertos," ID SA082P20, financed by Junta Castilla y León, Consejería de Educación, and FEDER funds. Their work contributed to the success of the BLOCKCHAIN'21 event and, finally, the Local Organization Members and the

Program Committee Members for their hard work, which was essential for the success of BLOCKCHAIN'21.

Javier Prieto
Alberto Partida
Paulo Leitão
António Pinto

Organization

General chair

Javier Prieto Tejedor	University of Salamanca, Spain, and AIR Institute, Spain

Advisory Board

Ashok Kumar Das	IIIT Hyderabad, India
Abdelhakim Hafid	Université de Montréal, Canada
António Pinto	Instituto Politécnico do Porto, Portugal

Program Committee Chairs

Alberto Partida	Universidad Rey Juan Carlos, Spain
Paulo Leitao	Technical Institute of Bragança, Portugal

Organizing Committee

Juan M. Corchado Rodríguez	University of Salamanca, Spain, and AIR Institute, Spain
Javier Prieto Tejedor	University of Salamanca, Spain, and AIR Institute, Spain
Roberto Casado Vara	University of Salamanca, Spain
Fernando De la Prieta	University of Salamanca, Spain
Sara Rodríguez González	University of Salamanca, Spain
Pablo Chamoso Santos	University of Salamanca, Spain
Belén Pérez Lancho	University of Salamanca, Spain
Ana Belén Gil González	University of Salamanca, Spain
Ana De Luis Reboredo	University of Salamanca, Spain
Angélica González Arrieta	University of Salamanca, Spain

Emilio S. Corchado Rodríguez	University of Salamanca, Spain
Angel Luis Sánchez Lázaro	University of Salamanca, Spain
Alfonso González Briones	University Complutense of Madrid, Spain
Yeray Mezquita Martín	University of Salamanca, Spain
Javier J. Martín Limorti	University of Salamanca, Spain
Alberto Rivas Camacho	University of Salamanca, Spain
Ines Sitton Candanedo	University of Salamanca, Spain
Elena Hernández Nieves	University of Salamanca, Spain
Beatriz Bellido	University of Salamanca, Spain
María Alonso	University of Salamanca, Spain
Diego Valdeolmillos	AIR Institute, Spain
Sergio Marquez	University of Salamanca, Spain
Marta Plaza Hernández	University of Salamanca, Spain
Guillermo Hernández González	AIR Institute, Spain
Ricardo S. Alonso Rincón	University of Salamanca, Spain
Javier Parra	University of Salamanca, Spain

Program Committee

Regio A. Michelin	University of New South Wales, Australia
Mo Adda	University of Portsmouth, UK
Rishav Agarwal	University of Waterloo, Canada
Imtiaz Ahmad Akhtar	Higher Colleges of Technology, Sweden
Sami Albouq	Islamic University of Madinah, Saudi Arabia
Ricardo Alonso	AIR Institute, Spain
Alejandro Alfonso Fernández	DigitelTS, Spain
Diego Andina	Universidad Politécnica de Madrid, Spain
Artem Barger	IBM, Israel
Francisco Luis Benítez Martínez	University of Granada, Spain
Javier Bermejo Higuera	Universidad Internacional de La Rioja, Spain
Bill Buchanan	Napier University, UK
Roben C. Lunardi	IFRS, Brazil
Roberto Casado Vara	University of Salamanca, Spain
Arnaud Castelltort	Montpellier, France
Giovanni Ciatto	University of Bologna, Italy
Victor Cook	University of Central Florida, EE.UU.
Manuel E. Correia	CRACS/INESC TEC; DCC/FCUP, Portugal
Gaby Dagher	Boise State University, Idaho, USA
Ashok Kumar Das	International Institute of Information Technology, India
Pankaj Dayama	IBM, India

Andreea-Elena Drăgnoiu	University of Bucharest, Romania
Enrique De la Cal Marín	University of Oviedo, Spain
Josep Lluis De La Rosa	TECNIO Centre EASY Innovation, UdG, Spain
Monika di Angelo	TU Wien, Austria
Roberto Di Pietro	Hamad Bin Khalifa University, College of Science and Engineering, Qatar
Joshua Ellul	University of Malta, Malta
Enes Erdin	Florida International University, EE.UU.
Xinxin Fan	IoTeX, India
Manuel J. Fernandez	University of Seville, Spain
Iñaki Fernández	Université de Lorraine, France
Christof Ferreira Torres	Interdisciplinary Centre for Security, Reliability and Trust (SnT), University of Luxembourg
Nikos Fotiou	Athens University of Economics and Business, Greece
Paula Fraga Lamas	University of A Coruña, Spain
Muriel Franco	University of Zurich, EE.UU.
Felix Freitag	Universitat Politècnica de Catalunya, Spain
Shahin Gheitanchi	Altera Corporation, EE.UU.
Raffaele Giaffreda	CREATE-NET, Italy
Mongetro Goint	Université Le Havre Normandie, France
Hélder Gomes	Escola Superior de Tecnologia e Gestão de Águeda, Universidade de Aveiro, Portugal
Dhrubajyoti Goswami	Concordia University, Canada
Ram Govind	Indian Computer Emergency Response Team, Ministry of Electronics and IT, India
Volker Gruhn	Universität Duisburg-Essen, Germany
Hao Guo	Northwestern Polytechnical University, China
Abdelatif Hafid	University of Montreal, Canada
Yahya Hassanzadeh-Nazarabadi	Ferdowsi University of Mashhad, Irán
Farzin Houshmand	University of California, EE.UU.
Qin Hu	George Washington University, EE.UU.
Maria Visitación Hurtado	University of Granada, Spain
Hans-Arno Jacobsen	University of Toronto, Canada
Marc Jansen	University of Applied Sciences Ruhr West, Germany
Eder John Scheid	University of Zurich, Switzerland
Raja Jurdak	QUT, Australia
Salil Kanhere	University of New South Wales, Australia
Christos Karapapas	Athens University of Economics and Business, Greece
Samuel Karumba	UNSW, Australia
Denisa Kera	Asia Research Institute (STS Cluster), Australia

Mohamed Laarabi	Mohammadia School of Engineering Rabat, Morocco
Oscar Lage	TECNALIA, Spain
Chhagan Lal	University of Padova, Italy
Xabier Larrucea	TECNALIA, Spain
Anne Laurent	LIRMM – UM, France
Xianfeng Li	Macau University of Science and Technology, China
Sotirios Liaskos	School of IT, York University, Canada
Alexander Lipton	Massachusetts Institute of Technology, EE.UU.
Fengji Luo	The University of Sydney, Australia
João Paulo Magalhaes	ESTGF, Porto Polytechnic Institute, Portugal
Yacov Manevich	IBM, Israel
Stefano Mariani	Università degli Studi di Modena e Reggio Emilia, Italy
Luis Martínez	University of Jaén, Spain
Imran Memon	Zhejiang University, China
Suat Mercan	Florida International University, EE.UU.
Yeray Mezquita	University of Salamanca, Spain
Juan Jose Morillas Guerrero	Universidad Politécnica de Madrid, Spain
Imtiaz Muhammad Anas	Boston University, EE.UU.
Daniel Jesus Munoz Guerra	University of Malaga, Spain
Ahmed Nadeem	King Saud University, Saudi Arabia
Malaw Ndiaye	Université Cheikh Anta Diop, Senegal
Mark Nejad	University of Delaware, EE.UU.
Mariusz Nowostawski	Norwegian University of Science and Technology, Norway
Andrea Omicini	Alma Mater Studiorum–Università di Bologna, Italy
Kazumasa Omote	University of Tsukuba, Japan
Arindam Pal	Data61, CSIRO, Australia
Rafael Pastor Vargas	UNED, Spain
Miguel Pincheira	Fondazione Bruno Kessler, Italy
António Pinto	ESTG, P.Porto, Portugal
Pedro Pinto	Polytechnic Institute of Viana do Castelo, Portugal
Steven Platt	Universitat Pompeu Fabra, Spain
Matthias Pohl	Otto-von-Guericke-Universität Magdeburg, Germany
Javier Prieto	University of Salamanca, Spain
Yuansong Qiao	Athlone Institute of Technology, Ireland
Bruno Rodrigues	University of Zurich, Switzerland
David Rosado	University of Castilla-La Mancha, Spain
Grigore Rosu	University of Illinois at Urbana-Champaign, EE. UU.

Gernot Salzer	Vienna University of Technology, Austria
Georgios Samakovitis	University of Greenwich, UK
Altino Sampaio	Instituto Politécnico do Porto, Escola Superior de Tecnologia e Gestão de Felgueiras, Portugal
Elio San Cristóbal Ruiz	UNED, Spain
Ricardo Santos	ESTG/IPP, Portugal
Nuria Serrano	AIR Institute, Spain
Wazen Shbair	University of Luxembourg, SnT, Luxembourg
Pradip Sharma	University of Aberdeen, UK
Chien-Chung Shen	University of Delaware, EE.UU.
Ajay Kumar Shrestha	University of Saskatchewan, Canada
Mark Staples	CSIRO, Australia
Radu State	University of Luxembourg, Luxembourg
Burkhard Stiller	University of Zurich, Switzerland
Wilhelm Stork	Karlsruhe Institute of Technology, Germany
Marko Suvajdzic	University of Florida, EE.UU.
Stefan Tai	TU Berlin, Germany
Chamseddine Talhi	École de Technologie Supérieure, Canada
Teik Guan Tan	Singapore University of Technology and Design, Malaysia
Ege Tekiner	Florida International University, EE.UU.
Subhasis Thakur	National University of Ireland, Galway, Ireland
Llanos Tobarra	UNED, Spain
Aitor Urbieta	IK4-Ikerlan Technology Research Centre, Spain
Julita Vassileva	University of Saskatchewan, Canada
Massimo Vecchio	Fondazione Bruno Kessler, Italy
Eduardo Vega-Fuentes	University of Glasgow, UK
Sebastián Ventura	University of Cordoba, Department of Computer Science and Numerical Analysis, Spain
Luigi Vigneri	IOTA Foundation, Germany
Roopa Vishwanathan	New Mexico State University, Mexico
Marco Vitale	Foodchain SpA, Italy
Chenggang Wang	University of Cincinnati, EE.UU.
Yawei Wang	George Washington University, EE.UU.
Komminist Weldemariam	IBM Research, Africa, and Queen's University, Canada
Amr Youssef	Concordia University, EE.UU.
Uwe Zdun	University of Vienna, Austria
Kaiwen Zhang	École de technologie supérieure de Montréal, Canada
Mirko Zichichi	Universidad Politécnica de Madrid, Spain
Avelino F. Zorzo	PUCRS, Brazil
André Zúquete	University of Aveiro, Portugal

BLOCKCHAIN'21 Sponsors

Support from National Associations

Contents

BLOCKCHAIN-Main Track

BLOCKCHAIN-Main Track

Formal Analysis of Smart Contracts: Model Impact Factor on Criminality

Malaw Ndiaye(✉) and Karim Konaté

UCAD, Dakar, Senegal
{malaw.ndiaye,karim.konate}@ucad.edu.sn

Abstract. Smart contracts certainly provide a powerful functional surplus for maintaining the consistency of transactions in applications governed by blockchain technology. However, the intended level of automation might cause cascading effects that have to be checked by formal methods of algorithmic proof. Our smart contract formal model analysis framework uses the Finite State Machine (FSM) theory which is a model of behavior composed of states, transitions and actions. In this model, a state stores information about the past, a transition indicates a state change and is described by a condition that would need to be fulfilled to enable the transition. An action is a description of an activity that is to be performed at a given moment. These conditions are properties that are checked during program execution. This formal analysis framework allows us to define a set of invariants on Finite State Machine behavior model and to propose an anomaly detection system based on the invariants of the smart contract.

1 Introduction

Ethereum, taken as a whole, can be viewed as a transaction-based state machine. The state can include such information as account balances, reputations, trust arrangements, data pertaining to information of the physical world; in short, anything that can currently be represented by a computer is admissible. Transactions thus represent a valid arc between two states; the 'valid' part is important there exist far more invalid state changes than valid state changes. Invalid state changes might, e.g., be things such as reducing an account balance without an equal and opposite increase elsewhere. A valid state transition is one that is produced by a transaction [36].

Smart contract is the program deployed in a distributed network that can acquire outside information via transations and update the internal state automatically. Majority of smart contract procedures are based on blockchain technology. Existing smart contracts control digital currencies principal. Whereas they were found having defects and deficiencies in the course of their operations, leading to more serious consequences, such as "The DAO" and Ethereum Parity wallet incident, which caused a large number of cryptocurrencies to be stolen, causing a large loss. If these program bug cannot be processed, smart contracts will be difficult to manipulate real assets [10].

J. Prieto et al. (Eds.): BLOCKCHAIN 2021, LNNS 320, pp. 3–13, 2022.
https://doi.org/10.1007/978-3-030-86162-9_1

1.1 Contributions

This paper makes the following contributions:

- We propose a framework for analyzing the smart contract model, we use an approach based on Finite State Machine theory to model the execution of smart contracts in the Ethereum environment.
- We will show the inadequacies of the model in relation to malicious smart contracts, i.e. show how the model facilitates attacks and steals capital.
- Proposal for an anomaly detection model based on smart contract invariants.

1.2 Organize

This paper is organized as follows:

- Section 2 presents background study and related work.
- Section 3 presents the study framework of the program as a whole and examines the flaws in the model of smart contract promoting crime.
- Section 4 describes our anomaly detection system based on the invariant calculation method in smart contract systems.
- Section 5 offers new research directions based on anomaly detection in smart contract systems.

2 Background Study and Related Work

2.1 Smart Contracts Operational Mechanism

Smart contracts generally have two attributes: value and state (Fig. 1). The triggering conditions and the corresponding response actions of the contract terms are preset using triggering condition statements such as "If-Then" statements. Smart contracts are agreed upon and signed by all parties and submitted in transactions to the blockchain network, then transactions are broadcasted via the P2P network, verified by the miners and stored in the specific block of the blockchain [22,34].

The creators of the contracts get the returned parameters (e.g., contract address), then users can invoke a contract by sending a transaction. Miners are motivated by the system's incentive mechanism and will contribute their computing resources to verify the transaction. More specially, after the miners receive the contract creation or invoking transaction, they create contract or execute contract code in their local Execution Environment. Based on the input of trusted data feeds and the system state, the contract determines whether the current scenario meets the triggering conditions. If the conditions are met, the response actions are strictly executed. After a transaction is validated, it is packaged into a new block. The new block is chained into the blockchain once the whole network reaches a consensus [10,34].

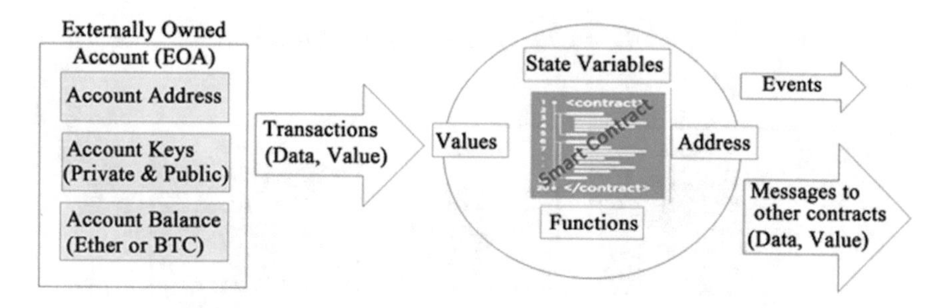

Fig. 1. Smart contracts operational mechanism

2.2 Smart Contracts Formal Model

A model M^* is a pair (Q, δ^*), where Q is a set of states and δ^* is a set of sequences of states satisfying property $\sum$ below. The set of states can be thought of as the set of all conceivable states of a program; i.e., all possible combinations of values of variables and "program counter" values. A sequence $q_1, q_2, \ldots \in \delta^*$ represents an execution that starts in state q_1, performs the first program step to reach state q_2, performs the next program step to reach state q_3 etc. The execution terminates if and only if the sequence is finite. The set δ^* represents all possible executions of the program, starting in any possible state [19].

Contract automata M^* is a quintuple:

$$\mathbf{M}^* = (\mathbf{Q}, \sum, \delta^*, s^*, F^*) \tag{1}$$

Among them:

- $\mathrm{Q} = \{q_1^*, q_2^*, \ldots, q_m^*\}$. Q is the set about all states of contract execution automata, q_i^* is contained in the state set of contract party, $q_i^* \in q_i$, (i=1, ... , m);
- $\sum$ is the set of all input events;
- δ^* is the set of all the transition functions,
  $\delta^* : Q \times \sum \rightarrow Q$
- s^* is the initial state, $s^* \in Q$
- F^* is the set of termination states, $F^* \subset Q$.

2.3 Transaction Formal Model

A transaction (formally, T) is a single cryptographically signed instruction constructed by an external actor [3,28,36,38]. While it is assumed that the ultimate external actor will be human in nature, software tools will be used in its construction and dissemination. There are two types of transactions: those which result in message calls and those which result in the creation of new accounts with associated code (known informally as "contract creation"). Both types specify a number of common fields: **nonce** (T_n), **gasPrice** (T_p), **gasLimit** (T_g), **to** (T_t), **value** (T_v), **init** (T_i), **data** (T_d) [30] (Fig. 2).

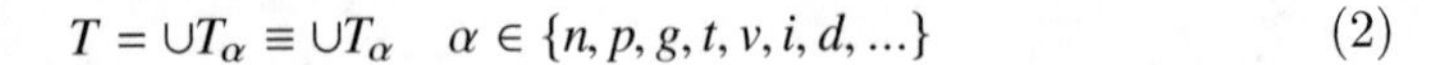

$$T = \cup T_\alpha \equiv \cup T_\alpha \quad \alpha \in \{n, p, g, t, v, i, d, ...\} \tag{2}$$

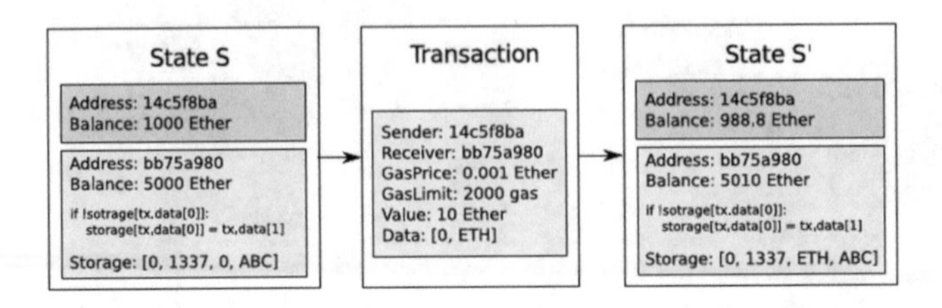

Fig. 2. Transaction model

3 Smart Contracts Sequential Execution Model

3.1 The Smart Contracts Sequential Programs

The specific model of deterministic sequential programs can be obtained by structuring the general state into [29]:

$$Q = (\pi, u) \tag{3}$$

π is the control component and assumes a finite number of values, taken to be labels or locations in the program. $L = \{l_0, l_1, ...l_n\}$

u is the data component and will usually range over an infinite domain. It be structured into state variables and data structures or functions (Fig. 1).

The transition relation δ can also be partitioned into a next-location function $N(\pi, u)$ and a data transformation function $D(\pi, u)$. $N(l, u)$ will actually depend on u only if the statement at l is a conditional.

We can thus express δ in terms of N and D:

$$\delta[(\pi, u), (\pi', u')] \iff \pi' = D(\pi, u) \& u' = N(\pi, u) \tag{4}$$

3.2 Smart Contracts Execution Model

An execution is a sequence of external transactions T each nesting one or more internal transactions (transitions T_α). Each transition starts with a message and proceeds in a sequence of commands. Commands may load or store data from and to the private storage, perform local computations (not affecting the storage), and initiate nested transitions [28].

a) Transition Invariant

   A transition invariant T is a superset of the transitive closure of the transition relation δ restricted to the accessible states Q [30]. Formally,

$$\delta^* \cap Q \times Q \subseteq T \tag{5}$$

Transition is valid

$$\Leftrightarrow \forall q_i \exists q_{i+1} \ / \ \delta(q_i, T_\alpha) = q_{i+1} \tag{6}$$

b) State Invariant
   A state invariant is a superset of Q. Given the transition invariant T and the set of starting states $I \in Q$, the set

$$I \cup \{q'|q \in I \ and \ (q', q) \in T\} \tag{7}$$

   is a state invariant. Conversely, a transition invariant can be strengthened by restricting it to a given state invariant [30].

In other words, an execution is normal if there is a finite sequence of consecutive valid transitions which begins with an initial state q_1 and ends with a final state q_n, without blocking any state (Fig. 3).

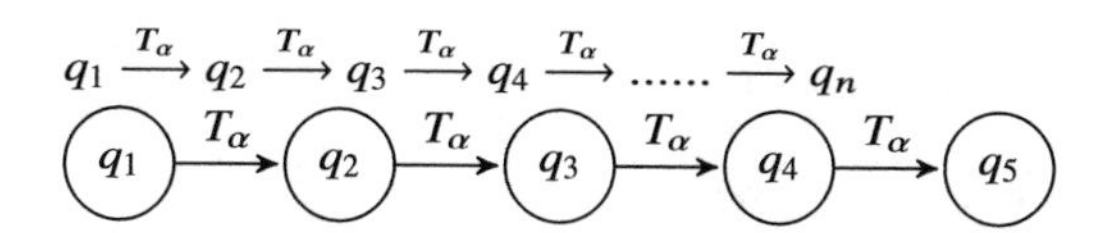

Fig. 3. Smart contracts execution model

3.3 Model and Vulnerable Smart Contracts Attack Vectors

In software systems, programming errors are usually the root cause leading to security breaches such as denial of service, buffer overflow, format string, code injection, etc. Coding errors may result either from defectively designed language features such as no built-in protection for accessing memory, or from invalid logic having high-level semantic error [12].

Smart contracts may contain vulnerabilities, which cause contracts to run on an unplanned scenario. However, these vulnerabilities are still harmless until an adversary takes advantage by exploiting them. Generally, he must send transactions, which are termed as attack vectors in security field, to exploit these vulnerabilities [23].

In formal verification (model verification, symbolic execution, theorem proof, translation and type verification) [1,2,4,5,14,18,21,22,28,31] as in the detection of vulnerabilities [6–9,11,13,15,20,24,25,27,32,33,37], all smart contracts which do not respect the properties of theorems 1 and 2, are carriers attack vectors. Table 1 represent the properties which are likely to be violated.

3.4 Model and Criminal Smart Contracts

We refer to smart contracts that facilitate crimes in distributed smart contract systems as criminal smart contracts (CSCs) [16]. In blockchain, the main activity of criminal smart contracts is based on Darkleaks, Generic public Leakage, Private Leakage, Key Theft, website defacement contract, Data Feed corruption,

Table 1. Ethereum application layer vulnerabilities attacks

Model impact factor		Properties Violation		
Attack name	Attack vectors	Theoreme 1	Theoreme 2	
		Termination	Fairness	Correctness
DAO ATTACK	Reentrancy	✗	✔	✗
Parity multisignature wallet	Delegate call injection	✔	✔	✗
	Erroneous visibility	✔	✔	✗
	Unprotected suicide	✔	✔	✗
	Frozen Ether	✔	✔	✗
BECToken attack	Integer overflow	✔	✗	✗
GovernMental attacks	DOS unbounded operations	✔	✔	✗
	Unchecked call return value	✗	✔	✗
	Call-stack depth limit	✗	✔	✗
	Transaction-ordering dependence	✔	✔	✗
	Timestamp dependence	✔	✔	✗
HYIP attack	DOS unexpected revert	✗	✔	✗
Fomo3D attacks	Generating randomness	✔	✔	✗
	DOS block stuffing	✗	✔	✔
ERC-20 signature replay	Insufficient signature information	✗	✔	✗
Rubixi attack	Erroneous constructor name	✔	✔	✗

Password theft. An example of a CSC is a smart contract for (private-)key theft. Such a CSC might pay a reward for delivery of a target key sk, such as a certificate authority's private digital signature key. The validity of criminal smart contracts is an indicator of criminals' success.

In their previous work, Juel and et al. [16,17] demonstrated that the execution of criminal smart contracts is always based on a time T_{end}. The time T_{end} marks the end of the execution whatever the state of the transaction i.e. the desirable terminal state is not always accessible therefore properties (4) and (7) are not always respected. This thesis is confirmed by the work of Yilei and et al. [35]. In their studies, they proposed a CSC based on PublicLeaks by formulating random factors such as the donation ratio. This contract is divided into five terminal states, one of which is unique in PublicLeaks because of its random nature (Table 2).

Table 2. Criminal smart contracts attacks

Model impact factor		Properties violation			
Attack name	Type of action	Correctness	Reachability	Real-time	Liveness
Leakage of secrets	Darkleaks	✓	✗	✗	✓
	Generic public leakage	✓	✗	✗	✓
	Private leakage	✓	✗	✗	✓
Key compromise	Key theft	✓	✗	✗	✓
Calling card crimes	Website defacement contract	✓	✗	✗	✓
	Data feed corruption	✓	✗	✗	✓
Password theft	Password theft	✓	✗	✗	✓

4 Proposition

It is important to understand that formal verification does not solve the problem of malicious smart contracts crime. To solve the problem of contracts while respecting the properties mentioned above, we propose an anomaly detection system on smart contracts. The idea is to create a consensus mechanism based on the behavior of smart contracts whose principle will be based on the prototype of normal behavior of the smart contract.

Anomaly detection overcomes the limitation of misuse detection by focusing on normal system behaviors, rather than attack behaviors. This approach is characterized by two phases: in the training phase, the behavior of the system is observed in the absence of attacks, and invariant calculation technique used to create a profile of such normal behavior. In the detection phase, this profile is compared against the current behavior of the system, and any deviations are flagged as potential attacks. Unfortunately, systems often exhibit legitimate but previously unseen behavior, which leads anomaly detection techniques to produce a high degree of false alarms. Moreover, the effectiveness of anomaly detection is affected greatly by what aspects of the system behavior are learnt.

Thus our model is composed of two modules: an invariant calculation algorithm to determine the normal profile of a smart contract, and an algorithm for monitoring the execution of the contract (Fig. 4).

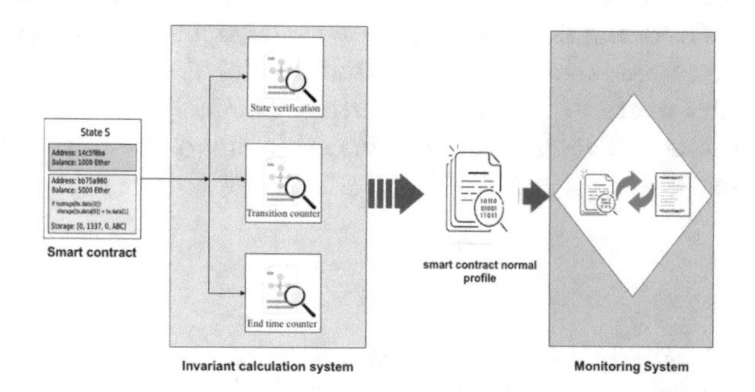

Fig. 4. Smart contracts anomalies detection system

5 Challenges

Given that all of the proposed solutions use techniques based on formal verification, it would be of great importance to orient the field of research towards the anomalies detection in smart contract systems. Anomaly detection is based on a program or host or network. Many distinct techniques are used based on type of processing related to behavioral model.

They are: Statistical based, Operational or threshold metric model, Markov Process or Marker Model, Statistical Moments or mean and standard deviation model, Invariant Model, Multivariate Model, Time series Model, Cognition based, Finite State Machine Model, Description script Model, Machine Learning based, Baysian Model, Genetic Algorithm model, Neural Network Model, Computer Immunology based.

The application of these detection techniques relating to the behavior model can solve the problem of attacks in smart contract systems.

Additional work could be carried out on smart contract anomaly detection techniques using artificial intelligence, ontology-based smart contracts for detecting malicious behavior, deep learning technique.

A future project could consist of working on a behavior detection model based on artificial intelligence because so far, cybersecurity systems using artificial intelligence have proven to be the most effective in protecting blockchain.

6 Conclusion

Termination, Fairness, Correctness, Reachability, Safety, Liveness and Real-time properties of a smart contract must be guaranteed in advance, before formal instantiation in a blockchain. This is certainly important for developers, as well as suppliers and consumers that rely on the soundness of a smart contract. Moreover, it furnishes a source of trust for users because trust is maintained by algorithmic concepts. For example, proving the correctness of smart contracts, a model of the actual correct behaviour of a contract is necessary in first place. Determining whether a contract reacts correctly is not always as trivial as it seems, and proving it (automatically) means that the behaviour must be defined as conditions in a formal notation, for instance (temporal) first order logics [26]. However, a good approach to smart contract model allows us to find a solution to the problems related to blockchain crime. Anomaly detection provides an answer to various problems such as vulnerable smart contracts and criminal smart contracts.

References

1. Abdellatif, T., Brousmiche, K.-L.: Formal verification of smart contracts based on users and blockchain behaviors models. In: 2018 9th IFIP International Conference on New Technologies, Mobility and Security (NTMS), pp. 1–5. IEEE (2018)

2. Amani, S., Bégel, M., Bortin, M., Staples, M.: Towards verifying ethereum smart contract bytecode in Isabelle/HOL. In: Proceedings of the 7th ACM SIGPLAN International Conference on Certified Programs and Proofs, pp. 66–77. ACM (2018)
3. Bartoletti, M., Galletta, L., Murgia, M.: A true concurrent model of smart contracts executions. In: International Conference on Coordination Languages and Models, pp. 243–260. Springer (2020)
4. Bhargavan, K., et al.: Short paper: formal verification of smart contracts. In: Proceedings of the 11th ACM Workshop on Programming Languages and Analysis for Security (PLAS), in Conjunction with ACM CCS, pp. 91–96 (2016)
5. Bigi, G., Bracciali, A., Meacci, G., Tuosto, E.: Validation of decentralised smart contracts through game theory and formal methods. In: Programming Languages with Applications to Biology and Security, pp. 142–161. Springer (2015)
6. Brent, L., et al.: Vandal: a scalable security analysis framework for smart contracts. arXiv preprint arXiv:1809.03981 (2018)
7. Chen, W., Zheng, Z., Cui, J., Ngai, E., Zheng, P., Zhou, Y.: Detecting ponzi schemes on ethereum: towards healthier blockchain technology. In: Proceedings of the 2018 World Wide Web Conference, pp. 1409–1418 (2018)
8. Di Angelo, M., Salzer, G.: A survey of tools for analyzing ethereum smart contracts. In: 2019 IEEE International Conference on Decentralized Applications and Infrastructures (DAPPCON). IEEE (2019)
9. Dika, A.: Ethereum smart contracts: Security vulnerabilities and security tools. Master's thesis, NTNU (2017)
10. Feng, T., Yu, X., Chai, Y., Liu, Y.: Smart contract model for complex reality transaction. Int. J. Crowd Sci. **3**(2), 184–197 (2019). https://doi.org/10.1108/IJCS-03-2019-0010
11. Fu, Y., et al.: EVMFuzzer: detect EVM vulnerabilities via fuzz testing. In Proceedings of the 2019 27th ACM Joint Meeting on European Software Engineering Conference and Symposium on the Foundations of Software Engineering, pp. 1110–1114. ACM (2019)
12. Hadjidj, R., Yang, X., Tlili, S., Debbabi, M.: Model-checking for software vulnerabilities detection with multi-language support. In: 2008 Sixth Annual Conference on Privacy, Security and Trust, pp. 133–142. IEEE (2008)
13. He, J., Balunović, M., Ambroladze, N., Tsankov, P., Vechev, M.: Learning to fuzz from symbolic execution with application to smart contracts. In: Proceedings of the 2019 ACM SIGSAC Conference on Computer and Communications Security, pp. 531–548 (2019)
14. Hirai, Y.: Formal verification of deed contract in ethereum name service, November 2016 (2016). https://yoichihirai.com/deed.pdf
15. Jiang, B., Liu, Y., Chan, W.K.: ContractFuzzer: fuzzing smart contracts for vulnerability detection. In: Proceedings of the 33rd ACM/IEEE International Conference on Automated Software Engineering, pp. 259–269. ACM (2018)
16. Juels, A., Kosba, A., Shi, E.: The ring of gyges: using smart contracts for crime. Aries **40**, 54 (2015)
17. Juels, A., Kosba, A., Shi, E.: The ring of gyges: investigating the future of criminal smart contracts. In: Proceedings of the 2016 ACM SIGSAC Conference on Computer and Communications Security, pp. 283–295 (2016)
18. Kalra, S., Goel, S., Dhawan, M., Sharma, S.: Analyzing safety of smart contracts. In: NDSS, Zeus (2018)
19. Lamport, L.: "Sometime" is sometimes "not never" on the temporal logic of programs. In: Proceedings of the 7th ACM SIGPLAN-SIGACT Symposium on Principles of Programming Languages, pp. 174–185 (1980)

20. Luu, L., Chu, D.-H., Olickel, H., Saxena, P., Hobor, A.: Making smart contracts smarter. In: Proceedings of the 2016 ACM SIGSAC Conference on Computer and Communications Security, pp. 254–269 (2016)
21. Murray, Y., Anisi, D.A.: Survey of formal verification methods for smart contracts on blockchain. In: 2019 10th IFIP International Conference on New Technologies, Mobility and Security (NTMS), pp. 1–6. IEEE (2019)
22. Nehai, Z., Piriou, P.-Y., Daumas, F.: Model-checking of smart contracts. In: 2018 IEEE International Conference on Internet of Things (iThings) and IEEE Green Computing and Communications (GreenCom) and IEEE Cyber, Physical and Social Computing (CPSCom) and IEEE Smart Data (SmartData), pp. 980–987. IEEE (2018)
23. Nguyen, Q.-B. Nguyen, A.-Q., Nguyen, V.-H., Nguyen-Le, T., Nguyen-An, K.: Detect abnormal behaviours in ethereum smart contracts using attack vectors. In: International Conference on Future Data and Security Engineering, pp. 485–505. Springer (2019)
24. Nguyen, T.D., Pham, L.H., Sun, J., Lin, Y., Minh, Q.T.: sFuzz: an efficient adaptive fuzzer for solidity smart contracts. arXiv preprint arXiv:2004.08563 (2020)
25. Nikolic, I., Kolluri, A., Sergey, I., Saxena, P., Hobor, A.: Finding the greedy, prodigal, and suicidal contracts at scale. In: Proceedings of the 34th Annual Computer Security Applications Conference, pp. 653–663 (2018)
26. Osterland, T., Rose, T.: Correctness of smart contracts for consistency enforcement. ERCIM NEWS **110**, 18–19 (2017)
27. Parizi, R.M., Dehghantanha, A., Choo, K.-K.R., Singh, A.: Empirical vulnerability analysis of automated smart contracts security testing on blockchains. In: Proceedings of the 28th Annual International Conference on Computer Science and Software Engineering, pp. 103–113. IBM Corp. (2018)
28. Permenev, A., Dimitrov, D., Tsankov, P., Drachsler-Cohen, D., Vechev, M.: VerX: safety verification of smart contracts. In: 2020 IEEE Symposium on Security and Privacy, SP, pp. 18–20 (2020)
29. Pnueli, A.: The temporal logic of programs. In: 18th Annual Symposium on Foundations of Computer Science (SFCS 1977), pp. 46–57. IEEE (1977)
30. Podelski, A., Rybalchenko, A.: Transition invariants. In: 2004 Proceedings of the 19th Annual IEEE Symposium on Logic in Computer Science, pp. 32–41. IEEE (2004)
31. So, S., Lee, M., Park, J., Lee, H., Oh, H.: VeriSmart: a highly precise safety verifier for ethereum smart contracts. In: 2020 IEEE Symposium on Security and Privacy (SP), pp. 1678–1694. IEEE (2020)
32. Torres, C.F., Schütte, J., State, R.: Osiris: hunting for integer bugs in ethereum smart contracts. In: Proceedings of the 34th Annual Computer Security Applications Conference, pp. 664–676 (2018)
33. Wang, H., Li, Y., Lin, S.-W., Ma, L., Liu, Y.: VULTRON: catching vulnerable smart contracts once and for all. In: Proceedings of the 41st International Conference on Software Engineering: New Ideas and Emerging Results, pp. 1–4. IEEE Press (2019)
34. Wang, S., Ouyang, L., Yuan, Y., Ni, X., Han, X., Wang, F.-Y.: Blockchain-enabled smart contracts: architecture, applications, and future trends. IEEE Trans. Syst. Man Cybern.: Syst. **49**(11), 2266–2277 (2019)
35. Wang, Y., Bracciali, A., Li, T., Li, F., Cui, X., Zhao, M.: Randomness invalidates criminal smart contracts. Inf. Sci. **477**, 291–301 (2019)
36. Wood, G.: Ethereum: a secure decentralized generalized transaction ledger (EIP-150 revision) (2020). http://gavwood.com/paper.pdf

37. Wustholz, V., Christakis, M.: Harvey: a greybox fuzzer for smart contracts. arXiv preprint arXiv:1905.06944 (2019)
38. Xia, X., Ji, Q., Le, J.: Research on transaction dependency mechanism of self-healing database system. In: 2012 International Conference on Systems and Informatics (ICSAI2012), pp. 2357–2360. IEEE (2012)

A Blockchain-Enabled Fog Computing Model for Peer-To-Peer Energy Trading in Smart Grid

Saurabh Shukla(✉), Subhasis Thakur, Shahid Hussain, and John G. Breslin

National University of Ireland Galway, Galway, Ireland
{saurabh.shukla,shahid.hussain,subhasis.thakur,
john.breslin}@nuigalway.ie

Abstract. The advancement in renewable energy sources (RESs) technology have changed the role of traditional consumers to prosumers. In contrast to the traditional power grid, the Smart Grid (SG) network provides a platform for peer-to-peer (P2P) energy trading between prosumers to buy or sell energy according to their requirements. The potential benefits of P2P energy trading can be realized through an efficient service provider of the communication network infrastructure. However, the current communication network is a trustless environment and thereby is unable to fully support the P2P energy trading requirements. Existing techniques in P2P energy trading with blockchain suffers from large network delay due to large network size; this further affects the network performance for P2P trading. In this paper, we present a novel Blockchain-Based Smart Energy Trading (BSET) algorithm along with a Blockchain-Enabled Fog Computing Model (BFCM) for P2P energy trading in Smart Grid. The proposed BSET algorithm provides a fully trusted minimum latency communication network that enables the prosumers to trade energy within their local premises. The algorithm was implemented using iFogSim, Truffle, ATOM, Anaconda, and Geth and evaluated against state-of-the-art communication network models for P2P energy trading. The simulation results revealed the effectiveness in terms of secure trading and network latency.

Keywords: Smart grid · Smart meter · Cyber-physical system · Fog computing · Blockchain · Cryptography · Cloud computing · Microgrid · Internet-of-Things

1 Introduction

Nowadays, the increasing demand for timely electrical energy consumption and monitoring has given rise to the role of a smart grid network over the traditional grid. The traditional grid is a centralized and one-way transmission of energy. Whereas the SG network is distributed two-way transmission of energy and information. Where prosumers and consumers have a major role to play in buying and selling energy in a decentralized manner. The current SG network extends the controlling, computation, monitoring and sensing of the information and electrical energy flow in a bidirectional way when compares to a traditional network where different buyers and sellers participate in the auction

J. Prieto et al. (Eds.): BLOCKCHAIN 2021, LNNS 320, pp. 14–23, 2022.
https://doi.org/10.1007/978-3-030-86162-9_2

and bidding. SG when enabled with Information and Communication Technology (ICT) and Cyber-Physical System (CPS) has become an autonomous and resilient energy trading system. The different and independent stakeholders can unite and build a trustworthy relationship to exchange the required electrical energy as per the end-user requirement in real-time. The existing state-of-the-art SG system requires a distributed platform for storing, processing, and controlling a large amount of information and energy to make the system reliable for P2P energy trading. This can be possible with the use of Fog Computing (FC) acting as the middleware layer at the edge of prosumers and consumers in a Neighbourhood Area Network (NAN), Building Area Network (BAN), Household Area Network (HAN), and Local Area Network (LAN) [1]. FC nodes can transmit the data in a real-time mode with a single hop count. This makes the SG network efficient for the performance related to energy trading and exchange of information between prosumers and consumers as they require minimum service latency, network usage and an efficient secure channel [2]. Next, to secure the energy trading transaction in the SG network blockchain can play a major role by acting as a decentralized system along with the FC environment. Recent studies have not evaluated the underlying blockchain performance for QoS requirement in P2P energy trading in an FC environment. Therefore, to meet the above-mentioned challenges we have proposed a novel paradigm using blockchain in the FC environment. The proposed work includes a novel BSET algorithm along with a BFCM for P2P energy trading in SG.

2 Background and Related Work

This section includes the research work related to P2P energy trading mechanism, auction of available energy between different distributed prosumers and consumers, along with blockchain-based techniques bidding techniques used in an SG network and CPS. Recent work related to energy trading shows that it is quite complex to design a decentralized P2P trading system that keeps an optimum balance between economic efficiency and data privacy [3, 4]. Network size, latency and security are the major features that affect the overall performance of P2P energy trading. Some of the recent works for energy trading have been highlighted under this section. In [5], the authors proposed a novel approach using a Hyperledger Fabric (HF) to strengthen the collaboration and coordination of the SG network with CPS. Furthermore, they designed a novel framework for peer-to-peer energy trading mechanism. In [2], the authors proposed a negotiation protocol for a double auction mechanism with additional security features for secure energy trading in the SG network between prosumers and consumers. The novel protocol is comparable with SG technology. Similarly, in [6], the authors proposed a peer-to-peer trading system method that minimizes the risk of energy loss and maximizes the profit and benefits for both buyers and sellers. The risk in energy distribution is measured using Markowitz portfolio theory and modified Sharpe ratio. The proposed work focuses on maximizing the benefit for prosumers and uses the particle swarm optimization technique.

In [7], the authors proposed a two-stage optimization model for maximizing the utilities with optimal strategies in peer-to-peer energy trading between prosumers and consumers in the power systems likes, microgrids, and SG network. The above model was implemented using a distributed algorithm. In [8], the authors conducted a detailed description of the problem related to peer-to-peer energy trading in a decentralized power system. The existing model for energy trading between prosumers and consumers lacks the real-world implementation of sharing energy using the model between different customers and sellers. In [9], the authors explained the potential of machine learning in the distributed environment to solve the issues involved in the energy trading problem between the different prosumers and consumers. They addressed several issues of the SG network such as resource allocation, communication schemes, the privacy of user data, and negotiation between the seller and buyers.

3 Conventional Energy Trading Model

This section discusses the conventional way of P2P energy sharing between the communities using smart meters in an SG network. See Fig. 1 for the conventional trading involving different communities.

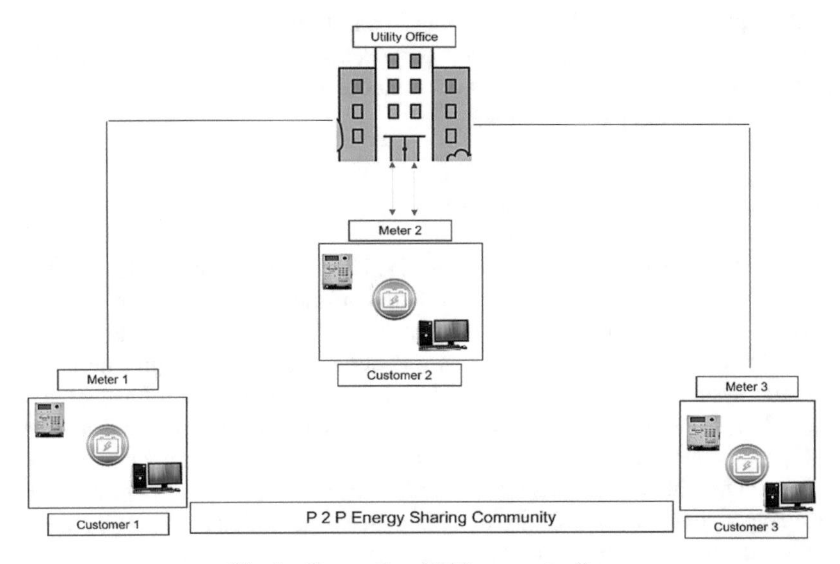

Fig. 1. Conventional P2P energy trading

Figure 1 shows the conventional way of P2P energy trading between the different communities i.e., the different distributed customers. Where one of the customers acts as prosumers and another one-act as consumers. The smart meters are further connected to utility offices. However, this conventional way of P2P energy trading suffers from the various QoS requirements i.e., service delay, and network usage for efficient P2P energy trading between the buyers and sellers of electrical energy in the SG network. In general, energy trading is conducted between different microgrids. Where both producers and consumers participate in using the energy supply from smart meters to measure consumption. The conventional energy trading model lacks secure private communication and energy transaction between prosumers related to P2P energy trading.

4 Blockchain-Enabled Fog Computing Model (BFCM)

In this section, we discussed the proposed advanced P2P energy trading system model which utilizes the concept of blockchain and FC to meet the network, latency and security requirement for energy trading involving smart meters. FC acting at the edge of networks transfers the electrical data in a single hop count Here in this system model, we have used blockchain technology to secure the user's anonymity and privacy for securing information during trading. See Fig. 2 for an advanced blockchain-based P2P energy trading system.

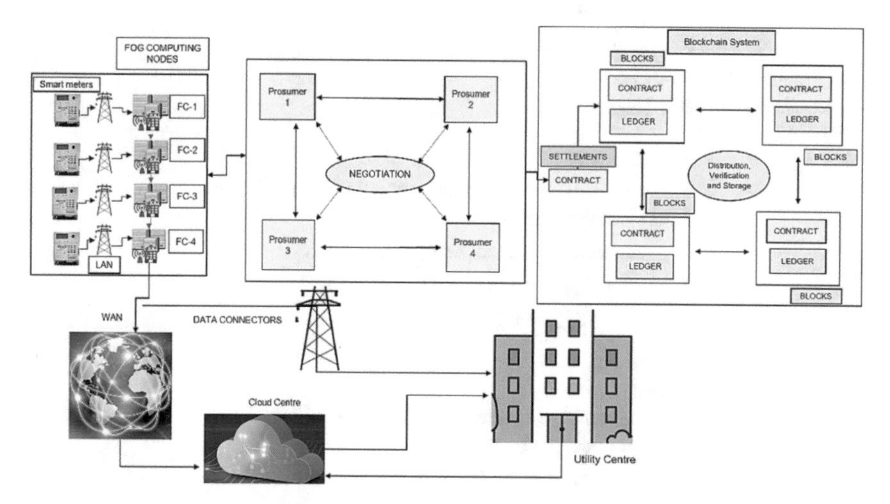

Fig. 2. Blockchain-enabled fog computing model for P2P energy trading in smart grid

Figure 2 shows the novel advanced blockchain-based P2P energy trading model. The model consists of a secure data transmission channel that uses blockchain-based technology. The blockchain here used for electrical data distribution, settlements of contracts, recording of transactions that occurred, data and user verification along with storage of blocks. Where FC nodes are used for minimizing the service latency and network usage at the edge of the smart meter deployed in the SG network. Next, the energy available for trading is transferred to prosumers and consumers in a real-time mode where negotiations are conducted for the auction of available energy to sell. The advanced model will help in achieving QoS requirement by minimizing the bottlenecks which affect the performance of the energy trading and achieve the best configuration for the system. Each prosumer can share his/her energy usage information which can be used by other neighbours with a secure channel using the blockchain model.

5 Blockchain-Based Smart Energy Trading (BSET) Algorithm

This section discusses the design and development of the BSET algorithm. The functioning of the proposed novel algorithm is divided into three different algorithms for the performance evaluation of efficient P2P energy trading between prosumers. Moreover, the time complexity of the proposed BSET algorithm for encryption function is $O(N)^2$ and for decryption function is $O(N)^3$.

Algorithm 1: Negotiation protocol of two users begin inside fog computing (FC) environment. In this part, the seller generated a contract and final contract based on the offer and counteroffer for negotiation with the buyer.

Algorithm symbols notations

$FCon$: Final contract
$Con_{i,j}^{QK}$: Contract
o' : Counteroffer
o: Offer
θ_k: Task

1. **Start**
2. For task θ_k , seller j generates an offer o to buyer i
3. **if** i accept o then
4. generate a contract $Con_{i,j}^{Q_k}$ based on o
5. $FCon_{i\theta_k,}^{tmp} = FCon_{i,\theta_k}^{tmp}$ o $Con_{i,j}^{\theta_k}$;
6. $FCon_{j,\theta_k}^{tmp} = FCon_{j,\theta_k}^{tmp}$ o $Con_{i,j}^{\theta_k}$;
7. return;
8. **else**
9. i generate a counteroffer o' to j
10. **if** j accepts o' then
11. generate contract $Con_{i,j}^{\theta_k}$ based on o
12. $FCon_{i,\theta_k}^{tmp} = FCon_{i,\theta_k}^{tmp}$ o $Con_{i,j}^{\theta_k}$;
13. $FCon_{j,\theta_k}^{tmp} = FCon_{j,\theta_k}^{tmp}$ o $Con_{i,j}^{\theta_k}$;
14. return
15. **else**
16. return null; /* no contract is formed */
17. **end if**
18. **end if**
19. **End**

Algorithm 2: Final contract determination begins between the peers in FC. This part of the algorithm deals with the final contract determination and contract determination deadline.

Algorithm symbol notations

$FCon_{i,\Theta_k}^{tmp}$: temporary contracts

$FCon_{i,\Theta_k}^{fn}$: Final contract determination using FC

$Dl'(\Theta_k)$: Contract determination deadline

$FCon$: Final contract

$Con_{i,j}^{QK}$: Contract

Θ_k: Task

FC : Fog computing

1. **Start**
2. for task Θ_k, buyer i sort temporary contracts in $FCon_{i,\Theta_k}^{tmp}$;
3. **while** t < $Dl'(\Theta_k)$ /*t is the real-time*/
4. **if** all temporary contracts in $FCon_{i,\Theta_k}^{tmp}$, have been processed then
5. **return**;
6. **else**
7. **if** $\sum_{c \epsilon FCon_{i,\Theta_k}^{fn}} c:cp > E_{\Theta_k}$ then
8. **return;**
9. **else**
10. get the next temporary contract C from $FCon_{i,\Theta_k}^{tmp}$;
11. i sends a transaction request message to j; U
12. i gets a reply message from j;
13. **if** j confirms the contract, then
14. $FCon_{i,\Theta_k}^{fn}$ <- $FCon_{i,\Theta_k}^{fn}$ U C; /* using FC */
15. $FCon_{j,\Theta_k}^{fn}$ <- $FCon_{j,\Theta_k}^{fn}$ U C; /* using FC */
16. **end if**
17. **end if**
18. **end while**
19. **End**

Algorithm 3: Encryption and decryption of electrical data using blockchain for secure communication between peers. The data is encrypted and then decrypted using a private blockchain and different cryptographic operations. The encryption algorithm is used to encrypt the meter data and readings from the outside world. The algorithm performs security of data transmission between smart meters, meter management system, and consumers using the blockchain. The electrical data is encrypted using private-public key arrangements the data is encrypted and then decrypted as per the user requirement.

Algorithm symbol notations
S_m : Smart Meters
K_{sym} : Symmetric key
K_{pub} : Public key
F_nK_{pub} : Fog node public key
$Encrypt_{Sym}$: Symmetric encryption
$Encrypt_{Asym}$: Asymmetric encryption
C: Ciphertext

C_K : Cipher key
F_nK_{prvt} : Fog node private key
C_s : Cloud server
ST_m : Smart Meters
FC_n: Fog computing nodes
E_{DP} : Electrical energy data packet
FC_n: Fog computing nodes
$SPARK$: Real-Time Analyzer (RTA)
S_m_D: Smart meter data

1. **Start**
2. (FC-based blockchain system is created)
3. Data classification Using 2-PCA Linear SVM
4. **if** (ST_m = $Malicious\ electric\ data$) then
5. get geo-location and send the data for verification to FC_n using SPARK
6. FC_n allocates the E_{DP} to F_s
7. **else if** (ST_m == $Authorized\ electric\ data$)
8. then
9. E_{DP} send to FC_n to C_s
10. **end if**
11. **function** Encryption (S_m_D)
12. **if** S_m confirms E_d storage over blockchain then
13. Generate a K_{sym}
14. $C <- Encrypt_{Sym}$ (S_m_D , K_{sym})
15. $C_k <- Encrypt_{Asym}(K_{sym}, F_nK_{pub})$
16. **else**
17. do no operation
18. **end if**
19. **end function**
20. **function** $DECRYPTION$ (C, C_k, F_nK_{prvt}, K_{sym})
21. K_{sym} <-$Decryption_{Asym}(C_k, F_nK_{prvt})$
22. S_m_D <- Decryption (C, K_{sym})
23. **end function**
24. **End**

6 Results and Discussion

This section discusses the results and simulation of the BSET algorithm in the iFogSim simulator. Next, we have used Anaconda (Python), Geth version 1.9.25, Ganache, Truffle (Compile) and ATOM as a text editor for creating smart contracts. iFogSim is an open-source simulator used for creating physical topology design, resource placement, and packet allocation by creating different edges, networks, nodes, and devices with cloud and fog sever. A detailed comparative analysis is conducted between FC and cloud-related latency, malicious node percentage, and network usage. The simulations are conducted between different prosumers at different physical topology configurations. Furthermore, the malicious node percentage increases with an increase in the number of prosumers. See Fig. 3 represent the physical topology configuration.

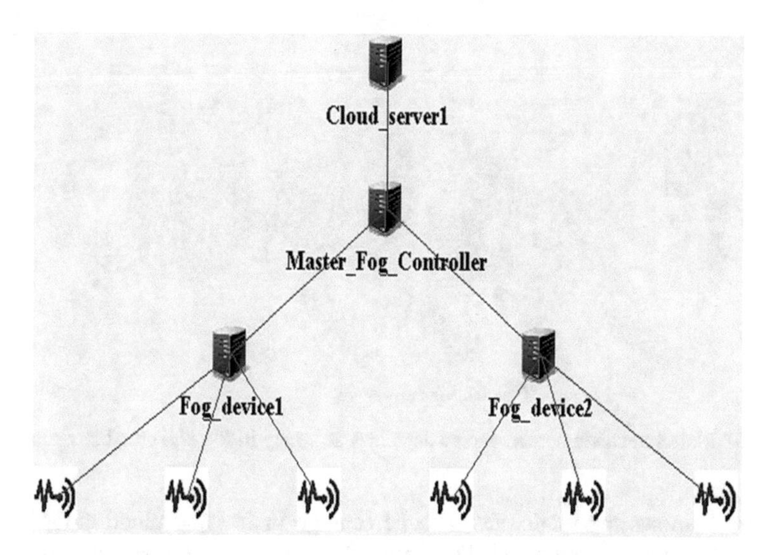

Fig. 3. iFogSim GUI configuration for physical configuration

Figure 3 shows the physical topology for configuration built in the iFogSim simulator. The configuration is solely based on the concept of a proposed system. The figure shows

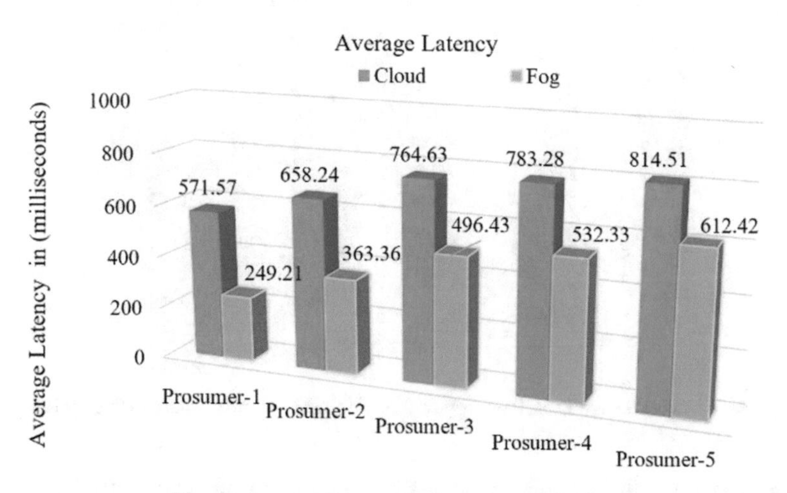

Fig. 4. Average latency performance evaluation

the fog devices connected with a cloud server and the smart meters embedded at the prosumers place. See Fig. 4 for average latency performance.

Figure 4 shows the average latency comparison between different prosumers during P2P energy trading in fog and cloud computing environment. The FC easily outperforms the cloud in terms of latency.

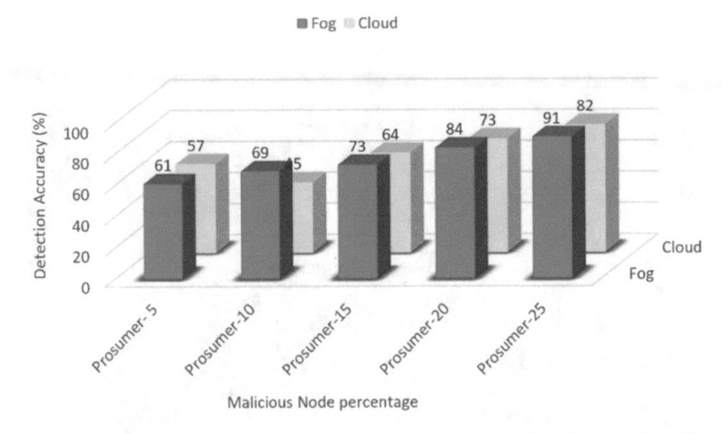

Fig. 5. Malicious node percentage vs detection accuracy in fog and cloud environment

Figure 5 shows the malicious node percentage in fog and cloud along with the detection accuracy in percentage. The figure shows that the detection accuracy in fog nodes is much greater when compared to the cloud for different distributed prosumers.

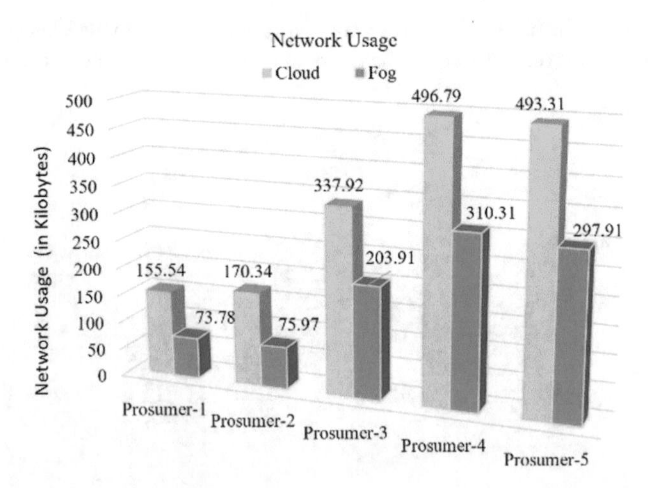

Fig. 6. Network usage in P2P energy trading for different prosumers in fog and cloud

Figure 6 shows the network usage by the different prosumers in P2P energy trading using fog and cloud. The FC easily outperforms the cloud in terms of network usage.

7 Conclusion

To overcome the limited communication, network efficiency, and data privacy issues of the SG network a fully distributed energy trading system using FC and blockchain is proposed. Next, we have proposed a novel Blockchain-Based Smart Energy Trading (BSET) algorithm. The algorithm is further divided into three other major sub-algorithms. 1) Negotiation protocols between two users 2) Final contract determination between the peers and 3) Encryption and decryption of electrical data using blockchain for secure communication. The main contribution of the proposed work is to improve P2P energy trading by providing a secure communication channel among prosumers along with determining the latency and network usage requirement.

Acknowledgements. This publication has emanated from research supported in part by a research grant from Cooperative Energy Trading System (CENTS) under Grant Number REI1633, and by a research grant from Science Foundation Ireland (SFI) under Grant Number SFI 12/RC/2289_P2 (Insight), co-funded by the European Regional Development Fund.

References

1. Ahsan, M.M., Ali, I., Imran, M., Idris, M.Y.I., Khan, S., Khan, A.: A fog-centric secure cloud storage scheme. IEEE Trans. Sustain. Comput. (2019). https://doi.org/10.1109/TSUSC.2019.2914954
2. Sarenche, R., Salmasizadeh, M., Ameri, M.H., Aref, M.R.: A secure and privacy-preserving protocol for holding double auctions in smart grid. Inf. Sci. **557**, 108–129 (2021)
3. Luo, F., Dong, Z.Y., Liang, G., Murata, J., Xu, Z.: A distributed electricity trading system in active distribution networks based on multi-agent coalition and blockchain. IEEE Trans. Power Syst. **34**(5), 4097–4108 (2018)
4. Sadhukhan, D., Ray, S., Obaidat, M.S., Dasgupta, M.: A secure and privacy preserving lightweight authentication scheme for smart-grid communication using elliptic curve cryptography. J. Syst. Arch. **114**, 101938 (2020)
5. Lohachab, A., Garg, S., Kang, B.H., Amin, M.B.: Performance evaluation of Hyperledger Fabric-enabled framework for pervasive peer-to-peer energy trading in smart Cyber-Physical Systems. Futur. Gener. Comput. Syst. **118**, 392–416 (2021)
6. Mohan, V., et al.: Realistic energy commitments in peer-to-peer transactive market with risk adjusted prosumer welfare maximization. Int. J. Electric. Power Energy Syst. **124**, 106377 (2021)
7. Jiang, A., Yuan, H., Li, D.: A two-stage optimization approach on the decisions for prosumers and consumers within a community in the Peer-to-peer energy sharing trading. Int. J. Electric. Power Energy Syst. **125**, 106527 (2021)
8. Tushar, W., et al.: Peer-to-peer energy systems for connected communities: a review of recent advances and emerging challenges. Appl. Energy **282**, 116131 (2021)
9. Wang, N., Li, J., Ho, S.-S., Qiu, C.: Distributed machine learning for energy trading in electric distribution system of the future. Electric. J. **34**(1), 106883 (2021)

Energy Trading Between Prosumers Based on Blockchain Technology

Susana M. Gutiérrez[1](✉), José L. Hernández[1], Alberto Navarro[2], and Rocío Viruega[2]

[1] Fundación CARTIF, 47151 Boecillo (Valladolid), Spain
susgut@cartif.es
[2] Alpha Syltec Ingeniería, SL., 47197 Valladolid, Spain

Abstract. Electrification of the energy systems and the current policies for sustainable development are increasing the potential for distributed energy resources (DERs). Within this trend, electricity grid generation and intermediary elements (e.g. batteries) are distributed, promoting a new business model where the consumers have become prosumers. That is to say, they can both generate and consume energy. The distributed energy generation is based on renewable (mainly, photovoltaics) and storage (i.e. batteries). However, the produced energy is unpredictable without forecasting tools and intermittent (i.e. dependent on the solar resources). Additionally, although the priority is the usage of this energy for self-consumption, the surplus of energy is usually sold to the DSO (Distribution System Operator). Then, the usage of Peer-to-Peer (P2P) energy trading fosters new business models and novel paradigms of energy system generation where people can generate their own energy and share it locally with each other. The aforementioned factors have encouraged a wider adoption of microgrids powered by renewable distributed energy sources and they provide a lot of economic and environmental benefits. For that end, this paper presents a P2P approach, combined with demand and production forecasting, to enable novel prosumers' business models and promoting renewable energy communities for energy transition.

Keywords: Blockchain · P2P · Smart contract · Energy trading · Microgrid · Energy transition

1 Introduction

Energy markets are evolving from centralized electricity generation to distributed concepts [1], thanks to the wide deployment of renewable self-consumption installations (e.g. photovoltaics in roofs). This aspect contributes to more efficient, environmentally-friendly and small-size generating units, as well as better demand management. This new trend is enhanced by the integration of new technologies, like blockchain and P2P (peer to peer). Under this approach, the new concept of microgrid arises where the distributed generation and consumption elements are integrated in the existing power system [2] and communicated though the aforementioned P2P technology. Thus, Distributed Ledger Technologies (DLTs) and, in particular, blockchain technology, have the

J. Prieto et al. (Eds.): BLOCKCHAIN 2021, LNNS 320, pp. 24–33, 2022.
https://doi.org/10.1007/978-3-030-86162-9_3

potential to transform the energy sector. The World Economic Forum, Stanford Woods Institute for the Environment, and PwC released a joint report [3] identifying more than 65 blockchain use-cases for the environment. These use cases include new business models for energy markets and, even more, moving carbon credits or renewable energy certificates onto the blockchain.

Since years ago, the penetration of renewables has been seen as a disruptive way to transform energy markets thanks to the integration of PV (photovoltaics) cells, wind turbines, Information and Communication technologies, among others [4]. Therefore, better balances of the energy flows are possible based on trading mechanisms [5], where blockchain is the enabler technology. Also, this blockchain technology breaks with the traditional role of the system aggregator by using the Smart Contracts concept as a virtual aggregator within the microgrid.

Thus, this paper presents a blockchain-based microgrid deployment so as to foster distributed energy resources within microgrids and energy trading mechanisms. Then, the implementation of this P2P concept aims to foster new electricity markets for energy transition. The implementation of blockchain plus Smart Contract allows the creation of transactions for energy sharing between agents in the microgrid. In this sense, local energy exchange is prioritized, reducing the grid stress (i.e. when supplying the excess of energy from the aforementioned renewables sources into the grid).

In order to detail the concept, Sect. 2 contains a very brief summary of the related work and how the aforementioned approach is sited in the current research context. Section 3 explains the approach and business model that applies for the energy trading solution proposed in this paper. Section 4 describes the system that has been developed and the technological aspects to be considered. Finally, Sect. 5 provides the real case demonstration and test results, and the set of conclusions is described in Sect. 6.

2 Related Work

In the last years, a lot of research works have been published in the field of blockchain, P2P and energy trading. One of the pioneers was the Brooklyn microgrid [6], where a distributed energy generation microgrid was set up to balance energy generation and consumption. The concept of the Brooklyn microgrid established self-produced energy trading mechanisms integrating both physical and virtual prosumers.

Other studies like [5] or [7] have been focused on the implementation of auction mechanisms based on blockchain and smart contracts for the optimization of the energy flows, allowing trading between peers. Additionally, blockchain has been deployed to enable a direct energy trading among consumers and facilitating the local power, such as the analysis in [8]. Finally, the authors in [9] propose a framework under where P2P trading is possible. Then, energy matching strategies are followed to match demand and renewable energy supply. Double auction principles are applied within the microgrid where the players count on renewable producers, consumers, and the distribution system operator (DSO).

However, it has to be highlighted that the previous researches, with the exception of the Brooklyn microgrid, lacks of real demonstrators and they are simulation-based in test environments, such as Ganache. Additionally, the results are focused on the prosumers energy trading, without considering the DSO, who plays a very important role, where only the research in [9] includes the DSO in the energy matching process. Finally, as stated before, the distributed energy generation, distribution and demand is intermittent, which is not considered within the previous researches and depends on the user's behavior. But, end-consumers are not considered as the core of the microgrid.

2.1 Contribution Beyond the State of the Art

The research presented in this paper goes beyond the state of the art in the axes that are described in the following bullets:

- First of all, to overcome the uncertainties in the available energy production, as well as demand uncertainties, combined blockchain and demand prediction has been implemented. While the blockchain technology enables the energy trading, machine learning for forecasting is applied on top of the blockchain to predict the production, demand and, thus, excess of energy. According to this, three stages in the energy trading are modelled: (1) predicted excess of energy to be sold and predicted energy demand to be covered (i.e. requirements for purchase); (2) execution of the trading according to the real condition; (3) cross-validation to double-check that the sold and purchased energy are the agreed ones. The advantage of this approach is the increased accuracy, based on predictions, but checked with real measurements of data from the smart meters.
- Secondly, the sales and purchases of energy are based on informed users, which are the core of the system. The predicted excess of production and the demand needs are provided to the prosumer through a user-friendly interface. Then, the end-consumer is empowered to decide about the proposals to sell and/or purchase energy. Moreover, the trading final decision is also possible thanks to the interface, where the end-consumer can make the decision about matching the sales-purchases, although the automatic auction is also possible for non-experience users.
- Next, the research presented in this paper goes a step forward as the test environment is a real, but controlled, scenario in contrast to Ganache simulation network.
- Finally, and related to the previous point, one of the actors within the system tested in this paper is the DSO or aggregator. Energy that has not been matched cannot be missed and, here, the DSO or aggregator plays the crucial role of being the buyer for the non-matched energy.

3 Peer to Peer Energy Trading and Business Model Approach

As mentioned before, the presented paper tests the P2P network in a real, but controlled, environment, where the energy that is produced should be consumed or dumped into the grid. Aligned with this, the role of the traditional consumers has also changed to become prosumers. Then, this drives to new business models where aggregators are not part of the energy market, while the Smart Contract acts as virtual aggregator. Thus, direct "communication channels" among end-consumers and DSOs are established through the P2P-blockchain network and Smart Contract. This new way, as stated in [10], benefits of using blockchain technology for a more transparent, distributed and secure transaction log. The first clear advantage is the confidence of the end-consumers, as well as their intervention in the energy markets, thus, empowering them and making them participants of the process.

According to that, this paper proposes a new paradigm in the energy trading market, where both sellers and buyers are able to exchange energy. The energy purchases and sales are guided by a predictive algorithm that always informs the user about the excess of energy that will be available to be sold, as well as the energy requirements according to the demand for the energy purchases. Thus, there are two ways of behaviour:

- *"User guided" energy trading*: This is provided for experienced and energy-aware users, empowering them in the final decisions. Both sellers and buyers are in charge of the sales and purchases proposals, the matching and the final validation. The transactions are "manually" generated and the developed smart contract provides the user with a method to generate such an energy transaction. That is to say, after having published the sales and purchases proposals, including the energy prices, the users are in charge of selecting the matches between sellers and buyers, being possible selling energy to multiple buyers and vice versa, but always specifying the total amount of energy and its price to sell/buy. Besides, the amount of energy not sold will be considered as transferred to the DSO (Distribution System Operator). Figure 1 shows an example of user guided energy trading in which the sequence of operations has been the following one:
  - Three "rounds" of energy transactions have been done. In the first round, the prosumer offered 4 energy units, seller1 bought 3 energy units and seller2 bought 1 (in this case, no energy has been transferred to the general grid). In the second round, the prosumer offered 3 energy units, seller1 bought 2.5, seller2 bought 0.5 and no energy was transferred to the general grid. During the third round, the prosumer put on sale 5 energy units, seller1 acquired 3 energy units, seller2 bought 1 energy unit and 1 energy unit was transferred to the general grid.

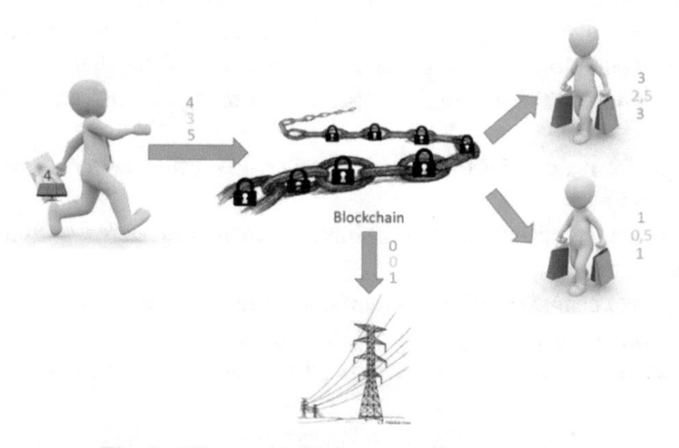

Fig. 1. "User guided" energy trading example

- *Auction energy trading*: This second approach is depicted for non-experienced users and the energy transactions are automatically generated by the smart contract once it is required to do it (after the expiration of the prediction horizon). The sales and purchases proposals are still manually published according to the information provided to the users (again, empowering users with information), but matching and transaction processes are automatically executed (including the ones linked to the DSO), sorting out the energy sales from the cheapest to the most expensive one and purchases in the reverse way.

Here, it is important to remark some important aspects related to the business models. First of all, the automatic auction is currently based on the average price where both the seller and the buyer benefit. The seller sells energy with a higher price than the offered by the DSO, while the buyer still buys energy with price below the DSO. Secondly, the manual validation is contrasted to the smart meter readings, which are currently not integrated in the system, but it will be done in the future steps to automatically detect malicious behaviours. Payment process is externalised in order not to create a token, so it is just energy power units trading.

4 System Overview

4.1 Selection of the Blockchain Technology

The selection of the blockchain technology has been made based on the described business models and the microgrid creation. It should be also considered the current Spanish legislation, which limits the creation of energy communities larger than 500 m [11]. This means that the possibility to set up energy trading mechanisms is reduced to local neighborhoods, where the players are well known.

Then, taking this into account, permissioned blockchain has been selected. Permissioned blockchains provide an additional level of security over typical blockchain systems like Bitcoin due to the fact that they require an access control layer. Then, only the administrator of the microgrid is able to accept new peers and, as stated above, the people taking part is perfectly known. Besides, it has to be considered that a physical connection is needed to allow the physical transmission of the energy from one user to another. Additionally, there are multiple benefits of using permissioned blockchains:

- Efficient performance: Permissioned blockchains offer better performance than permissionless ones.
- Proper governance structure: Permissioned networks do come with an appropriate governance structure.
- Cost-effective: Permissioned blockchains are more cost-effective when compared with the permissionless ones (e.g. proof of work algorithms (less computationally efficient) are not needed (a proof of authority algorithm is enough in this case).

As stated before, the payments are not performed within the blockchain, but with an external accounting software, that is not part of this paper.

According to the previous advantages, Hyperledger Besu[1] has been chosen in contrast to other possibilities such as Fabric or Quorum, among others. The consensus algorithm is PoA[2] (by using Clique). The reasons for selecting this technology are:

- It supports zero gas price transactions, then, no incurring extra costs.
- It supports private transactions to increase security.
- It supports Proof of Authority algorithm, reducing the gas usage.

4.2 Developed Smart Contract

Smart contracts provide the necessary means to design and implement fully automated contractual agreements among involved stakeholders. They mediate and monitor transactions, provide transparency and enforcement of contractual clauses. In this case, the smart contract has been developed using Solidity as programming language.

For multiple reasons, two versions of the smart contract have been developed. In one of them, the user is in charge of generate the transactions. The other version of the smart contract uses an automatic auction algorithm that makes it possible to automatically generate transactions between prosumers (as per Fig. 2). In both versions, both the seller and the purchaser have to validate the transaction: the purchaser has to validate that he/she has received the energy stated in the transaction, and the seller has to validate that he/she has generated and sent the amount of energy indicated in the transaction.

5 P2P Network Case Study Validation

5.1 Case Study Description

As aforementioned, in contrast to other researches, the validation has been made against a real, but controlled, environment, not just making use of simulated networks as Ganache. Thus, two scenarios have been deployed: (1) an initial test case with 4 Raspberry PI with the purpose of debugging Smart Contract and Hyperledger Besu configuration; (2) SYLTEC offices as real and controlled environment where nodes are the different instances of the offices, which are independently managed from the electrical perspective. Figure 3 shows the schema of the connected nodes, simulating a condominium where each office could be mirrored to a dwelling. In this configuration, the orange node is the one with PV plus batteries for storage, while the rest of nodes are just consumers. Each one of the rooms has its own energy meter, thus, independent consumption. The Besu network has been deployed in each one of the nodes, considering two bootnodes (or masters of the network). Moreover, to enable the transactions and creation of new blocks, CLIQUE protocol has been selected and, thus, 4 nodes are defined as signers (the two bootnodes that are by default and two additional ones to comply with more than half of the network).

[1] https://www.hyperledger.org/use/besu.
[2] Proof of Authority.

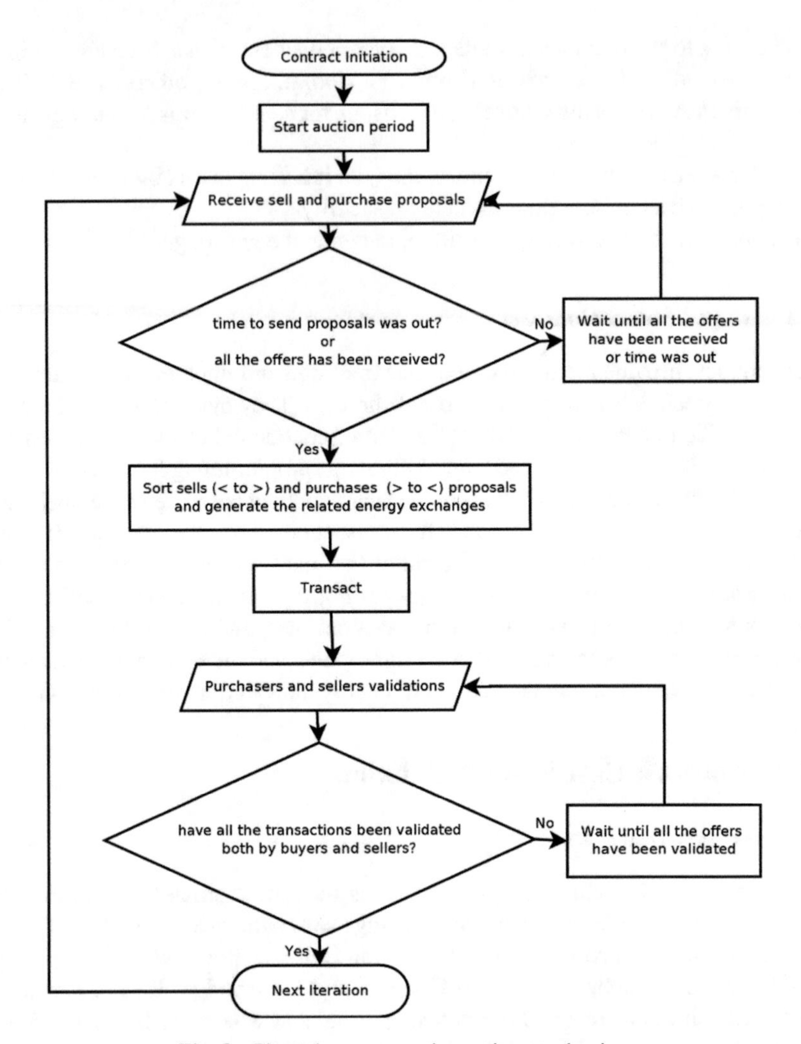

Fig. 2. Flow chart automatic auction mechanism

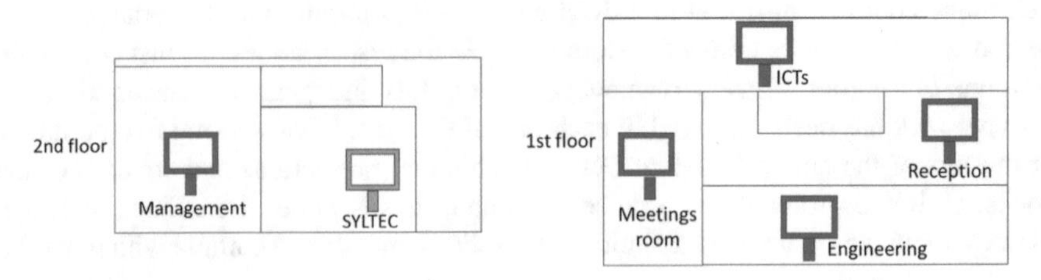

Fig. 3. SYLTEC demonstration environment

5.2 Demonstration and Validation

With the network and the Smart Contract deployed, the demonstration phase is characterized by the execution of the prediction algorithm on top of the blockchain network. The prediction algorithm determines the amount of energy from the batteries that will exceed from the forecasted demand. This is made within 24 h forecasted horizon and the user is informed about this energy. Then, the user has the possibility to "publish" this energy to be traded (note that the energy that is not traded and not used is directly dumped to the grid). For the demonstration, the user decides to sell this amount of energy. This is seller1, who determines 10 units of energy (kWh), to be sold at 0,07 €/kWh and expirationDate as a unix timestamp indicating 24 h after the publication. The output is a two-value vector indicating true as the sale has been created and the counter of total sales already published in the network.

From this sale, three buyers add the proposals of purchase with the unitsToBuy according to the predicted energy demand. As well, the price they are ready to pay and the expiration date as before. The summary of this data is depicted in Table 1.

Table 1. Validation tests for the energy trading

Who	Method	Input	Output
seller1	addSellProposal	unitsToSell: 10 unitEnergyPrice: 7 expirationDate: 1625129776	true (sale has been created) 1 (total amount of sales)
buyer1	addPurchaseProposal	unitsToBuy: 7 unitEnergyPrice: 8 expirationDate: 1625129776	true (purchase created) 1 (total amount of purchases)
buyer2	addPurchaseProposal	unitsToBuy: 4 unitEnergyPrice: 7 expirationDate: 1625129776	true (purchase created) 2 (total amount of purchases)
buyer3	addPurchaseProposal	unitsToBuy: 8 unitEnergyPrice: 6 expirationDate: 1625129776	true (purchase created) 3 (total amount of purchases)

For the validation, the "User guided" mechanism explained in Sect. 3 has been followed. Thus, the users are empowered to make the final decision. In this sense, the first transaction is the energy purchase from buyer1, with the corresponding address 0xBa84A879CB58AA0A93170C1b916b76871544B232, to the single seller, whose address is 0x7b937aaA1e2B4313d2ee8581A961146076869b99. As illustrated in Fig. 4, the event log on top demonstrates the trading has been performed. Here, the seller published energy at 0.07 €/kWh, while the buyer set a price of 0.08 €/kWh. However, within the manual trading, the users are informed to make the final decision. In this way, buyer is informed about the requested price of 0.07 €/kWh by the seller, therefore, the execution of the purchase is established to that price (see snapshot below) and 7 units of energy are bought to cover to whole demand of the buyer.

Next, it is the turn for buyer2, who is informed about the remaining energy from the seller, i.e. 3 units of energy from the original 10 subtracting the 7 aforementioned purchased. If buyer2 (address 0x8f28Dad50f69147254D37B46D9B49387042aF0be) is trying to buy additional units, e.g. 4, the transaction would not be executed by informing the user. In addition, the prediction is updated with real time data. Then, in this example, from the estimated 4 units of demand, the real energy requirement is just 2 units of energy (out of the original 4).

Last, there is one remaining unit for buyer3. However, in this version of the Smart Contract, if the purchase proposal contains a price that is lower than the final price, the transaction is not executed. The reason is to avoid malicious behaviors.

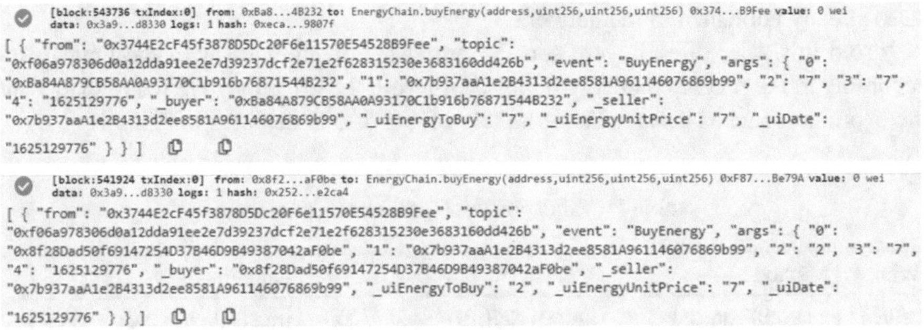

[block:543736 txIndex:0] from: 0xBa8...4B232 to: EnergyChain.buyEnergy(address,uint256,uint256,uint256) 0x374...B9Fee value: 0 wei
data: 0x3a9...d8330 logs: 1 hash: 0xeca...9807f
[{ "from": "0x3744E2cF45f3878D5Dc20F6e11570E54528B9Fee", "topic": "0xf06a978306d0a12dda91ee2e7d39237dcf2e71e2f628315230e3683160dd426b", "event": "BuyEnergy", "args": { "0": "0xBa84A879CB58AA0A93170C1b916b76871544B232", "1": "0x7b937aaA1e2B4313d2ee8581A961146076869b99", "2": "7", "3": "7", "4": "1625129776", "_buyer": "0xBa84A879CB58AA0A93170C1b916b76871544B232", "_seller": "0x7b937aaA1e2B4313d2ee8581A961146076869b99", "_uiEnergyToBuy": "7", "_uiEnergyUnitPrice": "7", "_uiDate": "1625129776" } }]

[block:541924 txIndex:0] from: 0x8f2...aF0be to: EnergyChain.buyEnergy(address,uint256,uint256,uint256) 0xF87...Be79A value: 0 wei
data: 0x3a9...d8330 logs: 1 hash: 0x252...e2ca4
[{ "from": "0x3744E2cF45f3878D5Dc20F6e11570E54528B9Fee", "topic": "0xf06a978306d0a12dda91ee2e7d39237dcf2e71e2f628315230e3683160dd426b", "event": "BuyEnergy", "args": { "0": "0x8f28Dad50f69147254D37B46D9B49387042aF0be", "1": "0x7b937aaA1e2B4313d2ee8581A961146076869b99", "2": "2", "3": "7", "4": "1625129776", "_buyer": "0x8f28Dad50f69147254D37B46D9B49387042aF0be", "_seller": "0x7b937aaA1e2B4313d2ee8581A961146076869b99", "_uiEnergyToBuy": "2", "_uiEnergyUnitPrice": "7", "_uiDate": "1625129776" } }]

Fig. 4. Transaction for the purchase confirmation

6 Conclusions and Future Work

This paper presents an approach for energy trading based on blockchain. Both the associated business model approach and the software that has been developed have been described. Besides, the selected blockchain technology has been detailed, and also the approach to the smart contract which is the core of the deployed system. The smart contract is the core of the solution, which allows the users to interact and exchange energy between prosumers.

The described system seeks to encourage the use of renewable energy sources, leading to prosumer proliferation and electricity cost reduction for consumers and this is achieved due to the fact that it makes easier the energy exchange between prosumers without the need of a central entity in charge of manage and authorize the transactions. Thanks to these P2P solutions, distributed energy generation and local energy communicates are fostered, contributing to flexibility, improved demand response capabilities and grid stress reduction.

Further studies and developments will focus on including payments and involving DSOs actively within the business models. At the moment, these payments are not integrated in the blockchain, where the possibility of tokenization the energy amounts, according to ERC20 standard, is a future work. Besides, other approaches to the auction mechanism (i.e. average price and considering the possibility of decimals of cents) would be tested to gather information about the best options in this sense. Finally, additional business models will be explored to allow multiple deployment options according to pre-agreements in the energy communities.

Acknowledgements. This research is being funded by "Instituto para la Competitividad Empresarial (ICE)" from Castille and Leon Community project entitled '€nergy Chain'.

References

1. Joint Research Centre (JRC): Distributed Power Generation in Europe: technical issues for further integration. https://ses.jrc.ec.europa.eu/publications/reports/distributed-power-generation-europe-technical-issues-further-integration. Accessed 23 Mar 2021
2. Pudjianto, D., Ramsay, C., Strbac, G.: Microgrids and virtual power plants: concepts to support the integration of distributed energy resources. Proc. Inst. Mech. Eng. Part A: J. Power Energy **222**(7), 731–741 (2008). https://doi.org/10.1243/09576509JPE556
3. World Economic Forum: Stanford Woods Institute for the Environment, PwC. Building Block(chain)s for a Better Planet. http://www3.weforum.org/docs/WEF_Building-Blockchains.pdf. Accessed 15 Mar 2021
4. Schleicher-Tappeser, R.: How renewables will change electricity markets in the next five years. Energy Policy **48**, 64–75 (2012). https://doi.org/10.1016/j.enpol.2012.04.042. ISSN 0301-4215
5. van Leeuwen, G., AlSkaif, T., Gibescu, M., van Sark, W.: An integrated blockchain-based energy management platform with bilateral trading for microgrid communities. Appl. Energy **263**, 114613 (2020). https://doi.org/10.1016/j.apenergy.2020.114613. ISSN 0306-2619
6. Mengelkamp, E., Gärttner, J., et al.: Designing microgrid energy markets: a case study: The Brooklyn Microgrid. Appl. Energy **210**, 870–880 (2018). https://doi.org/10.1016/j.apenergy.2017.06.054. ISSN 0306-2619
7. Lin, J., Pipattanasomporn, M., Rahman, S.: Comparative analysis of auction mechanisms and bidding strategies for P2P solar transactive energy markets. Appl. Energy **255**, 113687 (2019). https://doi.org/10.1016/j.apenergy.2019.113687. ISSN 0306-2619
8. Zhou, Y., Wu, J., Song, G., Long, C.: Framework design and optimal bidding strategy for ancillary service provision from a peer-to-peer energy trading community. Appl. Energy **278**, 115671 (2020). https://doi.org/10.1016/j.apenergy.2020.115671. ISSN 0306-2619
9. Han, D., Zhang, C., Ping, J., Yan, Z.: Smart contract architecture for decentralized energy trading and management based on blockchains. Energy **199**, 117417 (2020). https://doi.org/10.1016/j.energy.2020.117417. ISSN 0360-5442
10. Xu, X., et al.: The blockchain as a software connector. In: 2016 13th Working IEEE/IFIP Conference on Software Architecture (WICSA), pp. 182–191 (2016). https://doi.org/10.1109/WICSA.2016.21
11. Ecological Transition Spanish Ministry: Real Decreto 244/2019: Administrative, technical and economic conditions for electricity self-consumption. BOE-A-2019-5089, April 2019 (2019)

Protection Against Online Fraud Using Blockchain

Raphael Burkert, Jonas van Hagen, Maximilian Wehrmann, and Marc Jansen(✉)

Institute of Computer Science, Hochschule Ruhr West, Bottrop, Germany
{raphael.burkert1,jonas.van-hagen, maximilian.wehrmann}@stud.hs-ruhrwest.de, marc.jansen@hs-ruhrwest.de

Abstract. In this paper we present an application that, by using the blockchain, tries to inform the user about possible fraud attempts on the internet. The application presented is intended to prevent the user from landing on websites that have recently been the target of various attacks. Installed as a browser plugin it should alert the user upon visiting such a website. Alerts are triggered once a generated hash of the website no longer matches a hash that was stored on the blockchain. Included inside the hash is not only the HTML but also further website information such as used JavaScript and Cascading Style Sheets. By implementing a package into the build process the developer can easily publish the website hashes that are then stored on the blockchain for comparison. To see if our plugin can withstand common website attacks we compared it with the OWASP top 10 risks and check if an attack is mitigated or completely solved. We came to the conclusion that attacks directly targeting the front end of the website can be mitigated. Not included are attack vectors that directly target web servers and the data stored on them.

1 Introduction

Modern society is increasingly linked by the Internet. This linkage has simplified the lives of most people, but comes with its own dangers and problems. So-called cybercrime has risen steadily in recent years [1]. Here, in particular, the so-called computer fraud has been committed more frequently, in which the aim is "to obtain an illegal pecuniary advantage for oneself or a third party" [2]. However, not only this type of cybercrime has increased, but also spying and data extraction have resulted in more cases than in previous years. Therefore, it has become increasingly important to protect oneself properly on the Internet and to exercise an appropriate degree of caution. An active problem are manipulated websites, which pretend to fulfill a service but actually only aim to obtain the user's data to enrich themselves. Due to the increasingly sophisticated methods of fraud on the Internet, the following question arises:

Can websites and web applications displayed in the browser be made safer through the use of decentralized data storage?

J. Prieto et al. (Eds.): BLOCKCHAIN 2021, LNNS 320, pp. 34–43, 2022.
https://doi.org/10.1007/978-3-030-86162-9_4

We deal with a possible kind of protection against such manipulated websites in this paper. We present a plugin designed for the browser, which can be installed by the user and then alerts the user to problems with manipulated websites upon visiting them. The websites visited by the user are checked by the plugin against hashes generated directly in the browser, which are then cross-checked against hashes stored on the blockchain [3]. The generated hash consists not only of Hypertext Markup Language (HTML), but also of used JavaScript files, Cascading Style Sheets (CSS) and other statically linked elements. This should allow us to detect changes not only directly on the website, but also in used, perhaps manipulated, static resources of the website. In addition, the developer's side of the website will also be elaborated in this context to provide insight into both the advantages and disadvantages of our approach.

2 Related Work and Background

The most commonly used protocol for secure connection between server and client today is "HTTP over TLS". This protocol is used to encrypt the data exchanged between the server and the client. The TLS certificates used for this purpose serve, among other things, to authenticate the server, but do not offer the possibility of checking the integrity of the data sent by the server [4].

As already shown by Kalis and Belloum, a blockchain can be used to validate data. For this purpose, entries are created in a smart contract that consist of an identifier and a hash. Thus, a value can be stored on the blockchain for each file, which can then be validated by a third party [5].

In order to detect phishing for the user, the "AntiPhish" plugin was developed by Kirda and Kruegel in 2005. The presented approach is based on the functionality of auto-filling credentials included in many browsers. The plugin stores not only the credentials, but also in connection with them the domain where they were used. If a phishing attack redirects the user to an identical-looking website with a different domain, the plugin prevents the credentials from being filled in [6].

In another approach by Ross et al. the plugin "PwdHash" is presented. The plugin presented aims to prevent the theft of the real password for a website by having the plugin create a domain-specific password when an account is created on that site. This password is a hash calculated from the password entered by the user and the domain of the visited site. Thus, in the case that the user uses his password on a phishing site, the plugin generates a different hash than on the real site, consequently the transmitted hash cannot be used by the attackers as a password on the real site [7].

3 Architecture

For this paper, an approach was developed to address the problem of verifying the integrity of files served by a web server. For this purpose, a browser plugin and a command line tool were prototyped. To establish trust in the data used by the plugin, the blockchain is used as a secure public database. This is a community driven process to enable developers to publish website hashes, intentionally not creating a generalized whitelist managed by a central entity (Fig. 1).

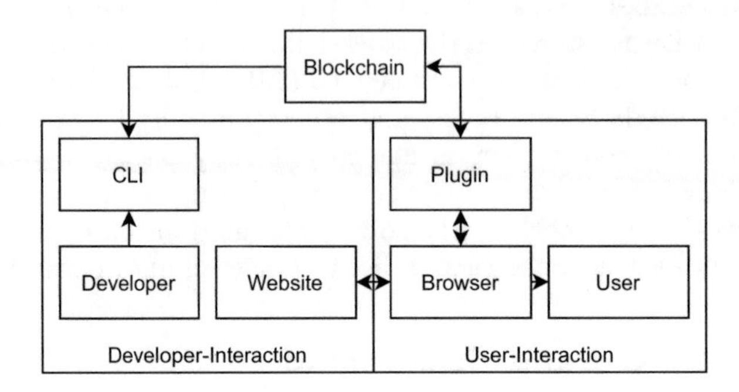

Fig. 1. Architecture overview

3.1 Browser Plugin

While designing the plugin, we did not plan any user interaction for the setup, it works right after installation and does not need to be configured. Since the plugin should support any website loaded in the browser, the flow of the plugin was standardized. It is composed of four individual, equally important, parts. When a page is loaded, the plugin's content script is executed for that page. The first task of the plugin is to look for all the required information and files of the page and pass them in a compact way to the subsequent tasks. In the next step, all the files of the website are individually loaded asynchronously from the server and hashed. While this is being executed, the plugin loads the JSON text transferred to the blockchain by the CLI tool, which is associated with the URL of the page. Now a comparison of the hashes generated by the plugin and loaded by the blockchain is pending. This is followed by feedback to the user of the plugin. The workflow is shown in Fig. 2 for clarification.

During the design of the presented workflow, special attention was paid to the speed and flexibility of the plugin, as well as to the variety of information for the user. In order to maintain a reasonable speed when using the plugin, as many steps as possible were parallelized. This was especially important with regard to loading and hashing the files of the website, since a website can contain any number of JavaScript and CSS files. By downloading and hashing the files in parallel, time is saved. A high flexibility in the plugin was important for us because especially in the area of web many innovations emerge, especially with regard to security and exploits one can never be sure to protect all possible corners. For this reason, a lot of attention was paid to the modularity of the plugin as well as the CLI, this should allow us to quickly respond to changes and new files that need to be hashed. The conceived modularity also helps us to increase the information variety for the user, as individual aspects and problems, which may

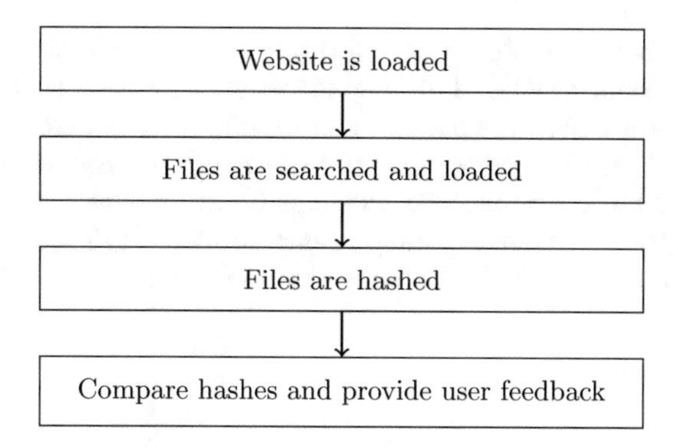

Fig. 2. Browser plugin workflow

be important for the user, can be quickly worked out and presented. An example of this would be the discovery of a new exploit in previously unrecorded files, which could then be hashed relatively quickly and thus appropriately recorded after only a minor update for both the CLI tool and the plugin.

3.2 CLI Tool and CI/CD

The command line tool is intended for use by the owner of a website and is available as a node package, so it can be used standalone as well as in a build process to create or update the entries on the blockchain. Through this tool, integration into a CI/CD pipeline is also possible. With this tool it should be possible for the owner of a website to register a valid hash that can then be checked by the browser plugin upon visits from the user. In case of a fraudulent website the hash would no longer match the one given by the owner and as such the user would be alerted by the browser plugin.

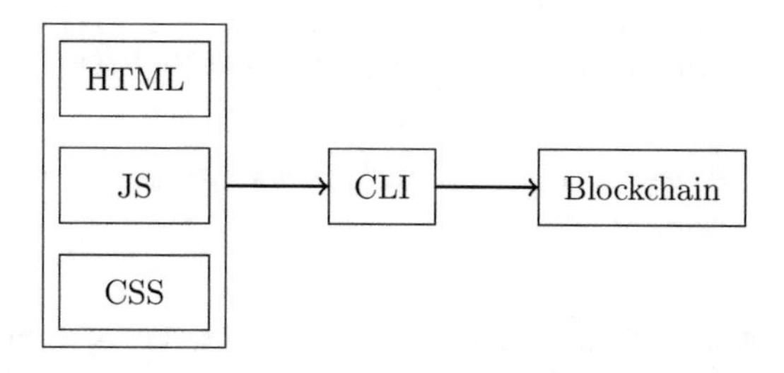

Fig. 3. Workflow of the CLI tool

As shown schematically in Fig. 3, the tool reads the files required for the webapp from the build directory. In the first step, the HTML files are loaded and the DOM is created. In the next step, the dependencies of the webapp to JavaScript, stylesheets and other statically included resources can be determined from the loaded DOM. This step is repeated for each HTML file until all dependencies have been determined. The hashing of the detected files is then performed based on the steps described in Sect. 4.1. As the hasing is generally quite fast it can be integrated into a default build pipeline. Intrusions or other malicious changes that trigger a plugin reaction by changing the hash can thus be eliminated quickly by just rebuilding the application.

4 Implementation

Selected points of the architecture described in the previous section, which are essential for our approach, will be discussed in the following.

4.1 Data Hashing and Storage

A central role in the approach presented in this work is the hashing of the server's files, which are then checked by the plugin during operation. In order to achieve both the desired modularity and information diversity, the structure was implemented as exemplified in Fig. 4. In order to reduce the number of entries stored on the blockchain for each supported website by the plugin, the storage of individual file hashes was intentionally omitted.

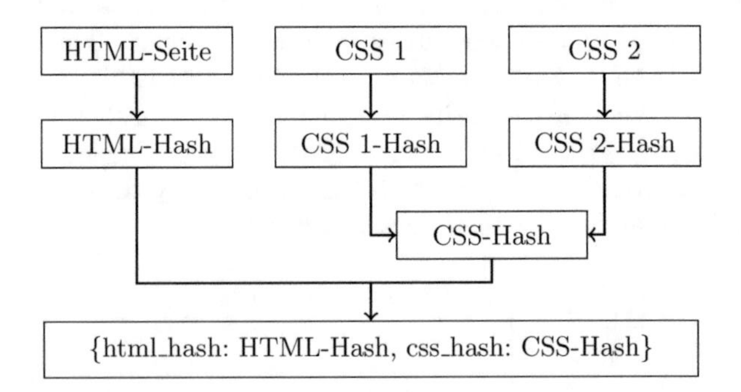

Fig. 4. Example overview of how individual hashes are merged and then stored as JSON on the blockchain.

In the first step, all files used statically by the website, such as JavaScript, Cascading Style Sheets, and images or media elements, are detected and individual file hashes are created. If multiple hashes of a file type exist, the hashes are concatenated and a new hash is generated from them. In this way, a hash is formed for each of the groupings. All these hashes are concatenated and hashed again in the next step to form the so-called "root hash", this is to be compared by the plugin as the first hash and can avoid unnecessary comparisons. If all hashes are available, they are transferred to the smart account as JavaScript object notation (JSON). Splitting the hashes by file type has the advantage that the plugin can superficially sort changes to files into threat levels based on these groupings. The described process is executed every time a website is opened by the browser plugin and the newly generated hashes are compared with those stored on the blockchain. For the hashing algorithm, the quasi-standard in blockchain, SHA-256, was chosen. Also playing a role here were the works of Wang and Yu [8] and Yie et al. [9], both of which demonstrated that the widely used MD5 algorithm is not collision-free.

4.2 Smart Contract

The data transferred from the CLI tool to the blockchain is processed by a smart contract. When data for a website is transmitted for the first time, the smart contract generates entries for the ownership assignment in addition to the entries for the transmitted data. These entries ensure that only the account that originally transmitted the entries can update them.

Algorithm 1. Smart Contract

```
procedure CREATEHASHENTRY(fileUrl, fileHash)
    hash ← getCurrentHash(fileUrl)
    if hash does not exist then
        addHashEntry(fileUrl, fileHash)
        addHashOwner(fileUrl, caller)
    else
        owner ← owner(fileUrl)
        if owner is caller then
            updateHashEntry(fileUrl, fileHash)
        end if
    end if
end procedure
```

5 Evaluation

For the evaluation of our project we decided to run different attack scenarios and examine how the use of our plugin can protect the user from the loss of their sensitive data. The different scenarios that will be evaluated are from the top ten list [10] published by OWASP. This list contains the top attack scenarios that a website should protect itself against. Since our plugin is designed for possible user related attack scenarios, this may mean that some of the items listed cannot be covered by our project. After a possible attack by such a scenario, it may still be possible for our plugin to prevent further consequences and problems for the user.

Table 1. OWASP top ten risks vs. plugin. *for details see related section

Problem	Mitigated	Solved
Injection	No	No
Broken authentication	Partial*	No
Sensitive data exposure	No	No
XML external entities	No	No
Broken access control	No	No
Security misconfiguration	Yes	Partial*
Cross-site scripting	Yes	Partial*
Insecure deserialization	Yes	Partial*
Using components with known vulnerabilities	No	No
Insufficient logging and monitoring	No	No

Injection. Injection attacks, such as SQL, NoSQL, OS, and LDAP injection, occur when untrusted data is sent to a database interpreter as part of a command or query. The attacker's hostile data can trick the interpreter into executing unintended commands or accessing data without proper authorization. Our plugin does not protect against this type of attack, since the data sent by the database in response is neither captured nor checked.

Broken Authentication. Application features related to authentication and session management are often implemented incorrectly, allowing attackers to compromise passwords, keys, or session tokens, or exploit other implementation flaws to temporarily or permanently assume the identity of other users. Again, our plugin does not intervene here, as incorrectly implemented authentication procedures remain in place. Only when referring to faulty authentication pages, in the case of an O-Auth procedure, our plugin could give the user the possibility to recognize that there is a problem, granted that the user is redirected towards the login-landing page that was hijacked.

Sensitive Data Exposure. Many web applications and APIs do not properly protect sensitive data such as financial, health, and personal information. Attackers can steal or modify such weakly protected data to conduct credit card fraud, identity theft, or other crimes. Sensitive data can be compromised without additional protection, such as encryption in memory or in transit, and requires special precautions when shared with the browser. Our approach does not provide any additional protection.

XML External Entities. Many older or poorly configured XML processors evaluate external entity references within XML documents. External entities can be used to expose internal files via the file URI handler, internal file shares, internal port scanning, remote code execution, and denial of service attacks. Since this is a server-side problem, our approach does not provide any additional protection.

Broken Access Control. Restrictions on what authenticated users are allowed to do are often not properly enforced. Attackers can exploit these vulnerabilities to access unauthorized functions and/or data, such as accessing other users' accounts, viewing sensitive files, modifying other users' data, changing access rights, etc. Since this problem is an implementation weakness of the website or the backend, our approach cannot provide security to the user.

Security Misconfiguration. Security misconfigurations are the most common problem. This is often a result of insecure default configurations, incomplete or ad-hoc configurations, open cloud storage, misconfigured HTTP headers, and cumbersome error messages containing sensitive information. All operating systems, frameworks, libraries, and applications must not only be securely configured, but also patched/updated in a timely manner. If there is any hijacking of the website due to these errors, our plugin can suitably protect the user in case malicious scripts or resources are injected.

Cross-Site Scripting (XSS). XSS errors occur whenever an application inserts untrusted data into a new website without proper validation or escaping, or updates an existing website with user-supplied data using a browser API that can create HTML or JavaScript. XSS allows attackers to execute scripts in the victim's browser that can hijack user sessions, deface websites, or redirect the user to malicious websites. If the XSS attack is heavy-handed and the attacker manages to embed JavaScripts one website, the plugin can warn about it upon visiting the page.

Insecure Deserialization. Insecure deserialization often leads to remote code execution. Even if deserialization errors do not lead to remote code execution, they can be used for attacks such as replay attacks, injection attacks, and privilege escalation attacks. Similar to *Security Misconfiguration*, if the website is hijacked as a result of the described weakness, the user can be protected on the next visit should the website be modified during the attack. The operator's backend is not protected.

Using Components with Known Vulnerabilities. Components, such as libraries, frameworks, and other software modules, run with the same privileges as the application. If a vulnerable component is exploited, such an attack can allow serious data loss or server takeover. Applications and APIs that use components with known vulnerabilities can subvert the application's defenses and enable various attacks and impacts. Since this usually involves an attack on internal data of the website, the plugin cannot actively protect against this issue.

Insufficient Logging and Monitoring. Inadequate logging and monitoring, coupled with lack of or ineffective integration with incident response, allows attackers to attack additional systems, maintain persistence, move on to additional systems, and manipulate, extract, or destroy data. Most security breach studies show that the time to discovery of a security breach is more than 200 days and is usually discovered by external parties rather than internal processes or monitoring. Since this is an internal problem of the website, the presented plugin cannot provide security to the user.

6 Conclusion and Future Work

For this paper, we successfully implemented the approach of a plugin for verifying the integrity of websites using decentralized data storage. However, the architecture presented here is still expandable, as frameworks that generate the website dynamically on client request are not supported by our CLI tool. These include e.g. *ASP.NET* and *Spring*, which are in the top 10 *Most Popular Technologies* in the Web Frameworks space according to StackOverflow's 2020 developer survey [11].

As shown in Table 1, the approach presented in this paper can mitigate and partially solve three of the presented risks. The risk "Broken Authentication" can also be partially mitigated. In total, four out of the ten risks can be mitigated or partially solved. While not overly great this needs to be taken with a grain of salt based on the fact that the risks presented by OWASP tend to be more about weaknesses in the code or backend of the website.

Another important point for future work on the approach is the verification of the checksums transmitted to the smart account. This was not addressed in detail in the approach presented, as it was not part of the objective of the work, but it is an important aspect, as the source of the transmitted checksums must be trusted to have transmitted the correct values. Since an automatic verification of the transmitted values within a smart contract is not possible, a community voting procedure could be used.

Another extension of the presented prototype would be an addition to the plugin. When visiting a manipulated website, the user is notified of changed files on the page with the help of an alert. In addition, a red exclamation mark appears next to the plugin's icon. If these measures are not sufficient, because the user does not perceive them properly, it would be conceivable to prevent the user from entering his username and password in the corresponding input fields of the website. But this would be a last step measure considering that it is a rather big intrusion and needs to always be handled with care.

As already shown in the evaluation, the goal of the plugin is to protect the user from possible attacks, as long as these change something on the website.

With Perl.com we can give a recent example where the presented approach could have warned users about a potential change to the website. In this case Perl.com was victim to domain hijacking [12]. If the hackers now modified resources used by the website, such as scripts, and it was secured by the actual owners using the presented approach, visitors would be immediately alerted to these changes. In this case, the previously described expansion to prevent the input of login information could have been helpful as well.

References

1. 2019 internet crime report. https://www.ic3.gov/Media/PDF/AnnualReport/2019_IC3Report.pdf. Accessed 03 Mar 2021
2. German criminal code section 263a computer fraud. https://www.gesetze-im-internet.de/englisch_stgb/englisch_stgb.html#p2385. Accessed 12 May 2020
3. Nakamoto, S.: Bitcoin: A peer-to-peer electronic cash system (2008). https://bitcoin.org/bitcoin.pdf. Accessed 12 Oct 2020
4. Felt, A.P., Barnes, R., King, A., Palmer, C., Bentzel, C., Tabriz, P.: Measuring https adoption on the web. In: Proceedings of the 26th USENIX Conference on Security Symposium, Ser. SEC 2017, , pp. 1323–1338. USENIX Association, Vancouver (2017). ISBN 9781931971409
5. Kalis, R., Belloum, A.: Validating data integrity with blockchain. In: 2018 IEEE International Conference on Cloud Computing Technology and Science (CloudCom), pp. 272–277 (2018). https://doi.org/10.1109/CloudCom2018.2018.00060
6. Kirda, E., Kruegel, C.: Protecting users against phishing attacks with antiphish. In: 29th Annual International Computer Software and Applications Conference (COMPSAC 2005), vol. 1, 2, pp. 517–524 (2005). https://doi.org/10.1109/COMPSAC.2005.126
7. Ross, B., Jackson, C., Miyake, N., Boneh, D., Mitchell, J.C.: Stronger password authentication using browser extensions (2005). https://crypto.stanford.edu/PwdHash/pwdhash.pdf. Accessed 04 Feb 2021
8. Wang, X., Yu, H.: How to break MD5 and other hash functions. In: Cramer, R. (ed.) Advances in Cryptology - EUROCRYPT 2005, pp. 19–35. Springer, Heidelberg (2005). https://doi.org/10.1007/11426639_2
9. Xie, T., Liu, F., Feng, D.: Fast collision attack on MD5. Cryptology ePrint Archive, Report 2013/170 (2013). https://eprint.iacr.org/2013/170. Accessed 03 Mar 2021
10. Owasp top ten. https://owasp.org/www-project-top-ten/. Accessed 18 Feb 2021
11. Stack overow developer survey 2020. https://insights.stackoverflow.com/survey/2020#technology-web-frameworks-professional-developers2. Accessed 02 Oct 2021
12. biran d foy, The hijacking of perl.com. https://www.perl.com/article/the-hijacking-of-perl-com/. Accessed 03 Mar 2021

Decentralized Online Multiplayer Game Based on Blockchains

Raphael Burkert, Philipp Horwat, Rico Lütsch, Natalie Roth, Dennis Stamm, Fabian Stamm, Jan Vogt, and Marc Jansen(✉)

Institute of Computer Science, Hochschule Ruhr West, Bottrop, Germany
marc.jansen@hs-ruhrwest.de

Abstract. Decentralized apps offer users significant advantages. One main advantage is that the content belongs to the users. Although decentralized apps have been developed, decentralized multiplayer games have still been a challenge. This research work is therefore based on the question: *Is it possible to create a turn-based real-time online multiplayer game based on decentralized applications on blockchain targeting the mobile market?* This question is answered by the successful development of a decentralized online multiplayer game based on a blockchain.

Keywords: dApp · Blockchain · Multiplayer · Gaming · Waves

1 Introduction

The gaming industry is towering over the global box office industry and music industry in terms of revenue, making it by far the most lucrative entertainment industry. In 2018 the global box office for films totaled US$41.7B in revenue, while global music revenues reached US$19.1B. In comparison the global game market alone generated US$152.1B in 2019 [1]. Out of all the segments that make up the gaming market the mobile game segment is by far the biggest, reaching a revenue of US$65.8B in 2019, thus being more than 43% of total revenue [2]. Reason being the huge availability of smartphones, reaching 3.2B worldwide smartphone users in 2019 and the accessibility of smartphone games mostly targeting casual gamers [3]. Additionally online multiplayer games are highly demanded by gamers and as such many developers are focusing on satisfying this demand. Though when it comes to multiplayer games the issue of connecting players arises. One solution to this problem is to decentralize servers using the blockchain. This comes with some major benefits such as cheaper server hosting and more security and privacy regarding user data [4]. All these aspects lead to the research question of this paper.

Is it possible to create a turn-based real-time online multiplayer game based on decentralized applications on blockchain targeting the mobile market?

J. Prieto et al. (Eds.): BLOCKCHAIN 2021, LNNS 320, pp. 44–53, 2022.
https://doi.org/10.1007/978-3-030-86162-9_5

This question is answered by creating a turn based multiplayer game based on the blockchain platform Waves as an easy to use, yet powerful enough technology. The steps necessary for achieving this dictate the structure of this paper which will be briefly described in the following. Firstly related work mainly focusing on blockchain and games based on it is being looked at. This is followed by the architecture and implementation of the proposed blockchain game. The last major chapter deals with the evaluation of the prior described implementation, which in closing leads to the conclusion and future work.

2 Related Work

The blockchain technology offers some advantages in the entertainment industry and especially in game development. The main advantage is that the blockchain technology "Offer[s] better controls for online players over virtual assets and allows them to use these assets across different gaming platforms" [5]. The disadvantages are also decisive for game development. In a blockchain random algorithms are difficult to program and transactions have to happen in a defined order, which makes real-time gaming basically impossible. Furthermore, smart contracts cost real money [6].

A good overview of current blockchain games is given in the study by Min et al. [7]. The authors created categories and assigned games to them based on four advantages of a blockchain for game development. The first category is "Rule Transparency" which mainly includes games of chance such as Satoshi Dice. The second category is "Asset Ownership", these are mainly collection games like CryptoKitties. The third category is "Asset Reusability" but this is not yet mature due to the relatively few games and the fourth is "User-Generated Content", where users can provide their own content via blockchain without commercial middlemen. In the paper by Du et al. [8], a distinction is made between two categories. One is the completely decentralized games, whereby these usually contain only simple game logic and low complexity. An example of this would be the aforementioned collection game CryptoKitties. The second category, a combination of the traditional game process with the blockchain technology, where part of the game process is executed in the blockchain. The paper describes the development of a poker game using the second category. The development of a completely decentralized app could not really be realized so far. The game CryptoKitties, for example, is equipped with a centralized server that stores special identifiers for cats or takes over administrative functions [6].

3 Architecture

For this paper a turn-based game inspired by the in 1980 released "Battlezone" was created. As a modern interpretation of this classic game we decided to integrate Augmented Reality (AR) aspects into our approach. Using AR our game does not depend on a generated game world and instead uses the GPS coordinates of the players making the entire globe the playing ground. In this chapter we will first explain how the game flow works and following that the required components for the game will be outlined.

3.1 Game Flow

Figure 1 shows the game sequence. First the player has to log in to his blockchain account. Since the game uses its own token instead of the blockchain currency, the player must first obtain a certain amount of it. To start the game the stake is transferred to the dApp and the player is placed in a queue. When another player is searching for a game, a new game instance is created with the searching player and the player from the queue. The player from the queue is the first in turn. After that the players take turns until a hit is scored and the game is finished. At the end of the game, the paid wager of both players is paid out to the winner.

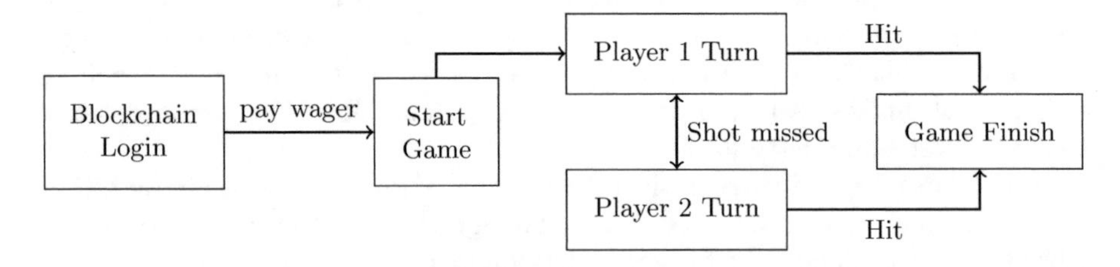

Fig. 1. Game flow

3.2 Components

The components required for the game are shown in Fig. 2. The game is provided to the player as a website, which is hosted on a web server and displayed in a browser on the device. Because of this the application is not completely decentralized. This could be solved through IPFS using a public gateway as access point [9]. The most important component is the dApp, which is hosted in the Waves network. Here the game logic is implemented and executed as well. This includes functions like the calculation of the impact position or the matchmaking. The information about whose turn it is and who the match winner is, if there already is one, is also stored here.

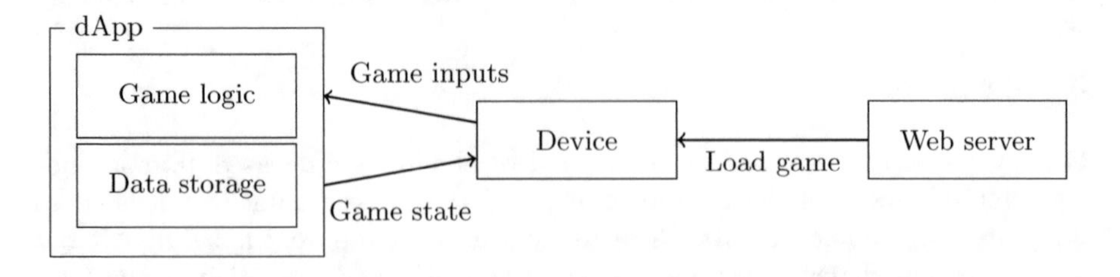

Fig. 2. Data flow between architecture components

4 Implementation

For the realization of the game flow, the most widely used web framework React.JS is used to provide a state of the art application [10]. Waves is used as blockchain technology. Waves is an open blockchain protocol and a development toolset for decentralized Web 3.0 solutions. The implementation provides an overview of basic React and Waves functionality and provides insight into vulnerabilities and hurdles identified during the development of the application.

4.1 ReactApp

The game sequence was implemented with a client-side JavaScript Web Application, which was build using React version 16.13.1, that can be accessed via a browser. The application is only available on mobile devices, so the built-in sensors, that are necessary for the game to run, can be used. The game follows an augmented reality approach, in which 3D-objects are fixed in the users field of vision and the environment is visible on the device in real-time. This is done by using the landscape camera of the device in combination with the JavaScript tool Aframe in version 1.0.4. Aframe is a framework for rendering 3D-objects on web browsers [11]. The augmented reality effect is enhanced through animated 3D-objects, which are projected into the environment and disappear after a few seconds.

Furthermore sensors are used to determine the players geographic location and thus the location of the device at the beginning of the game. Additionally the gyroscope is utilized to determine the angle and direction for the shot. By means of an integrated compass, the players receive an indication in which direction the shot is aimed, furthermore the compass includes a rough indication of the direction to the location of the opponent. To communicate with the blockchain, Signer was used, a framework developed by the Waves Community to simplify sign-in and communication with the blockchain. A more detailed description of which data is stored on the blockchain (smart account) and which functions (callables) are executed using Signers is given in Sect. 4.2. The stored data in the smart account is necessary to manage the course of the game. Changes on the smart account are retrieved and processed every 5 s using the Waves node API on the client side. During processing, a check is made whether an opponent has been found or if the player is still waiting for an opponent to join. Within the course of the game, the system checks which players turn it is and whether the game has been decided. If it is a player's turn, he is given the opportunity to adjust the direction, angle and force of the shot. The player whose turn it is not in turn receives feedback that he is waiting for the opponent to shoot. Once the game is decided, the game is ended and the players receive information about the game progress. Figure 3 offers a section of screens from the realized game.

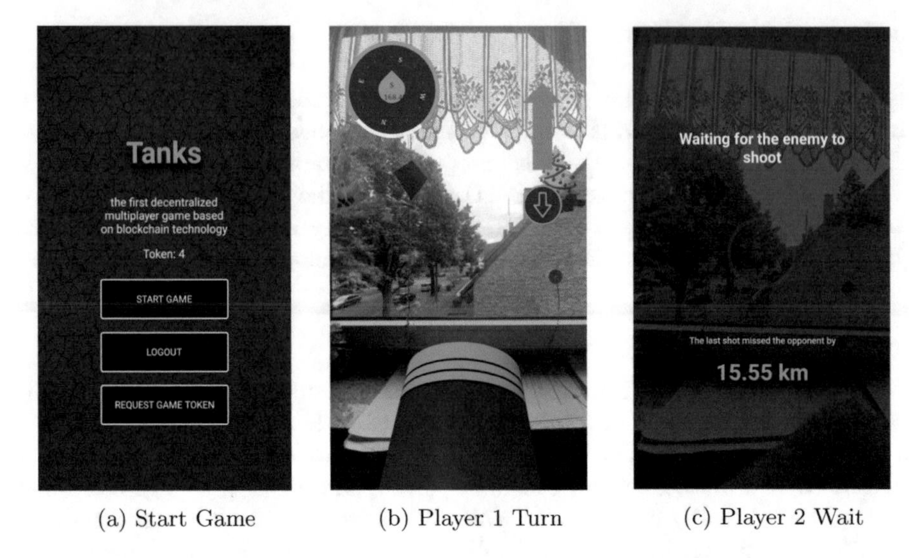

(a) Start Game (b) Player 1 Turn (c) Player 2 Wait

Fig. 3. Representation of the decentralized game based on react

4.2 Waves

Since most of the logic of the game is executed on the blockchain, the input from the players must be passed to the blockchain at the end of each turn. This transfer is done using callables, code functions that are directly executed on the blockchain. Only two callables are needed for the game we created. The first callable contains the logic for creating a game and accepting the wager. It is implemented in such a way that there is always a maximum of one player in the queue. In the second callable the input of the player is processed and it is determined whether the player has hit his opponent. Depending on the results of these calculations the game is ended and the winner is paid his winnings or it is the opponent's turn.

Limitation of Script Complexity. In Ride, to limit the execution time of account scripts, there is a value called complexity. Every line of code and every function call adds to this complexity. The complexity of the callables in a dApp script is limited to 4000 [12]. The final dApp Script of the presented approach has a complexity of 2584, what is below said limit. This was only achieved by optimizing the script with regards to script complexity, while reducing script understandability. However, games with more complex game mechanics would not be possible on the Waves blockchain because of this limitation. In the future this limitation could be bypassed by splitting code execution across multiple script calls.

Other blockchain networks like Ethereum do not have a limit of script complexity. Instead, the transaction fee paid for every script execution increases with the length of the executed code [13]. Developers have to make a trade-off between the complexity of the game mechanics and the transaction fees players are willing to pay.

Turing Completeness. The script language Ride does not contain while- or goto- operations, recursion, and for- operations are limited to a predefined number of rounds. Therefore this language is not Turing-complete. However, most smart contracts do not require Turing completeness [14], and the implemented functionalities did not require it either.

Trigonometric Functions. Ride does not - at least in Version 4 - support trigonometric functions like sin, cos and tan, so an approximation of them had to be implemented. However these could not be too complex because of the script complexity limitations. In following the approximations used for the trigonometric functions are described:

sin For the sin approximation, the formula invented by Bhaskara I. is implemented. For radians, it is stated in Eq. (1).

$$sin(x) \approx \frac{16x(\pi - x)}{5\pi^2 - 4x(\pi - x)} \tag{1}$$

In the utilized range of -3.15 to 4.72 the approximation has a maximum discrepancy of 0.00163.

cos The cosine function was not approximated itself but was implemented as $cos(x) = sin(\frac{\pi}{2} - x)$.

asin The arcus sinus was implemented as stated in [15]. This was also recommended as reference implementation for shader code by NVIDIA [16].

$$\begin{aligned} asin(x) &\approx sign(x) \cdot (\frac{\pi}{2} - \sqrt{1 - |x|} \cdot p(x)) \\ p(x) &= ((-0.01873 \cdot |x| + 0.07426) \cdot |x| - 0.21211) \cdot |x| + 1.57072 \end{aligned} \tag{2}$$

Inside the definition range from -1 to 1 this approximation has a maximum discrepancy of $9 \cdot 10^{-5}$. For values of x outside the definition range it was assumed that those were rounding errors and the values of $asin(-1)$ or $asin(1)$ were returned.

atan An approximation for arcus tangent was defined as in Eq. (3).

$$atan(x) \approx \begin{cases} -\frac{\pi}{2} - \frac{x}{x^2+0.28} & \text{, if } x < -1 \\ \frac{\pi}{2} - \frac{x}{x^2+0.28} & \text{, if } x > 1 \\ \frac{x}{1+0.28 \cdot x^2} & \text{, otherwise} \end{cases} \tag{3}$$

For any x, the approximation discrepancy does not exceed 0.00489.

atan2 The definition of atan2 uses atan, so this function is implemented as defined and the approximation takes place in atan.

5 Evaluation

Online multiplayer games have many different requirements compared to locally played single-player games especially when it comes to player connection and data storage. In the following sections we evaluated cheat safety in regards to stored data and real-time capabilities of our dApp game.

5.1 Cheat Safety

Because the implemented game is an online multiplayer game where two opponents battle each other about money, special attention has to be paid on the topic cheat safety.

One major strategy to avoid cheating is to implement the game mechanics server-side and to send the inputs directly to the server. Any calculation that is done client-side can be manipulated by cheat software. In contrast, manipulating the code behavior on the server-side is a lot harder. In this case, the server-side is the smart account with the Ride script attached to it. The function to execute a shot requires the aiming pitch and yaw angles as well as the shot force and calculates impact position, distance to the opponent and whether the opponent is hit completely on its own.

To calculate the distance to the opponent, the smart account has to know the exact position of the opponent. In contrast, the player himself must not know the opponents position. Otherwise he could calculate the perfect aiming direction and force and instantly hit his opponent. The problem is that everything the smart account stores can be accessed, even the player positions.

One approach solving this problem is to store the player positions encrypted and to only decrypt them when needed. Unfortunately the Ride scripts executed on a smart account do not have access to the smart account's private key or any other kind of secret [17]. This makes sense because the script execution is part of calculating new blocks on the blockchain, and this should be able to be done by any miner connected to the blockchain. For any miner to have access to the private key, the private key had to be published.

A technique to overcome this is the commit-reveal-scheme. It contains two phases: In the commit phase, every participant uploads a hash of his choice, so no other participant can see his choice during this phase. After every participant uploaded the hash, the reveal phase begins and every participant uploads his choice. The blockchain can check if this choice was not changed by hashing it and comparing to the hash uploaded in the commit phase [18]. After both phases, the results are determined without any participant taking other participants choice into account for his decision.

This technique might be good for elections or games that only take one round, but is not applicable for our use-case. To actually work, the "choice" must include the position of the player. The game takes place over multiple rounds, every one running a separate commit-reveal run. The opponent position would be known after the first round, cheaters would win in the second round.

5.2 Privacy

Similar problems to cheat safety arise in the player's privacy. Besides the aiming direction and power, the exact geographic location of each player is uploaded at the start of a game with precision in meters. To protect the player's privacy, at least this information should be encrypted and only be decryptable by the smart account. As already stated in Sect. 5.1, this is a major challenge.

5.3 Response Time Evaluation

For online multiplayer games it is important that the delay between turns is as low as possible. In regards to this we measured the response time of the blockchain between executing a shot in the game and getting the results of that shot from the blockchain. Due to the blockchain not being able to respond after executing scripts, it is necessary to permanently poll the blockchain. For testing purposes a poll rate of $t_{\text{poll}} = 50ms$ was chosen so that the time between poll events has negligible influence on the response time.

In Fig. 4 the results of the measurements are shown for a 3G connection, a 4G connection and a WLAN connection with an Android 10 device. The data consists of 40 data points each and has been measured on two different days.

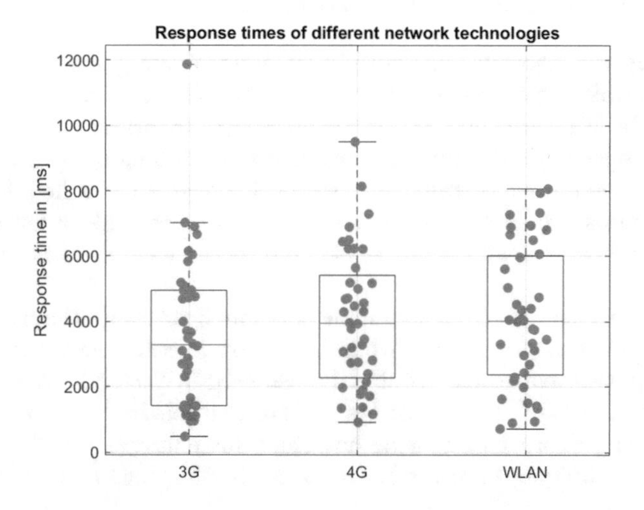

Fig. 4. Measured time between completing a shot and receiving a response from the Waves Testnet blockchain for different network technologies

The results show that there is a negligible difference in the response time between the different network connections. The averages of the different network types are within a 1 s window. The upper quartiles for 3G, 4G and WLAN are $t_{75} = 4950.15ms$, $t_{75} = 5414.05ms$ and $t_{75} = 6015ms$ respectively. Additionally, the measured worst-case has been recorded with a 3G connection with a response time of $t_{max} = 11872.70ms$. While this is an outlier, it is still an acceptable response time for a turn-based game as presented in this paper.

Surprisingly the fastest average response times were achieved with a 3G connection with an average response time of $t_{\text{avg}} = 3562.96ms$. Since the measurements have been taken in a rather small time frame, the results are prone to errors, like momentary blockchain traffic, which could cause this unexpected result. Furthermore we then performed a two-sample t-test on the data sets to determine if there is a significant difference in the data. The resulting p-values are as follows: $p_{\text{3G,4G}} = 0.30397$, $p_{\text{4G,WLAN}} = 0.91372$ and $p_{\text{WLAN,3G}} = 0.26351$. As a result of the t-tests we confirmed that there is no significant difference in the recorded samples at a 5% significance level.

For games requiring real-time updates in sub-second intervals the results are unsatisfying. While the minimum response times for all three technologies are less than 1 s, the lower quartiles are already above 1 s. On the other hand turn-based games can be implemented with minimal delay as in this type of game the time between moves is determined by the players choices and is not required to be in a fixed interval. Furthermore the results of the measurements indicate that the polling frequency can be lowered to reduce data traffic caused by polling the blockchain.

6 Conclusion and Future Work

For this paper a decentralized turn-based multiplayer game was successfully created. Possible improvements for the future are the complete decentralization, so that there is no central web server necessary, to provide the game to the players. Another possible improvement affects the polling of the smart account data, which would use less traffic, if it was pushed instead of pulled. Furthermore with this approach only games where the connection can take a few seconds are possible. Applications with a higher demand for connection speed need further research.

To increase cheat safety, a technique is required to secure the players data—in this case his exact location—while the dApp is still able to calculate with it. The commit-reveal-scheme is already discussed in Sect. 5.1, other techniques could be alterations of zero knowledge proof or multiple hashing procedures. The most promising approach might be homomorphic encryption [19]. However, all of the listed techniques need further research whether they could increase cheat safety or not.

References

1. Stewart, S.: Video game industry silently taking over entertainment world, 22 October 2019. https://www.ejinsight.com/eji/article/id/2280405/20191022-video-game-industry-silently-taking-over-entertainment-world
2. Jeni, C.: Mobile gaming industry analysis & trends in 2020 (June 2019). https://medium.com/@christinaalex/mobile-gaming-industry-analysis-trends-in-2020-c2a1b2df86d6

3. O'Dea, S.: Smartphone users worldwide 2016–2021, 20 August 2020. https://www.statista.com/statistics/330695/number-of-smartphone-users-worldwide/
4. riki28: Networkunits: The decentralized multiplayer platform, 03 January 2018. https://medium.com/@pusinggg22/networkunits-the-decentralized-multiplayer-platform-a81261ce76a9
5. Al-Jaroodi, J., Mohamed, N.: Blockchain in industries: a survey. IEEE Access **7**, 36500–36515 (2019). https://doi.org/10.1109/ACCESS.2019.2903554
6. Kingdoms Beyond: How decentralized does blockchain gaming need to be? 4 April 2019. https://medium.com/kingdomsbeyond/how-decentralized-does-blockchain-gaming-need-to-be-10862685b610
7. Min, T., Wang, H., Guo, Y., Cai, W.: Blockchain games: a survey (2019). https://ieeexplore.ieee.org/stamp/stamp.jsp?arnumber=8662573
8. Du, M., Chen, Q., Liu, L., Ma, X.: A blockchain-based random number generation algorithm and the application in blockchain games (2019). https://doi.org/10.1109/SMC.2019.8914618
9. Benet, J.: Ipfs - content addressed, versioned, p2p file system (2014)
10. Stackoverflow: 2020 developer survey, 10 October 2020. https://insights.stackoverflow.com/survey/2020
11. Kumar Ahir: How to create a virtual tour using a-frame—by kumar ahir—medium, 21 December 2018. https://medium.com/designerrs/how-to-create-a-virtual-tour-using-a-frame-164941fea573
12. Limitations—waves documentation, 24 September 2020. https://docs.waves.tech/en/ride/limits/
13. Valson, P.: Transaction fee estimations: How to save on gas? Part 1, 21 September 2020. https://upvest.co/blog/transaction-fee-estimations-how-to-save-on-gas
14. Jansen, M., Hdhili, F., Gouiaa, R., Qasem, Z.: Do smart contract languages need to be turing complete? In: J. Prieto, A.K. Das, S. Ferretti, A. Pinto, J.M. Corchado (eds.) Blockchain and Applications, Advances in Intelligent Systems and Computing, vol. 1010, pp. 19–26. Springer International Publishing, Cham (2020). https://doi.org/10.1007/978-3-030-23813-1_3
15. Abramowitz, M., Stegun, I.A. (eds.): Handbook of Mathematical Functions: With Formulas, Graphs and Mathematical Tables. Dover Publ., New York (1965). Page 81, equation 4.4.45 ; [outgrowth of a Conference on Mathematical Tables, Cambridge, Mass., Sept. 15–16, 1954, unabridged and unaltered republ. of the 1964 ed. edn. Dover books on intermediate and advanced mathematics
16. asin—nvidia developer zone, 03 May 2012. https://developer.download.nvidia.com/cg/asin.html
17. Mark Muskardin: Mastering the fundamentals of ethereum (for new blockchain devs) part iii – wallets, keys, and accounts—by mark muskardin—medium, 08 December2019. https://medium.com/@markmuskardin/mastering-the-fundamentals-of-ethereum-for-new-blockchain-devs-part-iii-wallets-keys-and-4cd3175b535b
18. Griffith, A.T.: Commit reveal scheme on ethereum—gitcoin—medium. Gitcoin, 14 December 2018. https://medium.com/gitcoin/commit-reveal-scheme-on-ethereum-25d1d1a25428
19. Armknecht, F., Boyd, C., Carr, C., Gjøsteen, K., Jäschke, A., Reuter, C.A., Strand, M.: A guide to fully homomorphic encryption. Cryptology ePrint Archive, Report 2015/1192 (2015). https://eprint.iacr.org/2015/1192

Parameter Identification for Malicious Transaction Detection in Blockchain Protocols

Vikram Kanth(✉), John McEachen, and Murali Tummala

Naval Postgraduate School, Monterey, CA 93943, USA
{vkkanth,mceachen,mtummala}@nps.edu

Abstract. As blockchain-based platforms become increasingly ubiquitous, malicious actors looking to either break the underlying platform or leverage it for nefarious purposes will also become more common. In order to combat these actors, robust mechanisms to detect and address illicit activities must be developed. Many of the current approaches to detecting abnormal activity in blockchain-based platforms are platform specific. In this paper we provide some generic parameters that should be valid for most permissionless blockchain platforms, particularly permissionless blockchain platforms that can be used for malicious transaction detection. We then analyze those parameters in the Ethereum cryptocurrency platform. These parameters include volumetric transaction rate and unique address activity.

1 Introduction and Background

In 2008, Satoshi Nakamoto released a whitepaper on Bitcoin [1]. That release ushered in the era of cryptocurrency, of which Bitcoin is the principal example. Its underlying technology blockchain, has been proposed for use or is used as a platform in a variety of applications including supply chain, medical, and cyber security amongst many others. As these types of platforms become more and more common, malicious entities looking to leverage them for illicit activities will also become more common. As a result, organizations looking to counter this activity must consider blockchain-based applications beyond cryptocurrencies, where much research has already been conducted.

There are several different types of illicit activity that have taken place on existing blockchain platforms (mostly cryptocurrencies). The Chainanalysis 2020 Crypto Crime Report lays out several categories of illegal activity including terrorism financing, stolen funds, scams, sanctions, ransomware, Darknet markets, and child abuse materials [2]. In some sense, these categories represent obvious financial crime of the sort most financial institutions have to deal with. One of the other major issues with blockchain platforms is that arbitrary information can be stored eternally due to the digital ledger. For example, 250,000 classified diplomatic cables (released by Wikileaks) are still recorded on the Bitcoin blockchain [3]. The 2.5 MB file was distributed across 130 Bitcoin transactions.

J. Prieto et al. (Eds.): BLOCKCHAIN 2021, LNNS 320, pp. 54–63, 2022.
https://doi.org/10.1007/978-3-030-86162-9_6

Whether it be to counter malicious financial activities in a cryptocurrency or to prevent illegal data from being stored on a blockchain platform, robust mechanisms to detect and address illicit activities must be developed.

As Sect. 1.1 will detail, there have been many proposed methods by which malicious blockchain activity can be detected. The purpose of this paper is to complement those approaches and to look to a future where blockchain-based applications are used for uses beyond cryptocurrencies. In that diverse environment, the establishment of broad parameters and methods that can identify potential nefarious or malicious activity will be particularly valuable. This paper details two such parameters, volumetric transaction rate and unique address activity that we believe will be applicable for use in most current and future blockchain protocols. To our knowledge, we are the first to explore volumetric transaction rate at the level of granularity presented in this paper. While unique address activity has been explored in past work, we believe we are the first to analyze short periods of time for the purposes of malicious actor/transaction detection. The remainder of this paper will be organized as follows: related work efforts, an analysis of our proposed parameters, and application of those parameters to the Ethereum platform.

1.1 Related Work

Recently, there have been several works that have provided different approaches to detect malicious activities in blockchain networks. Most of this work has been done in cryptocurrency platforms. Agarwal *et al.* detect malicious accounts in Ethereum using temporal graph properties [4]. Pham and Lee perform a graph based analysis to identify potential pernicious accounts in the Bitcoin dataset using *K-means* clustering. [5]. Zhang *et al.* proposed TXSPECTOR, a software framework that performs logic-driven analysis on Ethereum transactions and smart contracts to uncover attacks and vulnerabilities [6]. There are numerous other works detailing a wide range of potential options, including machine learning in detecting other forms of malicious activity, such as Ponzi schemes [7]. The majority of these efforts revolve around cryptocurrencies, specifically Bitcoin and Ethereum.

2 Parameters

The variety of approaches detailed in Sect. 1.1 leveraged a number of different aspects of specific cryptocurrency platforms. These aspects correspond to what would commonly be called parameters. For example, in Ethereum, transactions that interact with smart contracts can be isolated for further inspection. For most cryptocurrencies, the processing fee associated with transactions can be mapped to detect anomalies. The total amount transferred between accounts or the total number of coins (not specific to one cryptocurrency) are other parameters that can be considered. Unfortunately, most of these parameters are either specific to the use case, e.g. total amount transferred for cryptocurrencies or specific to platform, e.g. smart contract execution in Ethereum. As blockchain-based platforms become more common, these specific approaches, while valuable, will not be flexible enough to detect varied kinds of malicious activity. In order to bridge this gap, we present the following two parameters as examples of parameters that can be used in almost all blockchain platforms. The underlying assumption of our work is that creating a large malicious effect requires a large number of transactions.

2.1 Volumetric Transaction Rate

Regardless of what blockchain-based platform is used, the number of transactions that takes place in a specific time interval is a parameter that can be leveraged. As the hallmark of blockchain is the distributed ledger, a defender/investigator has the ability to both examine the total number of transactions and the specifics of any individual transaction. The goal from the defender perspective is to reduce the total number of transactions that must be investigated to a manageable number. One way to do this is to investigate periods of time in which a statistically large number of transactions takes place. In order to determine these time periods of interest, the underlying distribution of the transaction traffic must be understood. One of the fundamental contributions of this paper is an analysis of the distribution of Ethereum traffic over time. Not surprisingly, due to the relative youth of the platform, the transaction traffic distribution is non-stationary, or in other words, it changes over time. Based on the time period of interest, for example a single day, that day can be further subdivided into smaller time intervals (e.g. 10 min intervals) for more granular analysis. Once time intervals of interest are determined, the specific transactions within that interval can be examined using a variety of techniques or filtered even further using other parameters.

2.2 Unique Addresses Activity

As noted in Sect. 1.1, the detection of malicious addresses/accounts is an area of intense interest and scrutiny. Our analysis in this area tracks how many transactions are either initiated by or received by a specific entity. Due to the fact that all blockchain platforms will have to provide some sort of addressing for particular entities to carry out activities, this is a viable and far-reaching parameter. The idea behind any sort of address analysis is that addresses with high transaction volume are more likely to be malicious actors. Simply put, we care more about accounts moving millions of dollars a day than an account moving ten dollars a day. In a broader blockchain sense, the entities that attempt to transfer more data or are highly active are more able to perform nefarious activities. Unlike volumetric analysis, the distribution for this parameter is the well-known power-law distribution [5]. Our contribution in this area is to show how smaller time intervals in Ethereum traffic also follow this distribution type and to propose its use in conjunction with the earlier volumetric analysis as well as other parameters.

3 Application of Parameters to the Ethereum Platform

As noted in Sect. 2, while various parameters can be crafted to detect specific forms of malicious activity, it is unfeasible to apply all of those parameters to every single transaction that occurs. Instead, a more realistic approach is to identify specific time periods or particular entities that reduce the overall workload for the various detection algorithms. Compound this with the fact that individual blockchain protocols will have different mechanisms that allow for nefarious activity, and the range of possibilities for a malicious actor to abuse that protocol balloon. Thus, we proposed two parameters for permissionless blockchain protocols that can be used to cull the total number of suspicious time periods (volumetric analysis) or entities (addresses) of interest. In this section, we apply these ideas to the popular Ethereum platform [8]. We used the Google BigQuery ethereum_blockchain database for our data. Google runs a full Ethereum node and records the details of all transaction activity in the database.

3.1 Volumetric Transaction Rate Analysis

As noted in Sect. 2, one of the biggest issues with blockchain protocol analysis is the relative youth of the platforms. For example, Ethereum was released July 30, 2015. As with any new technology, Ethereum has seen interest wax and wane over the years. Figure 1 shows the number of transactions per day over the last 6 years.

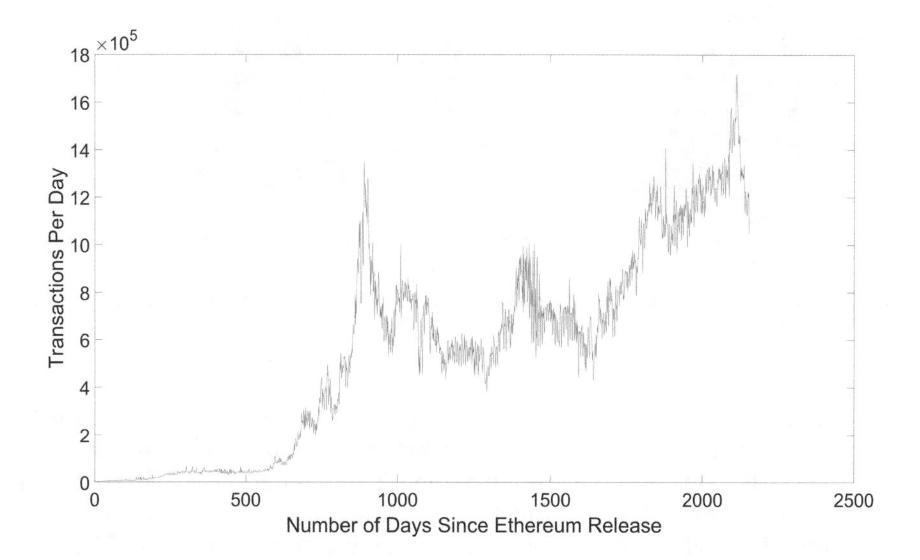

Fig. 1. Ethereum daily transactions chart from [9]

Clearly, an analysis of daily transaction data would not provide sufficient detail to suss out all but the most obvious nefarious activity. More importantly, the distribution of Ethereum transaction traffic appears to be shifting over time, which would be expected of a maturing platform. Ideally, we would want an idea of what typical activity looked like for a specific period of time in order to identify anomalous time periods. Our goal in this analysis is two-pronged. First, we want to determine what distribution type best captures transaction activity in more granular time intervals in the present moment. Secondly, we want to show that the distribution has changed over time. In order to accomplish this, we counted the number of transactions that occurred in 10 min intervals over several months. We then took those transaction counts and graphed them in a histogram. We chose time periods that occurred in the recent past as opposed to time periods corresponding to the release date of Ethereum. As Fig. 1 shows, very few transactions occurred during the first year post Ethereum release and we are interested in modern illicit activity. Figure 2 shows a comparison between January 2020 - April 2020 (left) and January 2021 - April 2021 (right).

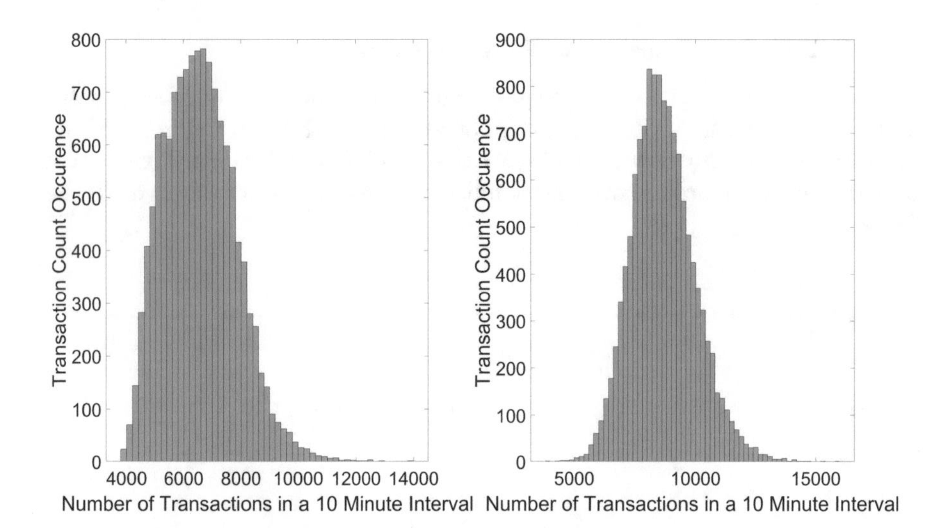

Fig. 2. Transaction Counts Comparison between Jan-Mar 2020 (Left) and Jan-Mar 2021 (Right)

Obviously, the number of transactions that occurred was much greater in 2021 than in 2020. The shape of the distributions also looks different. We then used MATLAB to fit various distribution models to this data. We tested different types of transformations on the number of transactions in a 10 min interval axis before fitting a model to our distribution. These transformations (see [10]) were a normalization (divide each x-axis value by the max x value), the log transform (each x value is replaced by $\log(x)$), and the reciprocal transform (each x value is replaced by $\frac{1}{x}$). A table of those results can be found in Table 1. We used a sliding window over three month intervals to better understand the effects of individual months on the aggregate distribution.

Table 1. Best-fit Distribution for Transaction Counts over Three Month Sliding Windows in 2020 and 2021

Aug-Oct 20	Norm	Log	Reciprocal	Sep-Nov 20	Norm	Log	Reciprocal
Distribution	gamma	normal	gev	Distribution	birnbaumsaunders	normal	inversegaussian
LL	1.25E+04	1.47E+04	1.22E+05	LL	1.29E+04	1.49E+04	1.21E+05
AIC	-2.50E+04	-2.94E+04	-2.44E+05	AIC	-2.58E+04	-2.98E+04	-2.46E+05
Oct-Dec 20	**Norm**	**Log**	**Reciprocal**	**Nov 20-Jan 21**	**Norm**	**Log**	**Reciprocal**
Distribution	lognormal	tlocationscale	lognormal	Distribution	lognormal	tlocationscale	lognormal
LL	1.50E+04	1.54E+04	1.23E+05	LL	1.52E+04	1.61E+04	1.24E+05
AIC	-3.00E+04	-3.08E+04	-2.46E+05	AIC	-3.03E+04	-3.22E+04	-2.48E+05
Dec 20-Feb 21	**Norm**	**Log**	**Reciprocal**	**Jan-Mar 21**	**Norm**	**Log**	**Reciprocal**
Distribution	lognormal	tlocationscale	lognormal	Distribution	lognormal	tlocationscale	lognormal
LL	1.45E+04	1.59E+04	1.21E+05	LL	1.40E+04	1.68E+04	1.22E+05
AIC	-2.89E+04	-3.18E+04	-2.42E+05	AIC	-2.79E+04	-3.35E+04	-2.44E+05
Feb-Apr 21	**Norm**	**Log**	**Reciprocal**	**Mar-May 21**	**Norm**	**Log**	**Reciprocal**
Distribution	lognormal	tlocationscale	lognormal	Distribution	lognormal	lognormal	gamma
LL	1.69E+04	1.60E+04	1.21E+05	LL	1.59E+04	1.56E+04	1.26E+05
AIC	-3.37E+04	-3.20E+04	-2.42E+05	AIC	-3.19E+04	-3.11E+04	-2.51E+05

Based on our testing, the reciprocal transform provided the best fit across all of our data. The criteria used were log-likelihood (LL) the Akaike information criterion (AIC), which is defined as:

$$\mathrm{AIC} = -2\log(\mathcal{L}(\theta|y)) + 2K \tag{1}$$

where $\log(\mathcal{L}(\theta|y))$ is the numerical value of the log-likelihood at its maximum point and K is the number of estimable parameters in the approximating model [11]. A more detailed description of these criteria and model selection can be found in [11]. Figure 3 shows the distribution models that best fits our data.

The distribution that best fit the transaction data changed over the last year. From an analysis perspective, having the most accurate distribution at the desired time of analysis is significant. For example, if we want to find a period of time where malicious actors took advantage of a platform vulnerability it would require understanding the distribution of transactions at a granular level. Also, the fact that the traffic distribution is non-stationary means that until the platform reaches a steady-state condition, we must continuously update and study these models. For our target time period (Jan-Mar 2021), the best fit distribution was log-normal. We can use the parameters of the distribution to highlight suspicious time intervals. For example, the probability density function (PDF) of the log-normal distribution is [12]:

$$f(x \mid \mu, \sigma) = \frac{1}{x\sigma\sqrt{2\pi}} \exp\left\{\frac{-(\log x - \mu)^2}{2\sigma^2}\right\}, \text{ for } x > 0 \tag{2}$$

The parameters of interest for this distribution are the mean μ, and the standard deviation σ. During the time period Jan-Apr 2021, $\mu = -9.0560$ and $\sigma = 0.1507$. Figure 4 shows the log-normal distribution fitted to the transaction

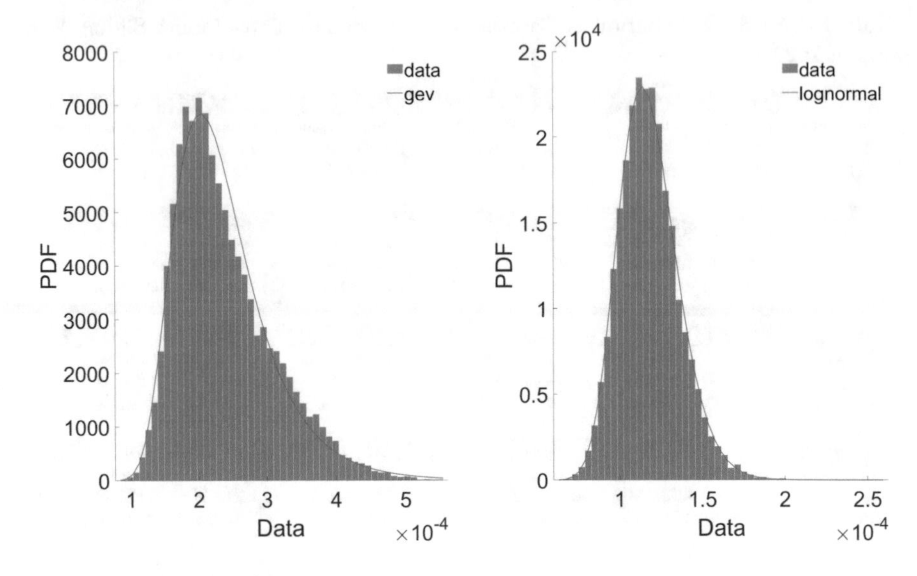

Fig. 3. Distribution Comparison between Jan-Mar 2020 (Left) and Jan-Mar 2021 (Right)

data from Jan-Apr 2021 with σ and $2 \cdot \sigma$ marked. These regions can be used to select time periods where the number of transactions falls well outside what would be expected based on the distribution. This culls down the total number of suspicious transactions that must be analyzed further. In this dataset, $2 \cdot \sigma$ corresponds to 10-minute intervals in which over 11590 transactions occurred. Any time interval with more transactions could be flagged as suspicious, greatly reducing the overall number of transactions that need to be deeply inspected. In our data, only 290 intervals out of 12816 total intervals or 2.26% of intervals had more than 11590 transactions occur.

3.2 Unique Address Activity Analysis

Another useful method to detect potential nefarious actors is to look at the distribution of entity activity. In the case of most blockchain protocols, an entity is represented by either a single address or a set of addresses. Assuming that transactions are the primary vehicle for malicious activity, looking at high traffic addresses or sets of addresses can be of great value. Pham and Lee [5] performed such an analysis on the Bitcoin dataset (from genesis up to April 7, 2013) and used k-means clustering as a method to detect outliers. While they focus on long term data, we specifically target more granular periods of time for our analysis. This facilitates capturing entities that only operate for short periods of time and are summarily discarded.

Addresses that initiated transactions were extracted from Ethereum transaction data between March 5, 2021 to March 15, 2021 using the Google BigQuery

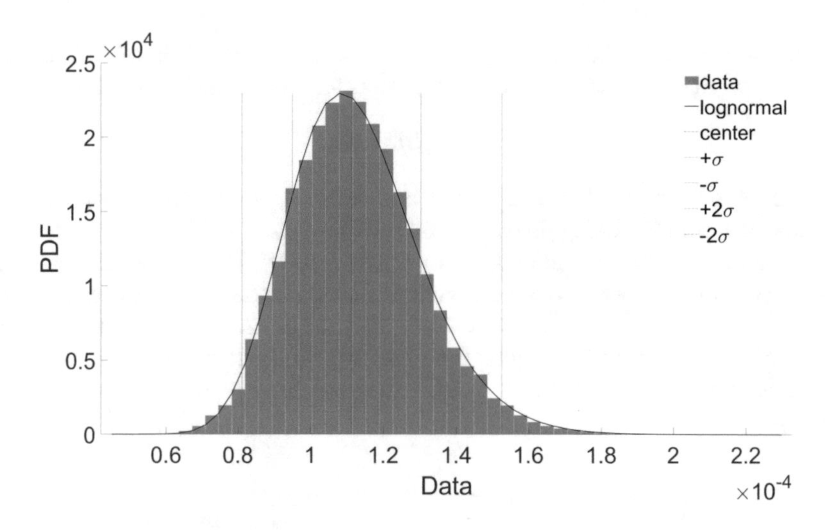

Fig. 4. Log-normal Distribution for Jan-Mar 2021

ethereum_blockchain database. Python3 was used for the visualizations and distribution fitting in this section. A histogram showing the number of transactions associated with the log (base 10) of the number of sending addresses for a ten day period between March 5-15, 2021 is shown in Fig. 5.

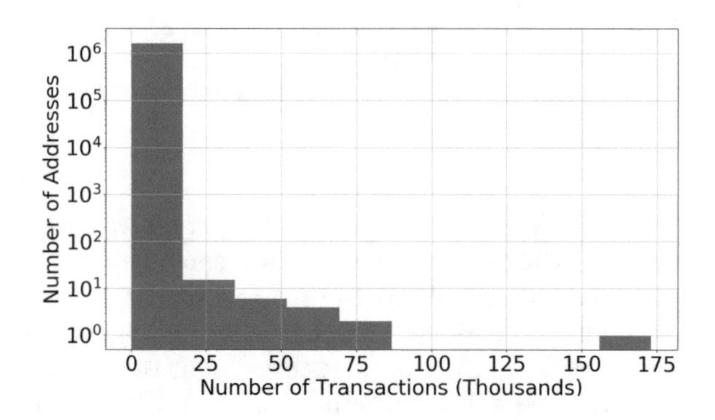

Fig. 5. Histogram of Number of Transactions vs Number of Addresses from March 5-15, 2021

As might be expected, the vast majority of addresses only initiate a few transactions. This means we can use this data to set a threshold of addresses whose transaction history might be suspicious. Furthermore, we can map a distribution to this data to further inform threshold selection. With this type of network data, logarithmic binning is often used to reduce the number of bins required for a histogram. Once that binning is complete, we used the center of the bin as a discrete data point for curve fitting [13]. Blockchain address data can often be fit to the power law distribution [4,5,14], which is defined as:

$$f(x \mid A, k) = Ax^k \tag{3}$$

For this distribution, A is amplitude and k is the exponent. Figure 6 shows a power law model being fitted to our 10 day period. The value of A was 6.985 ± 0.149 and the value of k was -0.139 ± 0.005.The R^2 value was 99.2% indicating that the model was an excellent fit for the data. With these parameters, a threshold can be set whereby accounts that initiate over a certain number of transactions during a particular time interval can be flagged. Then, the transactions initiated by that account can be analyzed further for malicious or illegal activity.

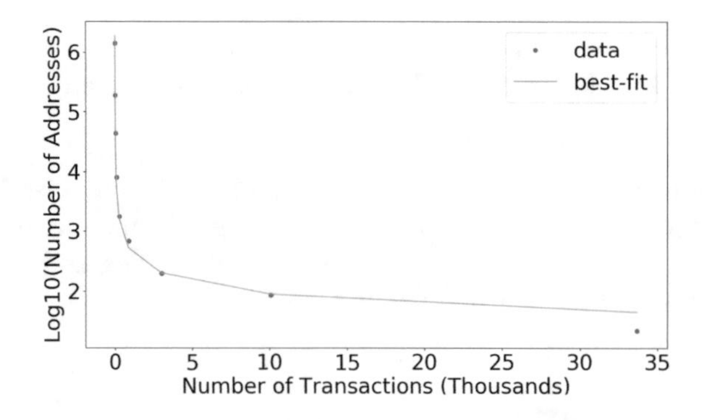

Fig. 6. Power law Fit of Histogram of Number of Transactions vs Number of Addresses from March 5-15, 2021

Not unexpectedly, looking at the same ten day period with respect to receiving address instead of sending addresses shows a similar distribution with different parameters. These distributions can help to reduce the total number of transactions that might be considered suspicious for analysis. Once these transactions have been identified, they can be used as input to a variety of follow-on detection algorithms (see Sect. 1.1 for further analysis). We stress that this trend exists for any individual day and thus can be used in conjunction with our volumetric analysis to further reduce the number of transactions of interest.

4 Conclusion

The problem of detecting malicious activity on blockchain platforms is fiendishly difficult. At the current time, most detection methods are ad hoc and platform specific. We have presented a set of parameters that can be used to evaluate any blockchain-based platform and reduce the total set of suspicious activity for further analysis. While both the address analysis and the volumetric analysis were conducted with the Ethereum dataset, they can be extended to most blockchain platforms. As they are general parameters, they can be used in conjunction with more sophisticated types of analysis, but provide an important baseline for each blockchain platform. Without a doubt, the usage of blockchain-based platforms will only increase. The security community's ability to detect and respond to malicious activity and actors will be integral in ensuring the safety of both these platforms and the users of these platforms.

References

1. Nakamoto, S.: Bitcoin: a peer-to-peer electronic cash system (2009). https://bitcoin.org/bitcoin.pdf
2. The Chainalysis 2020 Crypto Crime Report. https://go.chainalysis.com/2020-Crypto-Crime-Report.html
3. Editing the Uneditable Blockchain - Accenture. https://www.accenture.com/us-en/insight-editing-uneditable-blockchain
4. Agarwal, R., Barve, S., Shukla, S.K.: Applied Network Science **6**(1), 9 (2021). https://doi.org/10.1007/s41109-020-00338-3
5. Pham, T., Lee, S.: Anomaly detection in the bitcoin system - a network perspective (2017)
6. Zhang, M., Zhang, X., Zhang, Y., Lin, Z.: 29th USENIX Security Symposium (USENIX Security 20) (USENIX Association, 2020), pp. 2775–2792 (2020). https://www.usenix.org/conference/usenixsecurity20/presentation/zhang-mengya
7. Bartoletti, M., Pes, B., Serusi, S.: 2018 Crypto Valley Conference on Blockchain Technology (CVCBT), pp. 75–84 (2018). https://doi.org/10.1109/CVCBT.2018.00014
8. Vitalik, B.: A next-generation smart contract and decentralized application platform (2014). https://github.com/ethereum/wiki/wiki/White-Paper
9. Ethereum Daily Transactions Chart. https://etherscan.io/chart/tx
10. McDonald, J.: Data transformations - Handbook of Biological Statistics. http://www.biostathandbook.com/transformation.html
11. Burnham, K.P., Anderson, D.R.: Model Selection and Multimodel Inference?: A Practical Information-Theoretic Approach, 2nd edn. Springer, New York (2002)
12. Lognormal Distribution - MATLAB & Simulink. https://www.mathworks.com/help/stats/lognormal-distribution.html
13. Jurkiewicz, P., Rzym, G., Boryło, P.: Flow length and size distributions in campus Internet traffic. Comput. Commun. **167**, 15–30 (2021). https://doi.org/10.1016/j.comcom.2020.12.016
14. Adamic, L.A., Lukose, R.M., Puniyani, A.R., Huberman, B.A.: Search in power-law networks. Phys. Rev. E **64**(4), 046135 (2001). https://doi.org/10.1103/physreve.64.046135

Blockchain Validity Register for Healthcare Environments

Cinthia Paola Pascual Cáceres, José Vicente Berná Martinez(✉), Francisco Maciá Pérez, and Iren Lorenzo Fonseca

University of Alicante, San Vicente del Raspeig, Alicante, Spain
cppc2@alu.ua.es, {jvberna,pmacia}@dtic.ua.es, iren.fonseca@ua.es

Abstract. Currently, health information systems are decentralised and hyperconnected to third-party data sources and systems. For their proper use, they require adequate security measures to verify the validity of information, thus avoiding potential harm to patients using erroneous, corrupt or altered data. These systems must be able to avoid vulnerabilities to decentralisation, prevent record modification and data integrity problems, and facilitate fast and efficient verification operations, since it is very important that security mechanisms must not hinder the operation of these systems on which human lives depend. In this paper, we propose a decentralised registry system for healthcare environments based on Blockchain technology. This register can provide the same security as transaction registers in distributed databases, but allows for load balancing as it is a decentralised system, thus facilitating fast and efficient operations, and provides Blockchain mechanisms to maintain the integrity of the validation register.

Keywords: Security · Data veracity · Blockchain · Health system

1 Introduction

Today's healthcare systems have evolved from the electronic medical record, in which a patient's clinical information was stored digitally in a single medical centre, to the electronic health record, where all of a patient's medical information is stored in a distributed manner throughout the healthcare system so that it can be accessed and shared through different specialised tools [1]. Modern environments are made up of different medical centres such as hospitals, neighbourhood health centres, mobile systems, emergency departments and laboratories, among others. Each of them collects, stores and processes the information associated with the patients seen daily in these centres, and this information is stored in decentralised databases that ensure the availability, confidentiality and integrity of the data, while at the same time having to reconcile all this heterogeneous and distributed data flow [2].

In addition, new devices that collect data about patients and their environment through IoT-based technologies have been incorporated into the medical system [3]. The aim of these devices is to achieve more effective treatments and savings in healthcare through more comprehensive patient monitoring, but at the same time they lead to

J. Prieto et al. (Eds.): BLOCKCHAIN 2021, LNNS 320, pp. 64–73, 2022.
https://doi.org/10.1007/978-3-030-86162-9_7

more complex environments and turn the healthcare system into a huge, distributed and hyper-connected system with new and interesting security issues [4]. In healthcare environments, security threats have even more serious effects than in any other environment, as they directly threaten the lives of patients.

Cybersecurity is defined by the properties of data integrity, availability, confidentiality, traceability or authenticity, and in these super-distributed environments maintaining any of these properties becomes difficult. As far as the authenticity of the information itself is concerned, it can be seen as two problems. On the one hand on the validity of the information sources, checking and authorising only those sources and devices that are allowed to interact with the health system [5]. On the other hand, it is also necessary to verify the authenticity of the information, i.e., that the information to be used has not been damaged or modified, either accidentally or intentionally, since it was generated [6]. In this second aspect we can find various works that try to ensure the veracity of the information and where we can see that the use of Blockchain is very relevant [7] precisely because it provides attributes of immutability, transparency and traceability on the systems where it is applied.

In this paper we propose the use of Blockchain on decentralised systems to provide a layer of security on the veracity of the data so that any agent within the health system, human or not, can verify the authenticity of the information they are accessing, thus helping to preserve the quality of service and the life of the patient.

The following sections describe the main contents of the article: Sect. 2 will review how transactional systems generate security mechanisms, studying the models from the point of view of their general scheme; Sect. 3 will describe the proposal of this paper, detailing the proposed system, the processes it involves and the mechanisms it includes; Sect. 4 will review the results produced by the proposal and compare it with the schemes analysed in Sect. 2, extracting its advantages and shortcomings with respect to the current proposals; finally, Sect. 5 will draw the main conclusions of the proposal and indicate the future lines of work.

2 Analysis of the Problem of Data Immutability

Ensuring the veracity of information is a crucial issue to ensure a successful outcome while preserving patients' lives. As discussed above, an important issue is to ensure that those devices that can intervene in the system, generating and storing data, are duly authorised to do so, for which there are many authentication and certification proposals [7, 8]. But a medical prescription based on erroneous or inaccurate information can cause the treatment process to drag on or even worsen the patient's health instead of curing it. This is why we must provide mechanisms that ensure that the information used is that which was generated, and that it has not suffered alterations, accidental or malicious, during the processes of storage, treatment or access to the information that have been carried out from any point in the health network that the data have not been altered without the knowledge of the system [9, 10]. In addition, these mechanisms should take care not to affect the performance of systems that already have to cope with hundreds of thousands of transactions per day [11], because if performance is affected, the very functionality of the storage systems will be rendered unusable, even if security is very high.

Currently, most database management systems use security schemes based on storing transaction registers, where alterations to the data are noted [12]. These registers make it possible to know if a data has been altered and even return its value to its original state, undo operations, allow transactional replication and even high availability operations [13]. Thanks to these registries, it is possible to check the veracity of the data, determining if there is any transaction that affects the information, so that when accessing, for example, a registry in the database, it will be enough to check if there have been transactions that affect that registry to know if the data is or is not the one originally generated [14]. If we look at these schemes, we can basically find three transaction register generation strategies.

Centralised Transaction Register Scheme: In these schemes, a single centralised transaction register is maintained in which all actions performed on the data are registered in this other metadata store, the transaction store. This scheme allows maintaining a high performance of the databases, as any check on the veracity of the data is checked on a single register. However, on the downside, it is highly vulnerable, since in the event of damage or alteration of the register, there is no mechanism to check the validity of the register itself (Fig. 1).

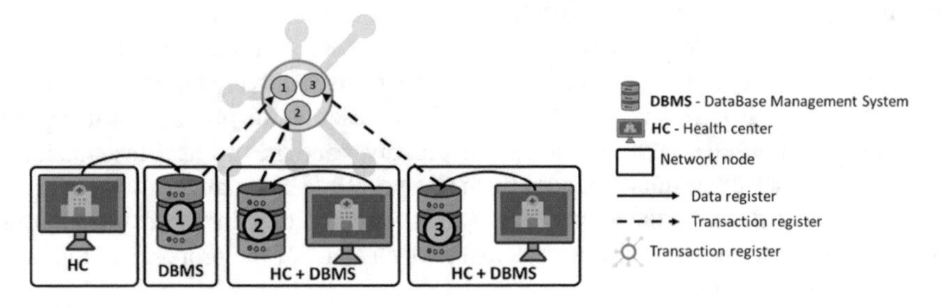

Fig. 1. Centralised transaction register.

Distributed Transaction Registry Scheme: Based on the above scheme, another model is to use a distributed scheme, Fig. 2. Each node stores a portion of the transaction log, usually the portion of the transaction log that relates to the data it stores. This scheme allows the registry to be less vulnerable to loss since it is divided into several nodes, and also allows high query performance to be maintained since the partial registers are smaller. However, on the downside, communication mechanisms are now necessary to know in which register it is possible that there is meta-information about the data and thus know its veracity, and the vulnerability to loss is also distributed, that is, now each register is vulnerable, although it is more difficult for all the registers to be lost at the same time.

Decentralised Transaction Registry Scheme: To minimise the risk of losing the activity registry, decentralised registries can also be created, where each node stores a copy of the activity registry, Fig. 3. With this scheme, it is not necessary to depend on a central node to make a query, and in the event that one of the registries is lost, it can be

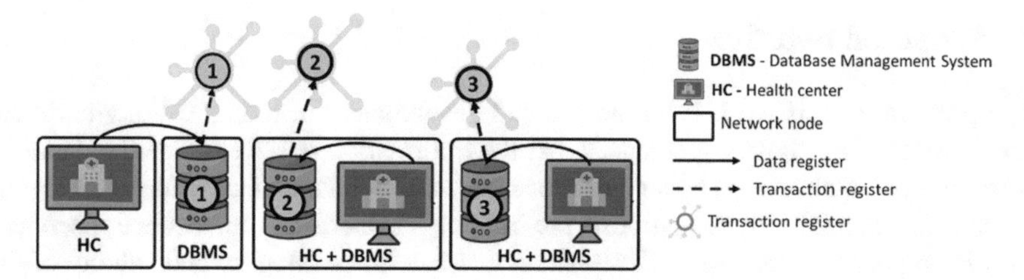

Fig. 2. Distributed transaction register scheme

easily recovered from another node. This scheme offers the advantage of both previous schemes, since the information is replicated and in case of loss it is easily replicable, and for queries it is also advantageous since it is not necessary to search for a specific registry, since the transactional validation information resides in all the registries. On the other hand, it presents a problem of integrity maintenance, since in case of alterations, manipulations or corruptions in several registries it is impossible to know which of all of them is valid and mechanisms must be enabled to discard altered registries. This can happen either through attacks or accidental modifications.

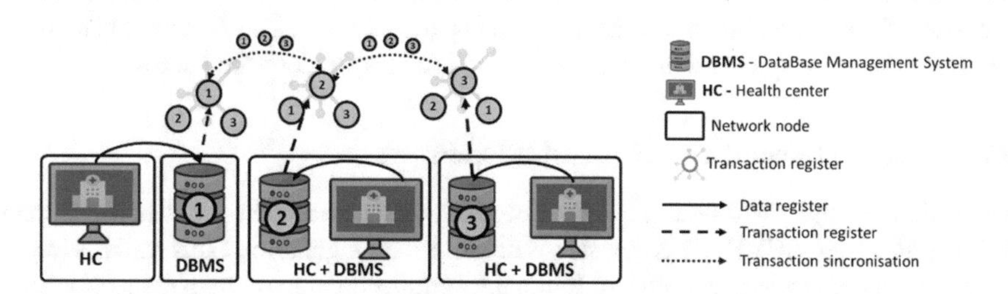

Fig. 3. Decentralised transaction *registry* scheme.

As can be deduced, decentralised schemes have the advantage of being less vulnerable to loss but imply the need to enable mechanisms to safeguard the integrity of the registry itself that is replicated across the network nodes [15].

Another important question is about the operation of the transaction registry itself. The purpose of this registry is to know all the operations performed on the data and to provide the ability to return the data to a previous state or its original state, undoing the changes [16]. Therefore, when we get a registry from our database, in order to check its validity, we should check if there are transactional registries that bring the data to the current state in which we found it, and this operation can greatly affect the performance of the validation operation, since it implies searching and checking all transactional registries on a data, and even more if we consider this highly distributed scenario that includes IoT devices [17] devices that can alternate along the nodes of the health network.

3 Proposed Solution

The proposal of this work is the design and development of a data validity certification register. This registry guarantees that the information retrieved from the database is truthful and authentic as created or processed, and in no case may it have undergone subsequent changes. For this purpose, the use of a Blockchain-based decentralised registry is proposed, as the use of Block-chain guarantees the integrity and validity of the distributed registry. This allows the advantages of a decentralised activity registry to be obtained, but mitigates the problem of discrepancies in the registry. The operation of this registry will be to store a token for each piece of data, so that it can be verified when a piece of data is retrieved from the information system as being authentic and unaltered. The validity register, unlike a transactional register, does not have the function of storing or identifying the operations performed on the data. This validity register only certifies that the accessed data is valid. The operation of the validity register is based on the fact that the data, when stored, store a signature which is generated from the validity register. This signature, in addition to the data, will be stored in the Blockchain, meaning that when a piece of data is retrieved, it will only be necessary to check that its signature is contained in the validity registry. If the signature doesn't exist, this will indicate data is not genuine and should be discarded. If the signature is found to exist and is correct, the system guarantees that the data is valid and true. The signature generation and registration process and the data validation process are described below.

3.1 Signature Generation and Registration Process

The validity register will be formed by a Blockchain and we will call it the Blockchain Validity Register BCVR. This register will store data signatures. This validity register will be a complementary system to the decentralised database management system (DBMS). The validity registry will reside in a decentralised manner on the nodes of the health system's network of nodes that are so determined. The idea is that only those nodes that are capable of processing Blockchain blocks will host a copy of the validity registry. Figure 4-a shows a schematic of what a possible health system network might look like. The main nodes with computational capacity large hospitals or health centres, will host an BCVR, which can coexist alongside the DBMS, although it is possible that the DBMS could exist on a node exclusively dedicated to this function. It is not mandatory that BCVR and DBMS coexist, the more nodes the BCVR has, the more decentralised it will be.

The Blockchain will follow the general scheme of this technology and each block will store (Fig. 4-b):

- *index:* the index of the block.
- *timestamp:* timestamp of when the block was mined.
- *signs:* will contain all the signatures that have been generated for the data. Each signature has its own timestamp and uid to generate a unique signature.
- *nonce:* the nonce of the block
- *blockHash:* the hash of the block
- *previousHash:* the hash of the previous block

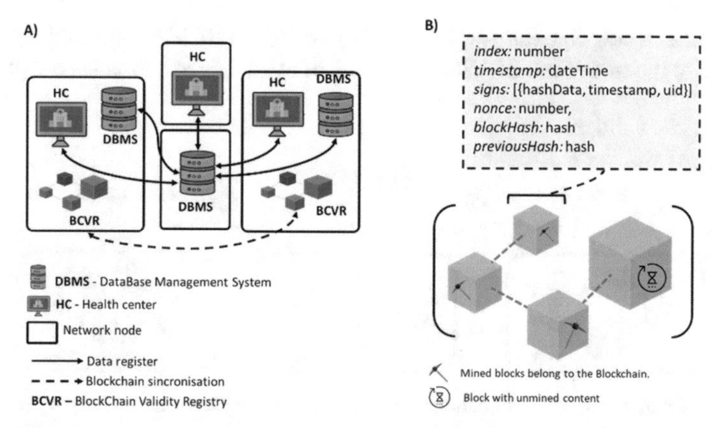

Fig. 4. **a)** Example of a network of nodes in the health system where some of them have BCVRs. Medical centres generate data that are stored in the DBMS. The DBMS nodes communicate with each other to synchronise information. BCVR nodes communicate with each other to synchronise the Blockchain. **b)** Description of the content of a block in the Blockchain.

In addition to the blocks that have been mined and are part of the BC, there will also be a block in which signatures pending signature will be stored. This block will store only the "signs" pending to be included in a new block.

When an agent or node wants to insert data into the DBMS, it must first obtain a signature from the BCVR. The steps for signature generation are (Fig. 5):

1. The agent that wants to insert data into the DBMS requests a new signature from the BCVR, for which it sends the data to be signed.
2. The BCVR generates a new signature. To generate this hash, the data, a timestamp and a uid are used to generate a hash that will serve as a signature. The uid makes the hash unique. The BCVR stores this data in the list of signatures to be mined.
3. The BCVR returns the generated hash to the agent together with the index of the last existing block in the Blockchain.
4. Finally, the agent inserts the data together with the signature and the index of the last block into the DBMS.

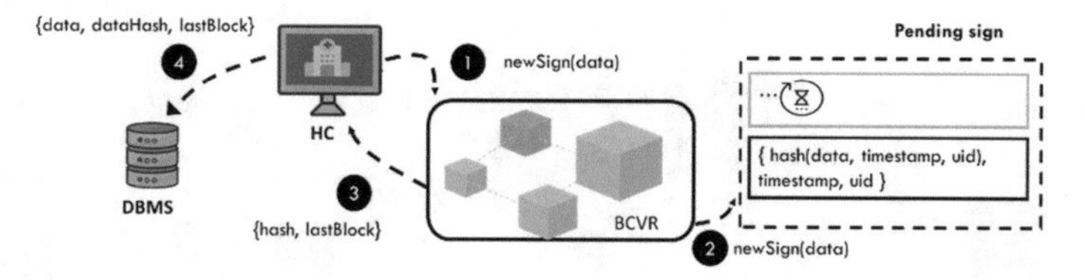

Fig. 5. Process of generating a new signature

As soon as a new signature is generated in one of the BCVR nodes, this signature is distributed to the rest of the BCVR nodes (Fig. 6) by means of a broadcast to the BCVR nodes.

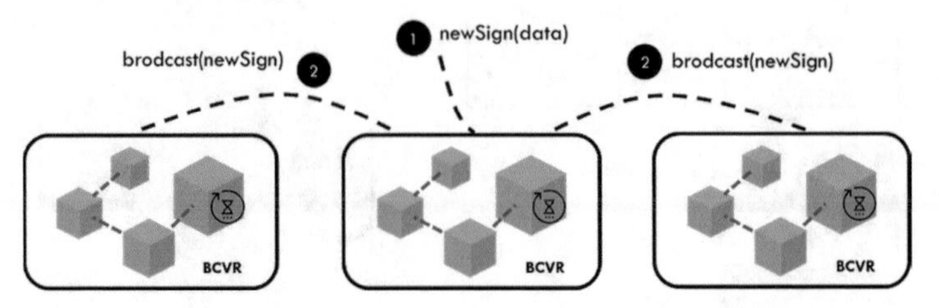

Fig. 6. Process in which a new signature is distributed to other *BCVR* nodes for the pending block to be stored.

When a certain number of signatures is reached in the pending signature block, a new block will be mined following the standard Blockchain procedure. This process will be carried out by the BCVR nodes and the first node that achieves the mining while complying with the restrictions imposed in the Blockchain (a certain number of '0' in the hash of the block) will communicate the new block to the rest of the nodes, thus incorporating the block into the chain.

3.2 Signature Verification Process

When accessing the DBMS data, it is possible to check whether the data is valid by means of its signature, for which purpose the process is as follows (Fig. 7).

1. The agent retrieving the data obtains from the BCMS the data, the hash of the signature and the lastblock index obtained during the signature.
2. The agent asks the BCVR if the signature is valid for that data, and indicates the lastblock.
3. The BCVR simply accesses the lastblock+1 and checks if there is any signature stored within this block that matches the signature obtained on the data.
4. Finally, the BCVR indicates whether the signature is valid or not.

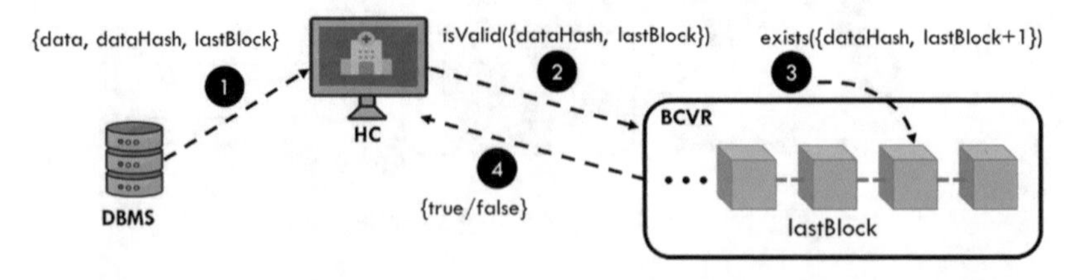

Fig. 7. Signature validation process

In this process, in addition to checking whether or not the signature exists in the block, it will also check the validity of the block itself (to ensure that it has not been tampered with). This block checking operation should only check that the hash of the previous block is correct and that the hash of the block itself is correct. If all conditions are met, and the signature exists within the block, then the data can be validated as correct. Otherwise, it shall be indicated as invalid.

4 Discussion of the Results

Our proposal manages in this way to provide a system capable of certifying the validity of data that is stored in a DBMS. The BCVR can be added to any DBMS system in such a way that only the signature information and the last block generated would need to be stored together with the data. By relying on Blockchain we extend the properties of this technology over the validity register. **Speed of verification**: the validation operation is based on the search of a hash stored in a specific block of the chain and is therefore a very fast operation. The data within the DBMS can be updated and each time it is updated, a new signature will be generated in the BCVR, leaving the previous one in disuse. Moreover, since the BCVR is decentralised, the signature validation operation can be requested to any node in the network, thus enabling load balancing. **Decentralised registry**: in the network of nodes that are part of the healthcare system, there may be one or more nodes that host the Blockchain. This allows the burden of operations to be distributed over the network and the registry to be protected against loss. Blockchain incorporates consensus mechanisms whereby nodes, in the event of multiple simultaneous mining or inconsistencies in the Blockchain, resolve the situation and maintain the integrity of the system. **Immutability of registry**: the use of Blockchain ensures that the validity registry is immutable. Each block contains within its data the hash of the previous block and its own hash, which prevents it from being altered or manipulated to forge a signature. This makes the registry highly secure. Moreover, as the chain is decentralised, if a block were to be altered (which is highly unlikely), the network of nodes itself, through the consensus of the Blockchain, would restore the legitimate chain.

If we compare our proposal with traditional transactional registry mechanisms, we find that the use of Blockchain is not able to provide traceability over data. Our proposal does not claim to do so either, it only requires the possibility to validate the data, but does not seek the ability to restore the data to its original state. Precisely our BCVR provides the speed to verify the data, but no other property related to traceability or non-repudiation of the information.

The creation of the BCVR implies the increase of transactions between the participants of the system, since it requires a signature request between the actor that stores data in the DBMS and the BCVR and also the communication between the nodes of the Blockchain to: first, keep the block chains synchronised both when mining and generating a new block; and second, when a new signature is generated in one of the nodes, transmit it to the other nodes to maintain the integrity of the list of signatures pending mining.

However, we can appreciate certain improvements with respect to the traditional schemes described in Sect. 2. With respect to a centralised or distributed registry scheme, our proposal allows load balancing on the execution of data verification requests, since it is possible to make the request on any node of the network. In the two previous schemes, requests must be made either to the central node that has the only transaction registry, or to the node where the part of the transaction registry that contains the information concerning the queried registry is stored. In our scheme, any node in the Blockchain network can be queried.

And finally, Blockchain technology completely avoids security and integrity issues about the registry. Due to the immutability property of the chain, it is virtually impossible for the validity registry to be modified. Furthermore, it must be taken into account that the registry aims to guarantee the quality of the data, preventing any unwanted alterations to the data. Unlike in a banking scenario, for example, where the attack is aimed more at financial reward, the main aim here is to prevent the unconscious degradation of data that could harm the health of patients being treated.

5 Conclusions

The use of Blockchain technology allows its properties to be extended to other contexts, in this proposal we are looking above all for the ability to guarantee that the data has not been accidentally or maliciously modified. The inclusion of new systems cannot generate worse performance of the main information systems, in this case the databases of the healthcare system. In a particularly sensitive scenario such as healthcare, where any alteration of the data can cause harm to the patient, being able to guarantee the veracity of the data and to do so in an agile way and with hardly any computational cost would provide the systems with new guarantees on the quality of the data. Our proposal can also be adapted to any DBMS system as it is independent of the technology used and complementary to any information system. Using a decentralised scheme for validation facilitates load balancing in highly distributed scenarios such as medical environments where the strong incursion of IoT systems to facilitate telemedicine must be taken into account. Currently our line of work pursues the possibility of incorporating the validity system in network nodes that can, without large computational requirements, generate a mining with an adequate level of security. Using these nodes, hybrid schemes between Blockchain and Edge Computing could be implemented to bring information processing closer to the nodes that will most commonly require it, although nodes containing the complete Blockchain will always be maintained.

References

1. Brighi, R.: The quality and veracity of digital data on health: from electronic health records to big data. Rev. Bioetica Derecho **42**, 163 (2018)
2. Halamka, J.D., Lippman, A., Ekblaw, A.: The potential for Blockchain to transform electronic health records. Harv. Bus. Rev. **3**(3), 2–5 (2017)
3. Swaroop, K.N., Chandu, K., Gorrepotu, R., Deb, S.: A health monitoring system for vital signs using IoT. Internet Things **5**, 116–129 (2019)
4. Alasmari, S., Anwar, M.: Security & privacy challenges in IoT-based health cloud. In: 2016 International Conference on Computational Science and Computational Intelligence (CSCI), pp. 198–201. IEEE (2016)

5. Xu, L., Chen, L., Gao, Z., Fan, X., Suh, T., Shi, W.: DIoTA: decentralized-ledger-based framework for data authenticity protection in IoT systems. IEEE Netw. **34**(1), 38–46 (2020)
6. Siddiqui, S.T., Alam, S., Ahmad, R., Shuaib, M.: Security threats, attacks, and possible countermeasures in Internet of Things. In: Kolhe, M.L., Tiwari, S., Trivedi, M.C., Mishra, K.K. (eds.) Advances in Data and Information Sciences. LNNS, vol. 94, pp. 35–46. Springer, Singapore (2020). https://doi.org/10.1007/978-981-15-0694-9_5
7. Shuaib, M., Alam, S., Daud, S.M.: Improving the authenticity of real estate land transaction data using blockchain-based security scheme. In: Anbar, M., Abdullah, N., Manickam, S. (eds.) ACeS 2020. CCIS, vol. 1347, pp. 3–10. Springer, Singapore (2021). https://doi.org/10.1007/978-981-33-6835-4_1
8. Alotaibi, B.: Utilizing blockchain to overcome cyber security concerns in the Internet of Things: a review. IEEE Sens. J. **19**(23), 10953–10971 (2019)
9. Krotofil, M., Larsen, J., Gollmann, D.: The process matters: ensuring data veracity in cyber-physical systems. In: Proceedings of the 10th ACM Symposium on Information, Computer and Communications Security, pp. 133–144 (2015)
10. Gollmann, D.: Veracity, plausibility, and reputation. In: Askoxylakis, I., Pöhls, H.C., Posegga, J. (eds.) WISTP 2012. LNCS, vol. 7322, pp. 20–28. Springer, Heidelberg (2012). https://doi.org/10.1007/978-3-642-30955-7_3
11. Kepner, J., et al.: Computing on masked data: a high performance method for improving big data veracity. In: 2014 IEEE High Performance Extreme Computing Conference (HPEC), pp. 1–6. IEEE (2014)
12. LeBlanc, P.: Microsoft SQL Server 2012 Step by Step: Micr SQL Serv 2012 Step. Pearson Education (2013)
13. Varga, S., Cherry, D., D'Antoni, J.: Introducing Microsoft SQL Server 2016: Mission-Critical Applications, Deeper Insights, Hyperscale Cloud. Microsoft Press, Redmond (2016)
14. Fowler, K., Gold, G.: SQL Server database forensics. Memory (2007)
15. Mooney, C.W., Jr.: The Cape Town Convention's improbable-but-possible progeny part one: an international secured transactions registry of general application. Va. J. Int'l L. **55**, 163 (2014)
16. Bernstein, P.A., Newcomer, E.: Principles of Transaction Processing. Morgan Kaufmann, Burlington (2009)
17. Rasolroveicy, M., Fokaefs, M.: Performance evaluation of distributed ledger technologies for IoT data registry: a comparative study. In: 2020 Fourth World Conference on Smart Trends in Systems, Security and Sustainability (WorldS4), pp. 137–144. IEEE (2020)

Implementation and Evaluation of a Visual Programming Language in the Context of Blockchain

Farouk Hdhili, Ramy Gouiaa, and Marc Jansen(✉)

Computer Science Institute, University of Applied Science Ruhr West, Bottrop, Germany
{farouk.hdhili,ramy.gouiaa,marc.jansen}@hs-ruhrwest.de

Abstract. Programming is not a simple task. Besides, the learning of syntax, the obstacles often seem insurmountable for the beginner. Syntax errors frustrate and distract attention from the actual algorithms and logic that need to be understood. With the help of visual programming, syntax errors can be avoided as much as possible by using predefined graphical boxes, which, due to their shape, can only be connected correctly to form a program. Moreover, Smart Contracts are mainly designed to exclude the need to trust a third party. Here using Visual programming makes perfect sense by making the creation of Smart Contracts much simpler without the help of a developer. Another advantage of Visual Programming Language (VPL) in Smart Contract creation is that the terms and conditions in Contracts can be read and understood from non-developers. In this research we aim to implement a Visual Smart Contract Creator and to conduct an empirical evaluation, which on the one hand, measures the semantic understandability of a visual representation of a Smart Contract created with our tool and compares it to its textual representation in Ride programming language. On the other hand, we evaluate the user acceptance of the tool.

Keywords: Visual programming · Blockchain · Smart contract · Waves blockchain

1 Introduction

Since the big hype of Bitcoin, Blockchain technology became the area of interest of many people with a different meaning. For developers, it represents a peer-to-peer system for ensuring trust among the network. Whereas, it means digital assets or rather cryptocurrencies for investors and in the financial industry. Additionally, it is the next internet generation from a technological perspective. Blockchain's potential goes far beyond powering cryptocurrencies as known and serves a much broader range of purposes; for instance, this technology could lower our dependency of intermediaries, such as banks or other companies. This technology offers a guarantee of trust and reduces the ability of data tampering.

After Satoshi's White paper [1] has been published, several Blockchains have emerged, and thus many cryptocurrencies have proliferated the market. Upon the idea of

J. Prieto et al. (Eds.): BLOCKCHAIN 2021, LNNS 320, pp. 74–82, 2022.
https://doi.org/10.1007/978-3-030-86162-9_8

decentralization, the second generation of the Blockchain builds a new concept, which is the integration of self-executing program code in the distributed network. Unfortunately, people outside the Information and communications technology area are unable to understand the code written in Smart Contract, since they are implemented in computer programming languages. Smart Contracts are public and accessible by everyone, but only experts can verify their contents. Moreover, the variety in the design of Smart Contracts could also be problematic for developers because it demands high requirements on knowledge and on time, as they have to learn a new programming language every time, they need to use features of a different Blockchain. In this work, we introduce a visual programming language that could solve the mentioned problems. We aim to create a user-friendly interface enabling essentially normal users to build their Smart Contracts by connecting graphical boxes generating the related Smart Contract in Ride, the programming language of the Waves Blockchain[1]. This interface will not only facilitate the creation of Smart Contracts, but it also permits to give a graphical representation to Smart Contracts, so that all users can interpret their semantics. As Napoleon said, "A good sketch is better than a long speech".

2 Background

This part will introduce Visual programming, Smart Contracts, and the Waves Blockchain, for which we intend to implement the Smart Contract creator tool.

2.1 Smart Contract

The definition of a contract is an agreement between two or more parties to perform (or not to perform) a specific action in return for remuneration [2]. Using classical contracts, each party must, therefore, trust the other for the execution of the agreement of their respective obligations [3]. As we mentioned in the last subsection, the second generation of Blockchain allows us to create programs in a trustworthy environment, which can replace or enhance classical contracts in many use cases. While a traditional legal contract defines the rules of an agreement between multiple entities, a Smart Contract goes further and ensures these rules in the Blockchain. Smart Contracts are IT (Information Technology) protocols that facilitate negotiation and allow verification of the validity and the execution of a contract.

Concretely, the contract of tomorrow will be computerized and automated, and many types of contractual clauses can thus be partially or completely self-executed. They work like any conditional statement of type "if-then" (if such condition is satisfied, then such consequence is executed).

A Smart Contract can be implemented in various programming languages, such as Solidity, C++, Golang, Ride, etc. The choice of the programming language is particularly important, as this is a potential target for malicious attack. If the selected language has gaps or flaws, attackers can take advantage of them and cause great harm. In addition, the program must have been implemented without errors, since an error in the code provides a loophole for attacks or can impair the function of the contract [4].

[1] https://docs.Wavesprotocol.org.

2.2 Visual Programming

Programming is usually known via typing line of codes or textual commands, which are difficult to learn and to use and it requires skills and knowledge to understand, that normal users do not have, however it can be also expressed as visual representations. The concept of visual programming languages (VPL) has been around for a long time, also sometimes referred to as graphical programming languages, and it is a language that enables the development of programs through graphical manipulation. All the commands are most often represented by colored boxes or puzzle pieces containing text or symbols. The graphical interface allows us to move the boxes or puzzle pieces to hang or nest them together to form a program such as Blockly[2]. In this way, the programming will be accessible to a wider range of people and even kids could draw their own programs. Furthermore, visual programming languages allow users to focus on learning computer-programming concepts by putting aside the aspects related to the language used. Another attraction of visual programming is the simplicity of the semantic, which could be very helpful for professional programmers, because it reduces both risk of making syntax errors and development time.

2.3 Flow-Based Programming

Flow-based programming [5] is a visual programming paradigm that allows developers to handle their programs like a map. The map represents an executable graph. Each node of the graph is a graphical Box containing inputs and output and can be connected with other Boxes. We could consider the data flow among the graph as stations that allow users to determine what is calculated next. Operations usually supply new data as a result, which is then used as input values for further operations. Furthermore, Ride programming language is based on the if-this-then-that rule.

2.4 Waves Blockchain

Waves Blockchain offers two types of Smart Contracts: Smart Accounts and Smart Assets. It is possible to convert any account into a smart account or dApp by adding a script written in Ride language that allows changing the account's behavior. It also offers several transaction types; for instance, data transactions allow users to store data in a Waves account or transfer transaction that permits to transfer tokens between accounts. This Blockchain technology adopts a non-Turing complete language for implementing Smart Contracts [6], which can be favorable for designing a VPL in terms of complexity and understandability.

The Ride programming language provides two types of functions: verifier and callable. Like Ethereum Smart Contracts, users can interact with deployed smart contacts by invoking certain callable functions. Callable functions are like normal functions that we use in any programming language but in a decentralized manner. It can store, e.g., new information in the key-value data storage of a smart account or perform transfer transactions. In contrast to Callable functions, the verifier function can appear only once in the attached script. This function is responsible for the validation of transactions and has no arguments.

[2] https://developers.google.com/blockly.

3 Research Methodology

Smart contracts contain agreements specified by multiple entities and they must be impartial and transparent. Unfortunately, these contracts are written in computer programming languages and only developers can implement and understand them.

Moreover, malware developers can falsify Smart Contracts' behavior so that end-users do not realize it since they cannot understand it. Making Smart Contracts understandable for non-developers can be considered one of the highest priorities in in bringing this kind of Blockchain usage to ordinary users, as we aim to make the use of Smart Contracts more approachable in all scenarios involving trust necessity.

Using a flow-based programming language in the Smart Contract creation could solve the understandability's problem for non-developers and maybe, in the future, eliminate the need for the It-specialist' intervention.

This work aims to implement and evaluate a visual programming environment, making smart contracts simpler for professional developers and allowing a wider range of users to take advantage of this technology independently of the native programming language of the Smart Contract.

This allows managers, decision-makers, and consumers to be directly involved in creating and understanding smart contracts' logic. The scope of this research is to implement a visual programming environment for smart contracts of the Waves platform and investigate three research questions:

- How effective/useful is a visual Smart Contract creator?
- How usable is the developed visual Smart Contract creator?
- Is the semantic understandability higher when using the visual Smart Contract creator tool?

For the sake of answering the research questions stated in the previous section, first-year students are invited to reply to various questions. This research's population is first-year students from different domains of study. Among the empirical research, participants need to answer two different kinds of questions. Questions related to our Smart Contract creator tool to evaluate the understandability of a visual representation of Smart Contracts, and on the other hand, questions based on the technology acceptance model have been asked to measure the usability and the ease of use of our tool. The first part of our questionnaire introduces the fundamentals of the Waves Blockchain, Smart Contracts, visual programming, and questions related to textual and visual Smart Contracts to examine visual Smart Contract comprehensibility compared to their textual representations. The second part presents questions to examine the acceptance of the tool, mainly usefulness and ease of use.

4 Implementation

This section will present the used visual programming paradigm and then introduce our Smart Contract Creator tool's functional aspects.

4.1 Smart Contract Creator

We adopted a Flow-based programming paradigm to implement our Tool. Flow-based programming is a visual programming technique that allows users to handle their programs like a map. The map represents an executable graph. Each node of the graph is a graphical Box containing inputs and output and can be connected with other Boxes. We could consider the graph's data flow as stations that allow users to determine what is calculated next. Furthermore, Ride programming language is based on a functional programming paradigm. For this reason, flow-based programming seems to be a suitable solution for our implementation [8], because users will be able to perceive the contract's logic in a flowing manner, in which functions are represented as nodes.

The presented Smart Contract Creator Tool is designed to be a single page application, which means that the user could perform all actions without reloading the page or the need to utilize a number of different pages.

We could split up a Smart Contract in Waves Blockchain into the three following components: verifiers, callables (see Sect. 2.3) and functions.

Upon this approach, we designed our interface so that each component of the Smart Contract will be created separately. The user will get a canvas to draw the desired logic for each component of the Smart Contract. We considered only the Verifier and callable components in the implementation, since we will use simple Smart Contract scenarios in the evaluation Evaluation results.

This part of the paper will deal with a quantitative research design. In which we present several statistical analyses of the collected data from our online survey. First of all, answers are investigated to determine how effective the use of a visual programming model in Smart Contract creation. Finally, we will study the user acceptance of the implemented tool using the technology acceptance model. To evaluate the user acceptance we used the following research model proposed by Money [9] (Fig. 1):

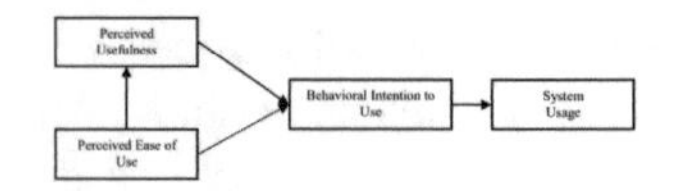

Fig. 1. Research model (Money 2004)

4.2 *Semantic* Investigation of a Visual Smart Contract

In order to answer our research question: "Is the semantic understandability higher when using the visual Smart Contract creator tool?" we compared answers about Smart Contracts of their visual representations to the same Smart Contracts in their textual representations.

In order to answer the first type of questions, we have included documentation accompanied by supplementary videos to introduce Smart Contracts and its advantages along with the basics of the Waves Blockchain.

To assess the understandability of the semantic related to a Smart Contract in its visual representation compared to its textual form, we have picked 6 different Smart Contract scenarios.

For each scenario, we presented a visual Smart Contract using our tool and its textual representation in the Ride programming language.

Participants got a random question associated with a visual or textual form of a Smart Contract, and they had to select the correct description from 3 given descriptions.

To guarantee the accuracy and the reliability of the data to analyze, we mixed visual and textual questions to compose two questionnaires.

The first questionnaire starts with a question related to a Smart Contract in its visual form and then will alternate with a question about the textual representation of a Smart Contract. In contrast, the second questionnaire starts with a textual representation.

The questions are sorted based on their difficulty, starting from easy to difficult.

As the difficulty of the questions increases, we classified the questions into 3 Groups for the statistical analysis:

Group 1: Easy (Questions_1, Question_2)
Group 2: Medium (Questions_3, Question_4)
Group 3: Hard (Questions_5, Question_6)

In this study 58 first-year students are involved. The statistical investigations will be depicted in the following sections (Table 1).

All correct answers (visual) versus correct answers (textual) from both questionnaires (174 questions)

Table 1. Correct answers of all questions

Questions	Wrong answer	Correct answer
Visual questions	22%	78%
Textual questions	34%	66%

The table above resumes the global results of both questionnaires. The correct answers about visual Smart Contracts are noteworthy higher than textual ones. More specifically the visual version of a Smart Contract is 12% better than the textual ones. The paired t-tests of above shown comparison ($p = 0.053$) says that the means of the two groups are not significantly different according to theory of statistical estimation [9].

All correct answers (visual) versus correct answers (textual) of experienced participants

As we asked participants about their experience in programming.

We considered participants that rated themselves from quite high to very high as experienced (Table 2).

The results shown above proves an advantage again in using visual Smart Contracts against traditional Smart Contracts also for experienced participants in programming.

The percentage of correct answers in visual questions is 6% better than the percentage of textual questions by subtracting the both percentage of correct answers (80%–74%).

Table 2. Correct answers of experienced participants

Questions	Wrong answer	Correct answer
Visual questions	20%	80%
Textual questions	26%	74%

The paired t-test indicates that the results of the above shown comparison are statistically not significant ($p = 0.426$).

All correct answers (visual) versus correct (textual) answers of inexperienced participants

Inexperienced participants are the ones having low or very low knowledge in programming (Table 3).

Table 3. Correct answers of unexperienced participants

Questions	Wrong answer	Correct answer
Visual questions	24%	76%
Textual questions	44%	56%

The results shown above proves again an advantage in using visual Smart Contracts against traditional Smart Contracts also for unexperienced participants in programming. As the unexperienced population is the aim of this paper, the results of this sections are considered more important than those of the experienced participants.

The percentage of correct answers in visual questions is 20% better than the percentage of textual questions. In contrast to the experienced participants, the visual representation of smart contact has 14% higher effectiveness in terms of understandably of a semantic.

The paired t-Test of the above presented comparison indicate a P-value ($p = 0.007$) less than 0.05. As a result, the difference between the visual and textual version of a Smart Contract for unexperienced participants is statistically significant.

Correct answers' analysis of experienced and unexperienced participants according to questions' difficulty

After analyzing all questions with respect to their difficulty, we noticed that results prove a significant advantage of using visual Smart Contracts over the textual ones for unexperienced participants ($p = 0.049$).

Our results also show that the percentage of correct answers increases proportionally with the difficulty of the questions. This means that a visual Smart Contract becomes more efficient the more complex it is, yet the paired t-test analysis of correct answers according to the difficulty groups for unexperienced participants does not show a statiscal significance ($p = 0.438$).

4.3 Usability and Ease of Use Analysis According to TAM

Besides the presented video of our tool in the documentation section, the first part of the questionnaire contains examination questions about visual and textual Smart Contracts. As a consequence, participants acquire more and more experience in our tool. The second part of the questionnaire includes 7-level Likert scale (1 = strongly disagree, 7 = strongly agree) questions about the usefulness and the ease of use of our tool. In this subsection, we analyze the collected rates and determine the relationships between our research model factors.

TAM's item and Cronbachs' Alpha analysis

The descriptive statistics of our factors are depicted in the following table. The means of the two first factors (Perceived ease of use, Perceived usefulness) are above 5. These results show that the average rate regarding the usefulness is "More or less agree." To measure the internal reliability of the questions asked in a questionnaire the coefficients Cronbach's Alpha have been calculated. Cronbach's Alpha coefficients are between .7 and .8 which is good [11] (Table 4).

Table 4. TAM' item analysis

Factors	Questions	Mean	Std dev	AVG	Cronbach's Alpha
Perceived ease of use	PEOU_01	5.21	1.33	5.26	0.858
	PEOU_02	5.14	1.34		
	PEOU_03	5.33	1.36		
	PEOU_04	5.39	1.47		
Perceived usefulness	PUSEF_01	5.52	1.08	5.39	0.824
	PUSEF_02	5.40	1.03		
	PUSEF_03	5.40	1.24		
	PUSEF_04	5.24	1.07		
Behavioral intention	BI_01	5.68	0.98	5.12	0.776
	BI_02	4.47	1.49		
	BI_03	5.21	1.24		
System usage	UI_01	3.49	1.59	3.85	0.852
	UI_01	4.22	1.86		

5 Conclusion and Future Work

In this research, we successfully designed and implemented a visual programming environment for Smart Contracts running on the Waves Blockchain. The collected data of our survey have shown promising results. Otherwise, as it is the first version of our tool, we can consider that a visual programming language can significantly improve Smart Contracts' understandability. Other research projects can also follow our methodology to consolidate our results by adopting, for example, other Blockchain implementations. Indeed, the Waves Blockchain provides an easy programming language for Smart Contract creation and does not support control-flow. Did that contribute to the success of our visual language?

Since control-flow mechanisms are mostly used in complex scenarios, we consider that end-users do not require control-flows in their most frequent demands. For this reason, to lean over non-Turing complete Smart Contract implementations to adopt visual language is noteworthy [7].

References

1. Nakamoto, S.: Bitcoin: a peer-to-peer electronic cash system (2008). https://Bitcoin.org/Bitcoin.pdf
2. Wright, A., De Filippi, P.: Decentralized Blockchain technology and the rise of lex cryptographia (2015)
3. Swan, M.: Blockchain: Blueprint for a New Economy. O'Reilly Media, Inc. (2015)
4. Luu, L., Teutsch, J., Kulkarni, R., Saxena, P.: Demystifying incentives in the consensus computer. In: Proceedings of the 22nd ACM SIGSAC Conference on Computer and Communications Security, pp. 706–719. ACM (2015)
5. Morrison, J.P.: Flow-Based Programming: A New Approach to Application Development. CreateSpace (2010)
6. Ivanov, S.: Waves platform—Whitepaper (2016). https://Wavesplatform.com/files/images/whitepaper_v0.pdf
7. Jansen, M., Hdhili, F., Gouiaa, R., Qasem, Z.: Do smart contract languages need to be turing complete? In: Prieto, J., Das, A.K., Ferretti, S., Pinto, A., Corchado, J.M. (eds.) BLOCKCHAIN 2019. AISC, vol. 1010, pp. 19–26. Springer, Cham (2020). https://doi.org/10.1007/978-3-030-23813-1_3
8. Mason, D., Dave, K.: Block-based versus flow-based programming for naive programmers. In: 2017 IEEE Blocks and Beyond Workshop (BB), pp. 25–28 (2017). https://doi.org/10.1109/BLOCKS.2017.8120405
9. Money, W., Turner, A.: Application of the technology acceptance model to a knowledge management system. In: Proceedings of the 37th Hawaii International Conference on System Sciences (HICSS 2004) - Track 8 (2004)
10. Fisher, R.A.: Theory of statistical estimation. In: Mathematical Proceedings of the Cambridge Philosophical Society, vol. 22. no. 5. Cambridge University Press (1925)
11. Gliem, J., Gliem, R.: Calculating, interpreting, and reporting Cronbach's Alpha reliability coefficient for likert-type scales. In: 2003 Midwest Research to Practice Conference in Adult, Continuing (2003)

Implementation and Evaluation of a Visual Query Language in the Context of Blockchain

Ramy Gouiaa, Farouk Hdhili, and Marc Jansen(✉)

Computer Science Institute, University of Applied Science Ruhr West, Bottrop, Germany
{ramy.gouiaa,farouk.hdhili,marc.jansen}@hs-ruhrwest.de

Abstract. Visual programming has become a term that stands for intuitive software development. It dispenses largely with textual notation and instead uses graphical program modules. The use of visual techniques is connected with the hope, among other things, to be able to create and understand programs more easily and better than before, which would open up access to software development for non-programmers. This could lead to a reduction of the enormous application backlog in many places. In this research, we design and build a web-based Visual Query builder tool to query the Waves blockchain database, using a flow-based programming language as well as an empirical evaluation of the tool that, on the one hand, measures the semantic understandability of a visual representation of a query, and compare it to its textual representation in SQL. On the other hand, we evaluate the user acceptance of the tool.

Keywords: Visual programming · Blockchain · BI · Waves

1 Introduction

More and more companies try to integrate blockchain technology in their business processes, gaining from the trust that those systems bring. On the other hand, process and higher management of those companies are used to real-time and up-to-date KPIs, generated by BI systems. Unfortunately, blockchains (due to their fundamental data structure of a linked list) do not allow for on-demand real-time queries against their data. Hence further research is necessary to provide possibilities to query blockchain-based data reasonably.

2 Background

This section will introduce visual programming languages, the Waves blockchain as a trading platform, and resume the WavesBI database architecture, which are the fundamentals in our research.

J. Prieto et al. (Eds.): BLOCKCHAIN 2021, LNNS 320, pp. 83–90, 2022.
https://doi.org/10.1007/978-3-030-86162-9_9

2.1 Visual Programming Language

Visual programming (VP) has grown in popularity in recent years, largely because of the promise that this new way of programming will significantly simplify software development. In general, a visual language is described being a formal language including visual syntax and semantics. This type of representation enables the visual language to be interpreted much more quickly and easily by humans. Visual languages are often referred to as flow-based because they represent complex structures as a flow of information [1]. There is also the definition as programmed attributed graph grammar (PAGG) [2].

The use of a visual programming language is controversial. One flagrant problem is that VPLs, generate programs which often do not satisfy the high requirements of a programming environment. It can also be more complex issues such as recursion, which are often not implemented. Still, that usually doesn't matter for SQL, since SQL (ANSI SQL92 standard) is not Turing Complete [3].

This is contrasted with the user friendliness of a VPL. Because of its abstract representation, it is easier to understand for people without programming knowledge and is therefore more quickly deployed. The reason for this, is that images can communicate things more simply and concisely, support understanding and remembering, have no language barriers and can therefore be understood by people of any language [4]. Evidence of this is provided by a study [5], in which the authors compared the user-friendliness of a visual query language with the use of SQL.

2.2 Waves Platform

The Waves project is a blockchain-based trading platform founded in 2015. Waves is the name of the coin for the corresponding Waves platform. The Waves blockchain allows users to perform various kinds of transactions, e.g., storing data in the Waves blockchain is called a 'Data transaction' or sending assets is registered as a 'Transfer transaction'. Precisely as the founders intended, it is effortless to trade or store cryptocurrencies (Bitcoin, Ethereum, Waves, Litecoin, etc.) as well as Fiat-coin (US-dollar, Euro) on the Waves platform. However, this also includes a quick and easy possibility to start crowdfunding or to invest in a foreign project. The Waves platform allows any user to create his custom digital token quickly and easily, and then start trading with it.

2.3 WavesBI Database Architecture

To request information from the Waves Network e.g. Transaction history, balances, etc., more flexibly than a traditional block explorer. We implemented a business intelligence system that aims to make blockchain data queryable and visualizable. We used a Python script (ETL) that extracts all valid blocks, transform and loads them into a relational database (MySQL) which represents our Operational data store.

3 Research Questions

As the demand for reporting and business intelligence has grown and become more and more noticeable, so needs business end-users to build their reports and visualizations.

Upon our stated above WavesBI Database, we added a reporting tool to visualize pieces of information about a given Token or asset. The aforementioned led us to write relatively complex SQL queries. Admitting that SQL-based Reporting Services opened up reporting to a whole new audience that hadn't used reporting tools before, in several companies, building reports is usually the developer's domain.

This research aims to implement a Visual SQL-query builder tool of a business intelligence system for the Waves platform and try to discover if, with the presented approach, usual users would be able to phrase and understand those kinds of queries.

For this purpose we conducted an empirical study and attempt to answer the three main questions:

- Is the semantic of a query understandable when using the Visual SQL-Query Builder Tool?
- How effective/useful is a Visual SQL-query Builder Tool?
- How usable is the Visual SQL-Query Builder Tool?

4 Research Methodology

To answer the research questions mentioned previously, we conducted an experiment with first-year students. We asked participants to answer an online survey in which we first introduce the topics that the study involves, as well as explanatory videos of the Visual SQL Query Builder tool. Second, participants are invited to answer various questions related to the tool.

5 Implementation

This part will introduce the visual programming method used to design and implement our tool and present its graphical user interface.

5.1 Visual Query Builder for SQL Query Creation

For the implementation of the Visual SQL query builder, some necessary conditions were specified. The Query Builder is designed to be a Web Application that gives users intuitive editing functionalities. The tool should be as simple as possible from the user's perspective by providing a user-friendly graphical interface to construct non-complex queries. A preliminary literature search was conducted to determine which type of VPL will be adopted and, secondly, which framework will be used to fulfill the previously stated criteria. So far, research has provided visualization techniques for queries [6, 7].

In our work, we used A Graph-based Model for SQL Constructs [8] in which the query is built using a network of visual boxes, interpreted as a series of functions that perform a specific task and have an input, output, or both.

Based on this approach, we designed the GUI as a Single Page Application running on the browser in which users can build visual queries then visualize retrieved data on the same page. The user interface is split into three distinct sections:

- The upper part provides a selectable Dropdown list in which the user selects the desired database fields to query.
- The middle section presents a canvas where the user builds the logic of the search request.
- In the bottom part, the user can visualize the query result from the MySQL database.

6 Evaluation Results

This study is conducted to delve into finding the effectiveness of using a visual query builder against traditional querying language (SQL). For this purpose, a quantitative data collection and analysis have been carried out. The results depicted below are divided into two parts. The first part will examine the correct answers of participants. On the other hand, the second part will investigate the user acceptance of our tool with respect to the research model proposed by Money [9] (Fig. 1):

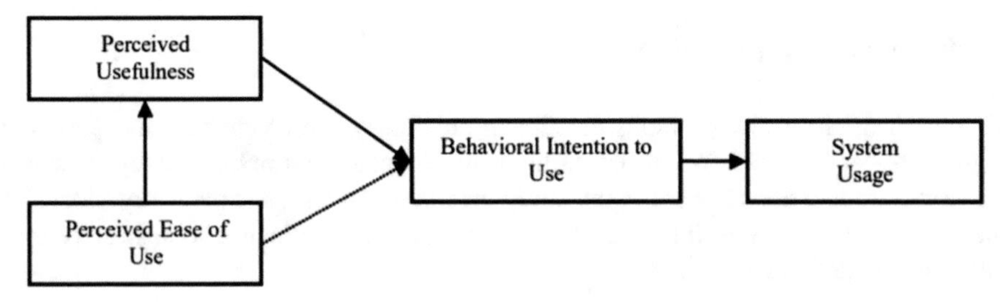

Fig. 1. Research model (Money 2004)

6.1 Investigation of the Comprehensibility of a Visual Query

This section studies the answers related to our Tool, to answer our research questions: "Is the semantic understandable when using the Visual SQL-query Builder tool?".

A total of 62 first-year students took part in the survey. The participants are categorized into two groups according to their overall knowledge in programming (experienced, unexperienced in programming). The goal of the classification is to examine the effectiveness of visual queries for each group.

The survey consisted of two questionnaires, including five questions. Each participant, will get randomly one version of the questionnaires.

The first questionnaire starts with a question about a visual query drawn with our Tool and then alternates with a question about a textual SQL query. On the other way around, the second questionnaire starts with a question about a textual SQL query. The complexity of the questions increases from the first one to the fifth one. Therefore, questions are as well divided into groups:

- Easy (Question_1)
- Medium (Question_2, Question_3)
- Hard (Question_4, Question_5)

The complexity groups enable us to interpret the difference between correct answers about visual queries and correct answers about textual queries. An ascending difference should mean that the significance of a visual representation of a query is growing. However, in this case, the correct answers about visual queries should be higher than the correct answers about textual queries.

The quantitative results are depicted in different comparisons levels in the following subsections.

All Correct Answers (Visual) Versus Correct Answers (Textual) from Both Questionnaires (174 Questions):

Table 1. All correct answers (visual) versus correct answers (textual) from both questionnaires.

Questions	Wrong answer	Correct answer
Visual questions	17%	83%
Textual questions	35%	65%

According to the table above (Table 1) the percentage of correct answers about visual queries are remarkably higher than textual ones. Otherwise, we have only half of the number of wrong answers with the visual approach. As mentioned above, the population can be regarded as experienced in SQL. Nevertheless, it did not change the fact that a visual query is more comprehensible than a textual query. The paired t-test shows a p-value ($p = 0.108$) greater than 0.05. Consequently, the difference between the correct answers about visual and textual SQL queries is statistically not significant.

All Correct Answers (Visual) Versus Correct Answers (Textual) of Experienced Participants:

Table 2. All correct answers (visual) versus correct answers (textual) of experienced participants.

Questions	Wrong answer	Correct answer
Visual questions	19%	81%
Textual questions	32%	68%

The table above (Table 2) shows one more time higher percentage of correct answers when using visual queries even when it comes to experienced participants in programming. Moreover, a 13% improvement of the correct answers in an experienced population greatly influences visual queries' effectiveness. The paired t-test of the visual version and the textual version in this context is statically not significant (p-value = 0.311).

All Correct Answers (Visual) Versus Correct (Textual) Answers of Unexperienced Participants:

Table 3. All correct answers (visual) versus correct (textual) answers of unexperienced participants.

Questions	Wrong answer	Correct answer
Visual questions	14%	86%
Textual questions:	42%	58%

We can notice better results than the experienced population, according to the table shown above (Table 3). In this population, the correct answers to questions about visual queries are 28% higher than the textual version. We also remark that the number of wrong answers for the textual query is 3 times higher than the visual ones the calculated (p-value = .005) of this comparison reveals that the difference between the correct answers of a visual query and a textual query is statistically significant.

Correct Answers' Analysis of Experienced and Unexperienced Participants According to Questions' Difficulty

According to the difficulty groups, correct answers were analyzed to look for more consistent results and study participants' behavior closer. The outcomes have proven that the visual version of a query is again advantageous. In addition, the percentage of correct answers about a visual version of a SQL query increases as the difficulty rises, i.e., the more complex the query, the more relevant the use of a visual language becomes. In contrast, experienced participants' results did not show significant growth in the percentage of correct answers.

The paired t-tests according to the stated analysis are statistically not significant.

6.2 Usability and Ease of Use Analysis According to TAM

As the participants will answer questions about visual and textual queries, and besides the presented video in the documentation part, participants will get enough knowledge about our tool. This part of the results examines the gathered ratings in order to determine how useful and how easy our system is. More specifically, the user acceptance of the tool will be discussed by investigating the relationships between the factors regarding our research model.

6.3 TAM Items and Cronbach's Alpha Analysis

After we analyzed TAM items and calculated the internal reliability (Cronbach's Alpha) coefficient, in this section we cover the participants' rate results. As shown in the following table, the average rate of the factors (Perceived ease of use, Perceived usefulness) is around five. These results show that the average rate regarding the usefulness is "More or less agree".

Table 4. TAM items & Cronbach's Alpha analysis.

Factors	Questions	Mean Rates	Std dev	AVG Rates	Cronbach's Alpha
Perceived ease of use	PEOU_01	4.93	1.42	4.90	0.887
	PEOU_02	4.77	1.50		
	PEOU_03	4.75	1.42		
	PEOU_04	5.17	1.42		
Perceived usefulness	PUSEF_01	5.33	1.27	5.34	0.838
	PUSEF_02	5.30	1.38		
	PUSEF_03	5.12	1.59		
	PUSEF_04	5.64	1.26		
Behavioral intention	BI_01	5.481	1.21	5.02	0.873
	BI_02	4.54	1.71		
	BI_03	5.04	1.63		
Usage intention	UI_01	3.64	1.70	3.8	0.878
	UI_01	3.96	1.98		

To resume the average rate is enough satisfying for a first version, knowing the usefulness's average rate is above 5.

Based on the answers from participants to the TAM questions in both questionnaires, all variables have a Cronbach's alpha coefficients above 0.8 (see Table 4), which is good [10].

7 Conclusion

There are numerous approaches, methods, and software packages for extracting data from systems to generate management and decision values. These have changed considerably over time and have adapted to new possibilities and challenges. Indeed scientific research in the field of reporting, data collection, and analysis is a significant factor for technological advancement.

This work tries to contribute to the enhancement of BI and querying technologies alongside blockchain and its applications by allowing a broader public to access this technology regardless of a person's technical abilities or knowledge background in computer science.

It should be noted that technological advancement in the field of visual programming greatly participates in this work's realization. To sum up, we have, in this context, joined a visual programming technique and a structured query language for querying databases in order to provide people with little or no knowledge in programming to build basic requests and visualize information related to blockchain data.

After the design and implementation of our Visual Query Builder tool, the data analysis of our online survey provided us great results, which is very encouraging in terms of advancement in database querying and data visualization of blockchain transactions from a non-It specialist perspective.

Integrating the presented approach in blockchain-based technologies could make the data stored on the blockchain easier to analyze and interpret, additionally allows the use of reporting technologies to the blockchain area.

References

1. Burnett, M., McIntyre, D.: Visual programming. Computer-Los Alamitos **28**, 14–14 (1995)
2. Göttler, H.: Graph grammars, a new paradigm for implementing visual languages. In: Ghezzi, C., McDermid, J. A. (eds.) ESEC 1989. LNCS, vol. 387, pp. 336–350. Springer, Heidelberg (1989). https://doi.org/10.1007/3-540-51635-2_48
3. Law, Y.-N., Wang, H., Zaniolo, C.: Query languages and data models for database sequences and data streams. In: 30th VLDB, September 2004
4. Shu, N.C.: Visual Programming. Van Nostrand Reinhold, New York (1988)
5. Catarci, T., Santucci, G.: Are visual query languages easier to use than traditional ones? An experimental proof. In: Kirby, M.A.R, Dix, A., Finlay, J.E. (eds.) People and Computers X: Proceedings of HCI 1995, Hudders-field, August 1995, Cambridge Programme on Human-Computer Interaction. Cambridge University Press (1995)
6. Balkir, N.H., Ozsoyoglu, G., Ozsoyoglu, Z.M.: A Graphical query language: Visual and its query processing. IEEE Trans. Knowl. Data Eng. **14**(5), 955–978 (2002)
7. Jaakkola, H., Thalheim, B.: Visual SQL – high-quality ER-based query treatment. In: Jeusfeld, M.A., Pastor, Ó. (eds.) ER 2003. LNCS, vol. 2814, pp. 129–139. Springer, Heidelberg (2003). https://doi.org/10.1007/978-3-540-39597-3_13
8. Papastefanatos, G., Kyzirakos, K., Vassiliadis, P. and Vassiliou, Y.: Hecataeus: a framework for representing SQL constructs as graphs. In: Proceedings of 10th International Workshop on Exploring Modeling Methods for Systems Analysis and Design-EMMSAD, vol. 5 (2005)
9. Money, W., Turner, A.: Application of the technology acceptance model to a knowledge management system. In: Proceedings of the 37th Hawaii International Conference on System Sciences (HICSS'04) - Track 8 (2004)
10. Gliem, J., Gliem, R.: Calculating, interpreting, and reporting Cronbach's alpha reliability coefficient for Likert-type scales. In: 2003 Midwest Research to Practice Conference in Adult, Continuing (2003)

On (Multi-stage) Proof-of-Works

Paolo D'Arco[1] and Francesco Mogavero[2(✉)]

[1] Dipartimento di Informatica, University of Salerno, Via Giovanni Paolo II, 132 I-84084 Fisciano, SA, Italy
pdarco@unisa.it

[2] Dipartimento di Ingegneria Informatica, Automatica e Gestionale (Antonio Ruberti), Sapienza University of Rome, Via Ariosto, 25 I-00185 Roma, Italy
mogavero@diag.uniroma1.it

Abstract. In this paper we analyze permissionless blockchain protocols, whose distributed consensus algorithm lies on a Proof-of-Work composed of $k > 1$ consecutive hash-puzzles, that have to be solved sequentially. Our contribution is twofold. First, under common assumptions in the literature, we provide a *closed-form* expression for the *mining probability* of a miner, that is, the probability that the miner completes the Proof-of-Work of the next block to be added to the blockchain before every other miner does. Second, we show that, contrary to single-stage Proof-of-Works (i.e., $k = 1$), in multi-stage Proof-of-Works, the mining probability *might not* be strictly related to the miner hash rate. This feature could be exploited by a smart miner, and could cause fairness and centralization issues in mining, which make the design of practical multi-stage Proof-of-Work blockchain protocols not trivial.

Keywords: Mining probability · Hypoexponential random variable · Proof-of-Work · Blockchain

1 Introduction

Single-stage Proof-of-Work blockchains. Bitcoin [1] was presented as a decentralized cryptocurrency working on top of a blockchain, which, in turn, was proposed as a public, tamper-resistant, distributed, and decentralized transaction ledger, maintained and replicated *entirely* and *consistently* by anonymous, unpermissioned, and trustless nodes, in a weakly synchronized [2] peer-to-peer network. The Bitcoin blockchain is structured as a chain of blocks. Each block has a set of block headers. Block headers include a hash pointer to a *Merkle tree* storing some transactions, and a hash pointer to the previous block in the chain [3]. Bitcoin transactions are mainly used for Bitcoin transfers between Bitcoin *addresses*. The only way to extend the Bitcoin blockchain, and to create new coins, is by *mining* a new block that accommodates a valid set of transactions. In order to mine a block, it is required to complete a Proof-of-Work (PoW, for short), which consists of a hash-puzzle to be solved. To do it, every competitor, called *miner*, is required to compose a new block containing a Merkle tree of valid transactions,

J. Prieto et al. (Eds.): BLOCKCHAIN 2021, LNNS 320, pp. 91–105, 2022.
https://doi.org/10.1007/978-3-030-86162-9_10

and to find a *nonce* to add to the block headers, in such a way that the double SHA-256 digest of the set of block headers is, in binary, lower than the *target* value of the hash-puzzle [3]. The first miner who finds a valid nonce for his block – the first miner who completes the PoW – is the winner. He broadcasts his block to the network, and he gets as a reward an amount of coins, given by the sum of the current *block reward* value and the *transaction fees* of every transaction in the mined block. At the same time, he starts mining the next block of the chain, following the one just mined. Any other *non-faulty* (i.e. *honest*) miner receives the block, and verifies that the block and its transactions are valid. If verifications are successful, he adds the block to his local copy of the blockchain, and then starts mining the next block in chain of the one received.

Bitcoin Scalability Problem and Multi-stage PoWs. To preserve security [4], Bitcoin blocks are limited to up 1 MB in size, and a new block is appended to the chain with an average latency of 10 min. Given these constraints, Bitcoin processes 7 transactions per second, making it unable to *scale* as a large-scale cryptocurrency exchange system. This limit appears particularly strict if compared to centralized currency exchange systems as VISA, which can process up to 20,000 transactions per second. Over the years, researchers have investigated how to improve Bitcoin's scalability without lowering its security and decentralization [5]. Sarkar has been the first researcher (and the last so far) to propose a multi-stage PoW blockchain protocol [6]. In a nutshell, a PoW is multi-stage if it is composed of $k \geq 1$ consecutive hash-puzzles, also called *stages*, that have to be solved sequentially. Every miner has to sequentially add k *nonces* to the headers of a block to complete his multi-stage PoW. Each miner can start the hash-puzzle number $s+1$ of his PoW *only after* he has found a valid nonce for the hash-puzzle number s. The first miner who completes the last hash-puzzle of his PoW is the winner. Besides this innovative multi-stage PoW mechanism, the author also proposed a pipeline-like mining architecture, in which multiple consecutive blocks are mined simultaneously. The goal was to obtain a lower inter-block latency, and, consequently, a higher transaction processing rate. However, Sarkar's work lacks a consistent and formal analysis of multi-stage PoWs, and this motivated us to provide it, in order to give scientific results that may facilitate further research and possibly the development of multi-stage blockchains in the future.

Organization of the Paper and Our Contribution. In Sect. 2, we describe Sarkar's *multi-stage PoW blockchain* protocol and point out some issues we found in its design. In Sect. 3, we recall a result of Houy [7], which shows that, assuming that number of miners, their hash rates, and the difficulty[1] are constant over time[2], in single-stage PoWs, as for the ones used in Bitcoin or Ethereum, the mining probability of a miner, that is, the probability that the miner completes the PoW of the next block to be added to the blockchain before every other miner does,

[1] Mathematically, it is the inverse of the hash puzzle target value.
[2] These assumptions are common in the blockchain literature [8–10].

coincides with the ratio of the global hash rate[3] the miner owns. As our first contribution, in Sect. 4 we generalize Houy's result to also include multi-stage PoWs, providing a *closed-form* expression for the mining probability valid in PoWs composed of $k \geq 1$ consecutive hash-puzzles. As our second contribution, in Sect. 5 we show that, if $k \geq 2$, then the mining probability *might not* be equal to the ratio of the global hash rate held by the miner. Finally, we conclude the paper in Sect. 6.

2 Multi-stage PoWs

The first multi-stage PoW protocol dates back to 2019 [6], and has been reviewed in 2020 with minor changes [12]. The author proposed an alternative mining game dividing the PoW in $k > 1$ consecutive hash-puzzles, that have to be solved sequentially. Let us describe the protocol more formally.

Hardware Incompatible Hash Functions. Two hash functions are hardware incompatible if, any kind of ASIC[4], or other special-purpose hardware, that allows computing the output of one function faster than general-purpose mining hardware, cannot be easily reconfigured to provide an advantage over general-purpose hardware in the computation of the other function. The protocol employs μ hardware incompatible hash functions, $G_0, \ldots G_{\mu-1}$. There is no correlation between k and μ. The purpose of using hardware incompatible hash functions is to make it harder for miners to obtain high hash rate values in multiple stages. The author suggested the NIST finalists for the SHA-3 competition as a valid set of hardware incompatible hash functions.

Genesis Blocks Composition and Their PoW. The first k blocks of the blockchain, $B_0, \ldots, B_{k-1}$ are the genesis blocks. They do not carry any transaction, and they must be mined to bootstrap the protocol and mine some initial coins. For each $i \in \{0, \ldots, k-1\}$, the structure of the i-th genesis block B_i is:

i,
bdigest$_i$,
t_i, η_i, τ_i, α_i, c_i

where

- i is the block number, such that $0 \leq i \leq k-1$
- $\text{bdigest}_0 = H_0\,(0, t_0, \eta_0, \tau_0, \alpha_0, c_0)$
- $\text{bdigest}_i = H_i\,(\text{bdigest}_{i-1}, t_i, \eta_i, \tau_i, \alpha_i, c_i)$, such that $1 \leq i \leq k-1$
- t_i is the target value of the hash puzzle of block i
- η_i is the nonce of block i

[3] The hash rate indicates the number of trials in the PoW game a miner performs per second, and is measured in *hash/s* [11]. By *global* we identify the number of trials performed per second by the entire network.

[4] Application specific integrated circuit.

- τ_i is the timestamp of the completion time of block i's PoW
- α_i is the address of the recipient of block i's reward
- c_i is the reward for mining block i

The PoW of the genesis block i is valid if $\text{bdigest}_i < t_i$, and is very similar to Bitcoin PoW.

General Blocks Composition and Their PoW. The PoW of a general block is divided into $k > 1$ consecutive stages. Given $s \in \{0, \ldots, k-1\}$, the hash function of the s-th stage is $H_s = G_{s \bmod \mu}$. Each stage target and difficulty parameters are set such that, *globally*, the expected time to solve each stage, denoted by T, is the same. There is no fixed amount of coins given to a single user as *block reward*, since the rewarding system is divided into *stage rewards*. A stage reward is a fixed amount of coins, given to the user that successfully completes a stage of a block. It consists of newly created coins, which will be effectively generated once the block has been fully mined and submitted to the network. Moreover, the winners of different stages of a block also have to divide the transaction fees obtained from the transactions in the block. The structure of a general block B_{bn} is:

$$\begin{array}{|l|}\hline \text{bn}, \\ \text{bdigest}, \\ \mathcal{L}, \\ t_0,\ \eta_0,\ \tau_0,\ \alpha_0,\ \mathsf{c}_0 \\ t_1,\ \eta_1,\ \tau_1,\ \alpha_1,\ \mathsf{c}_1 \\ \vdots \\ t_{k-1},\ \eta_{k-1},\ \tau_{k-1},\ \alpha_{k-1},\ \mathsf{c}_{k-1} \\ \hline \end{array}$$

where

- bn $\geq k$ is the block number
- bdigest is the digest of the block
- $\mathcal{L}$ is the possibly empty hash tree of transactions carried by the block
- t_s is the target value of the hash puzzle of stage s
- η_s is the nonce of stage s
- τ_s is the timestamp of the completion time of stage s's hash-puzzle
- α_s is the address of the recipient of stage s's reward
- c_s is the reward for completing stage s

Consider a general block B_{i+k}, with $i \geq 0$. The outputs of the stages, $g_{i+k,0}, \ldots g_{i+k,k-1}$, are

- $g_{i+k,0} = H_0\left(\text{bdigest}_i, i+k, \mathsf{RH}(\mathcal{L}_{i+k}), t_{i+k,0}, \alpha_{i+k,0}, c_{i+k,0}, \tau_{i+k,0}, \eta_{i+k,0}\right)$
- $g_{i+k,1} = H_1\left(\text{bdigest}_{i+1}, g_{i+k,0}, t_{i+k,1}, \alpha_{i+k,1}, c_{i+k,1}, \tau_{i+k,1}, \eta_{i+k,1}\right)$
- ...
- $g_{i+k,k-1} = H_{k-1}\big(\text{bdigest}_{i+k-1}, g_{i+k,k-2}, t_{i+k,k-1}, \alpha_{i+k,k-1}, c_{i+k,k-1}, \tau_{i+k,k-1},$ $\eta_{i+k,k-1}\big)$

where

- $\mathsf{RH}(\mathcal{L}_{i+k})$ is the root hash of the possibly empty hash tree that stores block's transactions
- $g_{i+k,s}$ is the output of stage s
  $t_{i+k,s}$ is the target value of the hash puzzle of stage s
- $\eta_{i+k,s}$ is the nonce of stage s
- $\tau_{i+k,s}$ is the timestamp of the completion time of stage s
- $\alpha_{i+k,s}$ is the address of the recipient of stage s's reward
- $\mathsf{c}_{i+k,s}$ is the reward for completing stage s

The PoW is valid if and only if $g_{i+k,s} < t_{i+k,s}$, for each $s \in \{0, \dots, k-1\}$. Finally, the value of $g_{i+k,k-1}$ is assigned to bdigest_{i+k}. For any $s \geqslant 1$, stage s of the PoW of block B_{i+k} requires two inputs: the output of stage $s-1$ and bdigest_{i+s}.

The second required input is always available in any stage of block B_k since the network has already mined blocks $B_0, \dots, B_{k-1}$ to bootstrap the protocol. Later, we will exploit this property in our analysis on the mining probability for block B_k. Due to the nature of the PoW, the author suggested a pipeline-like mining architecture, in which miners working in the same pipeline can freely partition themselves into k groups. The architecture may be realized by letting the k groups work in parallel and on different parts of the same task. If the number of the already mined general blocks in the blockchain is $i \geqslant 0$, then, for each $s \in \{0, \dots, k-1\}$, group s works on stage s of block B_{i+k-s}. For any s $\in \{0, \dots, k-1\}$, the hash rate of group s is equal to the sum of the hash rates of the group miners. Hence, miners belonging to different groups of the same pipeline mine collaboratively to complete the PoW before the other pipelines do and obtain the block reward. At the same time, miners in the same group mine *competitively*. Finally, separate pipelines compete against each other to mine the blocks. Generally, once a miner in a group completes a stage, he broadcasts to the network all the information necessary to start the successive stage of the same block and to prove that the previous stage was completed successfully. The pipeline-like mining architecture is shown in Fig. 1.

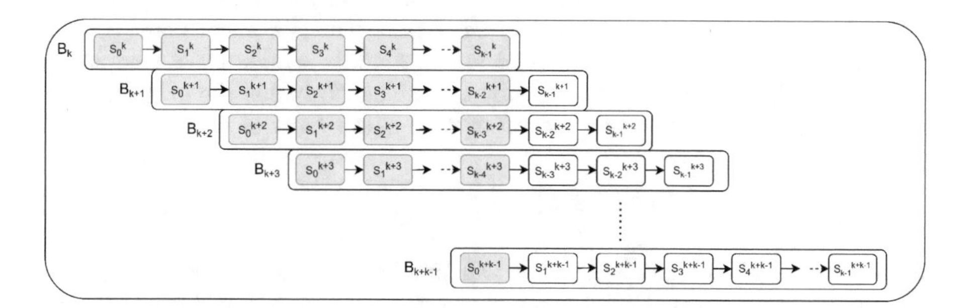

Fig. 1. The pipeline-like mining architecture with $i = 0$. The x axis denotes the time. The PoWs of at maximum k blocks can be active in the same moment.

Remarks. Notice that the design of the proposed protocol raises some issues.

1. Some *hardware incompatible* hash functions may be present in multiple stages. In this case, the advantage the miner has gained by buying an ASIC for those hash functions may be worthwhile in many stages.
2. A pipeline-like mining architecture cannot be easily constructed, since stage mining is a stochastic event. Therefore, the pipeline-like block mining architecture can hardly ever be perfectly synchronized among different stages. Considering a single pipeline, given a $i \geq 0$ and a $s \in \{1, \ldots, k-1\}$, it is easy to see that the probability that stage s of block B_{i+k} is completed exactly at the same time of completion of stage $s-1$ of block B_{i+k+1} is negligible.

Other possible security issues regarding this protocol have been raised in [13, 14].

3 Mining Probability in Single-Stage PoWs

A Mathematical Model for Individual Miners' PoWs. Let $h > 0$ be the hashing power of a miner, let $t > 0$ be the current target value of the hash-puzzle, and let $d = 2^{256}/t$ be the current difficulty of the hash-puzzle. The hash-puzzle output space is $\{0,1\}^{256}$, since the protocol uses a double SHA-256 hash function.

Using the *random oracle* assumption to model the hash function (assumption commonly used in the Bitcoin and the blockchain literature [15–17]), for any distinct input x, the output of the hash function can be considered as *uniformly* and *independently* distributed in the output space. It follows that, under the assumptions of constant hash-puzzle difficulty, every trial the miner makes to complete the PoW is an independent experiment, also known as *Bernoulli trial* [15,18], having probability of success $prob = t/2^{256} = 1/d$. Under the assumptions of constant hash-puzzle difficulty and constant hash rate, the number of successes *per unit of time* in a sequence of *identically distributed* Bernoulli trials is described by a *homogeneous Poisson point process* N, having *rate* parameter $\lambda = h/d$ [8,10]. The rate parameter indicates the average number of successes per unit of time[5]. The *homogeneous Poisson point process* is a stochastic process, which in our scenario describes the number of PoWs the miner completes per *interval of time*; a new point found in the homogeneous Poisson point process corresponds bijectively to a new valid *nonce* found – a new PoW completed, or, equivalently, a new block mined – by the miner. Formally, the probability of having $k \in \mathbb{N}_0$ successes in the interval of time long t units of time is [8]

$$P\big(N(t) = k\big) = \frac{(\lambda\, t)^k \, \exp(-\lambda\, t)}{k!} \, . \tag{1}$$

Block Inter-arrival Times. The inter-arrival time between two consecutive points in a homogeneous Poisson point process, as well as the inter-arrival time between the moment a miner starts the PoW of a block (i.e. "*point 0*" in his Poisson point

[5] In a homogeneous Poisson point process the average number of successes in t units of time is $\lambda\, t$ [19]. If we let t be a unit of time, i.e. $t = 1$, then the statement holds.

process) and the time the miner finds the first point (i.e. "*point 1*" in his Poisson point process), are described by an *exponential random variable* X having the same λ-parameter of the process [3,19].

Mining Probability. Consider $M \geq 2$ competing miners trying to mine the block with *block height* $\gamma \in \mathbb{N}_0$. The block height value of a block indicates the number of blocks in the blockchain preceding that block. For each $p \in \{1,\ldots,M\}$, let X_p be the exponential random variable describing the inter-arrival time between two consecutive blocks found by miner number p. The parameter of X_p is $\lambda_p = h_p/d$, where $h_p > 0$ is the hash rate of miner number p, and $d > 0$ denotes the difficulty of the hash-puzzle. Denote by $f_{X_p}(t) = P(X_p = t) = \lambda_p\, e^{-\lambda_p t}$ the *probability density function*, and by $R_{X_p}(t) = P(X_p > t) = e^{-\lambda_p t}$ the *survival function* relative to X_p. If each miner starts mining block γ at the same time, then the winner is the miner whose Poisson point process first finds a point. Notice that the processes are independent from each other. Therefore, for each $p \in \{1,\ldots,M\}$, the probability that miner p mines the block, is given by:

$$\begin{aligned} P\,(p \text{ mines block } \gamma) &= \int_0^{\infty} f_{X_p}(t) \prod_{\substack{z=1 \\ z\neq p}}^{M} R_{X_z}(t)\, dt = \int_0^{\infty} \lambda_p \exp(-\lambda_p\, t) \prod_{\substack{z=1 \\ z\neq p}}^{M} \exp(-\lambda_z\, t)\, dt \\ &= \int_0^{\infty} \lambda_p \prod_{z=1}^{M} \exp(-\lambda_z\, t)\, dt = \int_0^{\infty} \lambda_p \exp(-\sum_{z=1}^{M} \lambda_z\, t)\, dt \\ &= \lambda_p \int_0^{\infty} \exp(-\sum_{z=1}^{M} \lambda_z\, t)\, dt = \lambda_p \frac{-1}{\sum_{z=1}^{M} \lambda_z} \times \left[\exp(-\sum_{j=0}^{M} \lambda_j\, t)\right]_0^{\infty} = \frac{h_p}{\sum_{z=1}^{M} h_z}\,. \end{aligned} \tag{2}$$

Hence, the mining probability is independent of the hash-puzzle difficulty and target values, and is equal to the ratio of the global hash rate the miner holds.

Global Mining. A new PoW is *globally* completed every time one of the M miners "finds a point" in his Poisson point process. Thus, the global block mining event can be mathematically represented as the *superposition*[6] of the M homogeneous Poisson point processes. Due to the mutual independence of the processes, their superposition is still a homogeneous Poisson point process having *rate* parameter $\lambda_{glob} = \sum_{p=1}^{M} \lambda_p = \sum_{p=1}^{M} h_p/d$. Hence, the waiting time until a new PoW is globally completed is given by an exponential random variable with rate parameter λ_{glob}. The expected value of the exponential random variable – the expected time until a new block is mined –, λ_{glob}^{-1}, should be set to be $\lambda_{glob}^{-1} = 10\,\text{min}$ [3,18].

4 Mining Probability in Multi-stage PoWs

In this Section, we extend the result of expression (2) and propose a closed-form[7] expression for the mining probability. The closed-form expression works for every blockchain protocol whose PoW is composed of $k \geqslant 1$ consecutive hash-puzzles, under the assumptions that the number of miners, their hash rates, and the difficulty values of the hash-puzzles are constant throughout time.

[6] The merge of the N processes by cumulating their respective points.

[7] Expression computable in a finite number of standard operations.

As announced in Sect. 2, in the protocol proposed by Sarkar, the expression is applicable to the PoW of the first general block B_k. Indeed, all the genesis blocks have already been mined before the PoW of block B_k had started. Consequently, as soon as a miner completes stage number $s \leq k-2$ of the PoW of block B_k, he can immediately start stage $s+1$. Conversely, given a block B_{i+k}, such that $i \geqslant 1$ and $s \in \{0,\ldots,k-2\}$, when a miner completes stage s of block B_{i+k}, he may need to wait until block B_{i+s+1} is mined and $bdigest_{i+s+1}$ becomes available.

4.1 Notation

We use the following notation:

- the integer $k \geq 1$ denotes the number of consecutive hash-puzzles a PoW is composed of.
- the integer $M \geq 2$ denoted the number of competing miners involved in the PoW of a block.
- $d_0,\ldots,d_{k-1}$ are the positive real values representing hash-puzzles difficulties.
- $h_{p,s}$ is a positive real value, denoting the hash rate of miner number p on stage s, for a given $p \in \{1,\ldots,M\}$ and a given $s \in \{0,\ldots,k-1\}$.
- $X_{p,s}$ is the exponential random variable describing the time miner number p takes to complete the stage s of a block, for a given $p \in \{1,\ldots,M\}$ and a given $s \in \{0,\ldots,k-1\}$.
- $\lambda_{p,s} = h_{p,s}/d_s$ is the positive real parameter of $X_{p,s}$, for a given $p \in \{1,\ldots,M\}$ and a given $s \in \{0,\ldots,k-1\}$.

Let $p \in \{1,\ldots,M\}$. The time miner number p takes to complete the entire PoW of a block is given by the *hypoexponential random variable* $X_{p,+_k} = \sum_{s=0}^{k-1} X_{p,s}$. The random variable $X_{p,+_k}$ is simply the sum of the time miner number p takes to complete every stage of his PoW. We can represent the time to complete a PoW of a block this way, since every single stage can be described by its own independent Poisson point process. The Poisson point process related to the first stage starts at the same time the PoW of the block does. Completing a stage is equal to finding the first point into its Poisson point process. As soon as the first point in a Poisson point process is found, the stage process is not relevant anymore, and the next stage (i.e. the next Poisson point process) starts. Therefore, the time to complete every single stage is described by an exponential random variable, while the time to complete the PoW is the sum of the individual exponential random variables corresponding to the stages.

4.2 The Hypoexponential Random Variable

Let us describe the hypoexponential random variable, to provide the basis for our computation of the closed-form expression for the mining probability. The *probability density function* and the *survival function* of the hypoexponential random variable have been presented by Scheuer [20], and, as *closed-form* expressions, by Amari and Misra [21].

Literature Results on the Hypoexponential Random Variable. Consider a miner number $p \in \{1, \ldots, M\}$. As explained in [20], the literature studies on the hypoexponential random variable have been divided into three major subcases, as follows:

1. In the first subcase, $\lambda_{p,s} = \lambda_{p,s'}$, for every pair $s, s' \in \{0, \ldots, k-1\}$, with $s' \neq s$. The hypoexponential random variable is reduced to an *Erlang random variable* with *shape* parameter k and *rate* parameter λ_p, with $\lambda_p = \lambda_{p,0} = \ldots = \lambda_{p,k-1}$.
2. In the second subcase, $\lambda_{p,s} \neq \lambda_{p,s'}$, for every pair $s, s' \in \{0, \ldots, k-1\}$, with $s' \neq s$.
3. In the third subcase, $\lambda_{p,s}$ can be either equal or not equal to $\lambda_{p,s'}$, for every pair $s, s' \in \{0, \ldots, k-1\}$, with $s' \neq s$.

The first two cases have been deeply studied in mathematics over the years, and we will not focus on them. Scheuer, Amari, and Misra have been the first researchers focusing on the third, and most general, scenario.

4.3 Probability Density Function and Survival Function

In this Subsection we recall the Amari and Misra's *closed-form* expressions for the *probability density function* and the *survival function* of a hypoexponential random variable [21]. We omit the majority of the low-level technical details that have led to the composition of the two expressions, which are not necessary for our goals, and refer the readers to the original papers for the complete mathematical analysis.

Let $p \in \{1, \ldots, M\}$. Let $f_{X_{p,+k}}(t) = P(X_{p,+k} = t)$ be the *probability density function*, and let $R_{X_{p,+k}}(t) = P(X_{p,+k} > t)$ be the *survival function* of $X_{p,+k}$. In order to compute $f_{X_{p,+k}}(t)$ and $R_{X_{p,+k}}(t)$, the exponential random variables $X_{p,s}$ and $X_{p,s'}$ (such that $s' \neq s$, and $s, s' \in \{0, \ldots, k-1\}$) which satisfy the condition $\lambda_{p,s} = \lambda_{p,s'}$ must be grouped together. Their common λ value must be denoted with a new parameter. For example, if $k = 4$, $\lambda_{p,0} = \lambda_{p,3}$, and $\lambda_{p,1} = \lambda_{p,2}$, then two new parameters, $\beta_{p,1}$ and $\beta_{p,2}$ are:

$$\lambda_{p,0} = \lambda_{p,3} = \beta_{p,1}; \quad \lambda_{p,1} = \lambda_{p,2} = \beta_{p,2} \, .$$

Denote by $r_{p,1}$ and $r_{p,2}$ the number of λ-parameters having value equal to $\beta_{p,1}$ and $\beta_{p,2}$, respectively. If the number of different λ values is a_p, then a_p pairs are obtained: $(\beta_{p,1}, r_{p,1}), \ldots, (\beta_{p,a_p}, r_{p,a_p})$. It holds that $\sum_{q_p=1}^{a_p} r_{p,q_p} = k$. The probability density function $f_{X_{p,+k}}(t)$ is ([20], Eq. 13)

$$f_{X_{p,+k}}(t) \stackrel{\text{def}}{=} B_p \sum_{q_p=1}^{a_p} \sum_{l_p=1}^{r_{p,q_p}} \frac{\Phi_{p,q_p,l_p}(-\beta_{p,q_p})}{(r_{p,q_p} - l_p)! \, (l_p - 1)!} \, t^{r_{p,q_p} - l_p} \exp(-\beta_{p,q_p} \, t) \, , \quad (3)$$

where

- $B_p = \left(\prod_{q_p=1}^{a_p} (\beta_{p,q_p})^{r_{p,q_p}}\right)$, (see [21], *Notation* Paragraph))
- $\Phi_{p,q_p,l_p}(t) = (-1)^{l_p-1} \times (l_p-1)! \times \sum_{\Omega_{2_p}(1)} \prod_{\substack{j_p=1 \\ j_p \neq q_p}}^{a_p} \binom{i_{j_p}+r_{p,j_p}-1}{i_{j_p}} \times \tau_{j_p}$ (see [21], Eq. 4)
- $\tau_{j_p} = (\beta_{p,j_p}+t)^{-(r_{p,j_p}+i_{j_p})}$ (see [21], between Eqs. 3 and 4)
- $\Omega_{2_p}(1) = \{i_{j_p} \in \mathbb{N}_0,\ \text{for each}\ j_p \in \{1,\ldots,a_p\},\ \text{s.t.}\ \sum_{\substack{j_p=1 \\ j_p \neq q_p}}^{a_p} i_{j_p} = l_p - 1\}$ (see [21], *Notation* Paragraph)

The survival function is ([21], Eq. 3)

$$R_{X_{p,+k}}(t) \stackrel{\text{def}}{=} B_p \sum_{q_p=1}^{a_p} \sum_{l_p=1}^{r_{p,q_p}} \frac{\Psi_{p,q_p,l_p}(-\beta_{p,q_p})}{(r_{p,q_p}-l_p)!\,(l_p-1)!}\, t^{r_{p,q_p}-l_p} \exp(-\beta_{p,q_p}\, t)\,, \quad (4)$$

where B_p is defined as above, and

- $\Psi_{p,q_p,l_p}(t) = -(-1)^{l_p-1} \times (l_p-1)! \times \sum_{\Omega_{2_p}(0)} \prod_{\substack{j_p=0 \\ j_p \neq q_p}}^{a_p} \binom{i_{j_p}+r_{p,j_p}-1}{i_{j_p}} \times \tau_{j_p}$ (see [21], Eq. 5. See [21], *Notation* Paragraph for the "−" sign)
- τ_{j_p} as above
- $\Omega_{2_p}(0) = \{i_{j_p} \in \mathbb{N}_0,\ \text{for each}\ j_p \in \{0,\ldots,a_p\},\ \text{s.t.}\ \sum_{\substack{j_p=0 \\ j_p \neq q_p}}^{a_p} i_{j_p} = l_p - 1\}$ (see [21], *Notation* Paragraph)
- $\beta_{p,0} = 0,\ r_{p,0} = 1$, (see [21], between Eqs. 2 and 3)

4.4 A *Closed-form* Expression for the Mining Probability

Let $p \in \{1,\ldots,M\}$ be a miner, and γ a non-negative integer. The probability that miner number p is the first one in completing the PoW of block γ is

$$P(p \text{ mines block } \gamma) = \int_0^\infty f_{X_{p,+k}}(t) \prod_{\substack{z=1 \\ z \neq p}}^{M} R_{X_{z,+k}}(t)\, dt\,. \quad (5)$$

By substituting expressions (3) and (4), the integration above is equal to

$$\int_0^\infty \left(B_p \sum_{q_p=1}^{a_p} \sum_{l_p=1}^{r_{p,q_p}} \frac{\Phi_{p,q_p,l_p}(-\beta_{p,q_p})}{(r_{p,q_p}-l_p)!\,(l_p-1)!}\, t^{r_{p,q_p}-l_p} \exp(-\beta_{p,q_p}\, t) \right) \times \left(\prod_{\substack{z=1 \\ z \neq p}}^{M} B_z \left(\sum_{q_z=1}^{a_z} \sum_{l_z=1}^{r_{z,q_z}} \frac{\Psi_{z,q_z,l_z}(-\beta_{z,q_z})}{(r_{z,q_z}-l_z)!\,(l_z-1)!}\, t^{r_{z,q_z}-l_z} \exp(\beta_{z,q_z})\, t \right) \right) dt\,.$$

Due to the linearity of integration, the expression above is equal to

$$\Big(\prod_{z=1}^{M} B_z\Big) \int_0^{\infty} \Big(\prod_{z=1}^{M} \big(\sum_{q_z=1}^{a_z} \sum_{l_z=1}^{r_{z,q_z}} \frac{\Theta_{z,q_z,l_z}\,(-\beta_{z,q_z})}{(r_{z,q_z}-l_z)!\,(l_z-1)!}\, t^{r_{z,q_z}-l_z}\, \exp(\beta_{z,q_z})\, t\big)\Big)\, dt\,,$$

where

$$\Theta_{z,q_z,l_z}\,(-\beta_{z,q_z}) = \left\{ \begin{array}{ll} \Phi_{z,q_z,l_z}\,(-\beta_{z,q_z}), & \text{iff } z = p \\ \Psi_{z,q_z,l_z}\,(-\beta_{z,q_z}), & \text{otherwise} \end{array} \right\}$$

Due to the distributive property of multiplication over addition, and the linearity of integration, the expression above is equal to

$$\Big(\prod_{z=1}^{M} B_z\Big) \times \sum_{q_1=1}^{a_1} \sum_{l_1=1}^{r_{1,q_1}} \cdots \sum_{q_M=1}^{a_M} \sum_{l_M=1}^{r_{M,q_M}} \Big(\Big(\prod_{z=1}^{M} \frac{\Theta_{z,q_z,l_z}\,(-\beta_{z,q_z})}{(r_{z,q_z}-l_z)!\,(l_z-1)!}\Big) \times \int_0^{\infty} t^{\sum_{z=1}^{M}(r_{z,q_z}-l_z)}\, e^{-\sum_{z=1}^{M}\beta_{z,q_z}\, t}\, dt\Big)\,. \tag{6}$$

A corollary of the Gamma Function definition.
With $t_2 = \eta t$, for each $\alpha \in \mathbb{N}_0$, and $\eta \in \mathbb{R}$ such that $\eta \neq 0$, it holds that

$$\int_0^{\infty} t^{\alpha} \times \exp(-\eta\, t) dt \;=\; \int_0^{\infty} \frac{1}{\eta^{\alpha}} {t_2}^{\alpha} \times \exp(-t_2) \times \frac{1}{\eta} dt_2 \;=\; \frac{1}{\eta^{\alpha+1}} \int_0^{\infty} {t_2}^{\alpha} \times \exp(-t_2)\, dt_2\,.$$

By definition of Γ, the *Gamma function* [22], it holds that $\int_0^{\infty} {t_2}^{\alpha} \times \exp(-t_2)\, dt_2 = \Gamma(\alpha+1) = \alpha!$; and therefore, $\frac{1}{\eta^{\alpha+1}} \int_0^{\infty} {t_2}^{\alpha} \times \exp(-t_2)\, dt_2 = \frac{\alpha!}{\eta^{\alpha+1}}$.

We can apply the corollary to expression (6). Indeed, for each $z \in \{1, \ldots M\}$, we have that $l_z, r_{z,q_z} \in \mathbb{N}$ and $l_z \leq r_{z,q_z}$, and hence, $\sum_{z=1}^{M}(r_{z,q_z} - l_z) \in \mathbb{N}_0$. Moreover, for each $z \in \{1, \ldots M\}$, we have that $\beta_{z,q_z} > 0$, and hence $\sum_{z=1}^{M} \beta_{z,q_z} \neq 0$. Therefore, by setting $\alpha = \sum_{z=1}^{M}(r_{z,q_z} - l_z)$ and $\eta = \sum_{z=1}^{M} \beta_{z,q_z}$, applying the corollary it holds that

$$\int_0^{\infty} t^{\sum_{z=1}^{M}(r_{z,q_z}-l_z)}\, e^{-\sum_{z=1}^{M}\beta_{z,q_z}\, t}\, dt \;=\; \int_0^{\infty} t^{\alpha} \times \exp(-\eta\, t) dt \;=\; \frac{\alpha!}{\eta^{\alpha+1}}$$
$$= \big(\sum_{z=1}^{M} \beta_{z,q_z}\big)^{-1-\sum_{z=1}^{M}(r_{z,q_z}-l_z)} \times \Big(\big(\sum_{z=1}^{M}(r_{z,q_z} - l_z)\big)!\Big)\,.$$

Hence, expression (6) is equal to

$$\Big(\prod_{z=1}^{M} B_z\Big) \times \sum_{q_1=1}^{a_1} \sum_{l_1=1}^{r_{1,q_1}} \cdots \sum_{q_M=1}^{a_M} \sum_{l_M=1}^{r_{M,q_M}} \Big(\Big(\prod_{z=1}^{M} \frac{\Theta_{z,q_z,l_z}\,(-\beta_{z,q_z})}{(r_{z,q_z}-l_z)!\,(l_z-1)!}\Big) \times \big(\textstyle\sum_{z=1}^{M}\beta_{z,q_z}\big)^{-1-\sum_{z=1}^{M}(r_{z,q_z}-l_z)} \times$$
$$\times \Big(\big(\textstyle\sum_{z=1}^{M}(r_{z,q_z} - l_z)\big)!\Big) = \Big(\prod_{z=1}^{M} B_z\Big) \times$$
$$\times \sum_{q_1=1}^{a_1} \sum_{l_1=1}^{r_{1,q_1}} \cdots \sum_{q_M=1}^{a_M} \sum_{l_M=1}^{r_{M,q_M}} \left(\frac{\Big(\prod_{z=1}^{M} \frac{\Theta_{z,q_z,l_z}\,(-\beta_{z,q_z})}{(l_z-1)!}\Big) \times Multinomial_{(\sum_{z=1}^{M}(r_{z,q_z}-l_z));\,(r_{1,q_1}-l_1),\,\ldots,\,(r_{M,q_M}-l_M)}}{\big(\sum_{z=1}^{M}\beta_{z,q_z}\big)^{1+\sum_{z=1}^{M}(r_{z,q_z}-l_z)}}\right),$$

where

$$Multinomial_{\left(\sum_{z=1}^{M}(r_{z,q_z}-l_z)\right);\,(r_{1,q_1}-l_1),\,\ldots\,,(r_{M,q_M}-l_M)} = \frac{\left(\sum_{z=1}^{M}(r_{z,q_z}-l_z)\right)!}{\prod_{z=1}^{M}\left((r_{z,q_z}-l_z)!\right)}$$

Is a multinomial coefficient. Letting

- $\Phi'_{z,q_z,l_z}(t) = \frac{\Phi_{z,q_z,l_z}(t)}{(l_z-1)!} = (-1)^{l_z-1} \times \sum_{\Omega_{2_z}(1)} \prod_{j_z} \binom{i_{j_z}+r_{z,j_z}-1}{i_{j_z}} \times \tau_{j_z}$
- $\Psi'_{z,q_z,l_z}(t) = \frac{\Psi_{z,q_z,l_z}(t)}{(l_z-1)!} = -(-1)^{l_z-1} \times \sum_{\Omega_{2_z}(0)} \prod_{j_z} \binom{i_{j_z}+r_{z,j_z}-1}{i_{j_z}} \times \tau_{j_z}$
- $\Theta'_{z,q_z,l_z}(-\beta_{z,q_z}) = \begin{cases} \Phi'_{z,q_z,l_z}(-\beta_{z,q_z}), & \text{iff } z=p \\ \Psi'_{z,q_z,l_z}(-\beta_{z,q_z}), & \text{otherwise} \end{cases}$

then, the mining probability of miner p is

$$P(p \text{ mines block } \gamma) = \left(\prod_{z=1}^{M} B_z\right) \times \sum_{q_1=1}^{a_1}\sum_{l_1=1}^{r_{1,q_1}} \cdots \sum_{q_M=1}^{a_M}\sum_{l_M=1}^{r_{M,q_M}}$$
$$\left(\frac{\left(\prod_{z=1}^{M}\Theta'_{z,q_z,l_z}(-\beta_{z,q_z})\right) \times Multinomial_{\left(\sum_{z=1}^{M}(r_{z,q_z}-l_z)\right);\,(r_{1,q_1}-l_1),\,\ldots\,,(r_{M,q_M}-l_M)}}{\left(\sum_{z=1}^{M}\beta_{z,q_z}\right)^{1+\sum_{z=1}^{M}(r_{z,q_z}-l_z)}}\right). \tag{7}$$

To the best of our knowledge, expression (7) is no further simplifiable. In an extended version of this abstract, available at [23], we show that, when $k = 1$, expression (7) reduces to expression (2).

5 Relation Between Hash Rate and Mining Probability

The mining probability in a multi-stage PoW can be practically computed with several tools, such as Matlab or Wolfram Mathematica. We used the *hypoexponential random variable* library of Wolfram Mathematica [24] to compute the mining probability through expression (5). Alternatively, we also implemented a prototypical Mathematica Library, which computes the mining probability directly through expression (7). Benchmarking results prove that, with $M \leq 5$ and $k \leq 5$, our implementation of expression (7) is faster than the built-in Mathematica library[8]. Therefore, one practical future application of expression (7) may be a time-efficient computation of the mining probability. At the moment, however, our prototypical implementation is slower than the built-in Mathematica library with higher values for M and k. Optimized implementations of expression (7) may significantly improve the time performances[9].

[8] The benchmarking was performed on a HP Pavilion Laptop 15-cs2023nl, equipped with a quad-core CPU Intel Core™ i7-8565U 1.80 GHz CPU, 8 MB cache, and 16 GB (2 × 8) SO-DIMM SDRAM 2400MHz DDR4.

[9] Our source code, its comprehensive documentation, and the information regarding the benchmark results are available on GitHub: https://github.com/FraMog/MiningProbabilityMultiStageProof-of-Work.

In the following, we describe the relationship between the ratio of the global hash rate a miner holds and his mining probability. In particular, we prove that if $k > 1$, then the ratio of the global hash rate owned by a miner and his mining probability are *not necessarily* equal.

Example 1. Let $M = 2$, $k = 2$, and $d_0 = 2^8$, $d_1 = 2^{12}$. If $h_{1,0} = 1053.3420821484203\, hash/s$, $h_{1,1} = 3350.877902092879\, hash/s$, $h_{2,0} = 388.6077318015238\, hash/s$, $h_{2,1} = 6217.723708824381\, hash/s$, then the first miner possesses the 39.99% of the global hash rate[10] and obtains a mining probability of 0.49100464.

Example 2. Let $M = 4$, $k = 4$, and $d_0 = 2^8$, $d_1 = 2^{12}$, $d_2 = 2^{16}$, $d_3 = 2^{10}$. If $h_{1,0} = 145.199661766661\, hash/s$, $h_{1,1} = 591.7504823551661\, hash/s$, $h_{1,2} = 2583.430837941575\, hash/s$, $h_{1,3} = 292.1757055733901\, hash/s$, $h_{2,0} = 1.0007894670483253\, hash/s$, $h_{2,1} = 16.012631472773204\, hash/s$, $h_{2,2} = 256.20210356437127\, hash/s$, $h_{2,3} = 4.003157868193301\, hash/s$, $h_{3,0} = 8090.186554101536\, hash/s$, $h_{3,1} = 4183.275450846497\, hash/s$, $h_{3,2} = 1.0005803101702 99\, hash/s$, $h_{3,3} = 5168.46086015704\, hash/s$, $h_{4,0} = 5434.391620198674\, hash/s$, $h_{4,1} = 4195.160557464889\, hash/s$, $h_{4,2} = 1.0005808334883484\, hash/s$, $h_{4,3} = 5169.58494402513\, hash/s$, then the first miner possesses the 9.99% of the hash rate and obtains a mining probability of 0.9987.

Such cases can occur in the PoW of the first *general block*, B_k, using the protocol proposed by Sarkar. Indeed, let us focus on *Example* 1 and suppose that the second miner is initially the only miner in the blockchain network, having just mined the *k genesis blocks* to bootstrap the blockchain protocol. Let the hash functions used in the two stages be *hardware incompatible*. Following the protocol, the difficulties of the hash-puzzles are set and updated at repeated intervals to let the expected time to complete the two stages be the same. It means that $\mathbb{E}[X_{2,0}] \stackrel{\text{def}}{=} 1/\lambda_{2,0}$ is equal to $\mathbb{E}[X_{2,1}] \stackrel{\text{def}}{=} 1/\lambda_{2,1}$. This constraint is satisfied in the example. Right before the second miner starts mining block B_k, the first miner joins the mining game. Similar cases to the one described in *Examples* 1 and 2 might advantage a clever miner, who may optimally divide his hash rate among the different hash-puzzles and obtain a mining probability value greater than the ratio of the global hash rate he holds. This inequality between ratio of the global hash rate and mining probability also causes the mining process to be *unfair*. For instance, in *Example* 2 the first miner has the 90% probability to win the PoW of every block, and thus he expects to mine averagely the 90% of the blocks in the long-term, by owning only the 10% of the global hash rate.

Moreover, mining fairness is a requirement to keep PoW blockchains *decentralized.* In particular, Bano et al. stated that, to mitigate centralization risks, the number of valid blocks mined by a miner should be proportional to the ratio of the global hash rate he owns [25]. Therefore, in order to implement a

[10] The ratio of the global hash rate the first miner possesses is $(h_{1,0} + h_{1,1})/(h_{1,0} + h_{1,1} + h_{2,0} + h_{2,1})$.

truly *decentralized* multi-stage PoW blockchain, further investigations for a fair multi-stage PoW mining design are required.

Individual and Pooled Mining. It is also worth noting that, if $k = 1$, then the mining probability of a miner is always the same whether his opponents are competing against each other or cooperating and joining their forces inside a mining pool because the ratio of the global hash rate the miner holds is the same in both cases. The equivalence does not hold if $k > 1$.

Example 3. Let $M = 3$, $d_0 = 2^8$, $d_1 = 2^{12}$, $h_{1,0} = h_{2,0} = h_{3,0} = 100\, hash/s$, and let $h_{1,1} = h_{2,1} = h_{3,1} = 50\, hash/s$. In this setting, the ratio of the global hash rate of the first miner and his mining probability amount to $\frac{1}{3}$. Instead, if the second and the third miner collaborate in a mining pool p', so that $h_{p',0} = 200\, hash/s$ and $h_{p',1} = 100\, hash/s$, then the mining probability of the first miner is equal to $\frac{1073}{3315} < \frac{1}{3}$.

6 Conclusion

In this paper we obtained a *closed-form* expression for the *mining probability*, which is valid, under assumptions commonly used in the blockchain literature, in blockchains whose PoW is composed of $k \geq 1$ consecutive hash-puzzles. We proved that, if $k \geq 2$, then the ratio of the global hash rate held by a miner and his mining probability are *not necessarily* equal. This might advantage a clever miner, who may optimally divide his hash rate among the different hash-puzzles, and obtain a mining probability value higher than the ratio of the global hash rate he holds. Such a possibility can open up centralization issues. Multi-stage PoW blockchain deserve further and careful investigations to be practically implemented.

Acknowledgements. Francesco Mogavero's work was partially funded by Sapienza's *Progetto di Ateneo 2020: La disintermediazione della Pubblica Amministrazione: il ruolo della tecnologia blockchain e le sue implicazioni nei processi e nei ruoli della PA.*

References

1. Nakamoto, S.: Bitcoin: a peer-to-peer electronic cash system (2008). http://bitcoin.org/bitcoin.pdf
2. Wang, W., et al.: A survey on consensus mechanisms and mining strategy management in blockchain networks. IEEE Access **7**, 22328–22370 (2019)
3. Narayanan, A., et al.: Bitcoin and Cryptocurrency Technologies: A Comprehensive Introduction, 1st edn. Princeton University Press, Priceton, New Jersey, USA (2016)
4. Sompolinsky, Y., Zohar, A.: Secure High-Rate Transaction Processing in Bitcoin. In: Böhme, R., Okamoto, T. (eds.) FC 2015. LNCS, vol. 8975, pp. 507–527. Springer, Heidelberg (2015). https://doi.org/10.1007/978-3-662-47854-7_32

5. Croman, K., et al.: On Scaling Decentralized Blockchains. In: Clark, J., Meiklejohn, S., Ryan, P., Wallach, D., Brenner, M., Rohloff, K. (eds.) FC 2016. LNCS, vol. 9604. Springer, Heidelberg (2016). https://doi.org/10.1007/978-3-662-53357-4_8
6. Sarkar, P.: Multi-Stage Proof-of-Work Blockchain. Cryptology ePrintArchive, Report 2019/162 (2019). https://eprint.iacr.org/2019/162
7. Houy, N.: The Bitcoin Mining Game. Ledger **1**, 53–68 (2016)
8. Bowden, R., et al.: Block arrivals in the Bitcoin blockchain (2018). arXiv: 1801.07447 [cs.CR]
9. Kraft, D.: Difficulty control for blockchain-based consensus systems. In: Peer-to-Peer Networking and Applications 9 (2015)
10. Rosenfeld, M.: Analysis of Bitcoin Pooled Mining Reward Systems (2011). arXiv: 1112.4980 [cs.DC]
11. https://en.bitcoinwiki.org/wiki/Hashrate
12. Sarkar, P.: A New Blockchain Proposal Supporting Multi-Stage Proof-of-Work. Cryptology ePrint Archive, Report 2019/162 (2020). https://eprint.iacr.org/2019/162
13. Chang, D., et al.: Spy based analysis of selfish mining attack on multi-stage blockchain. Cryptology ePrint Archive, Report 2019/1327 (2019). https://eprint.iacr.org/2019/1327
14. D'Arco, P., Ansaroudi, Z.E.: Security attacks on multi-stage proof-of-work. In: SPT-IoT 2021, pp. 698–703. IEEE Xplore (2021)
15. Garay, J., et al.: The Bitcoin Backbone Protocol: Analysis and Applications. In: Oswald, E., Fischlin, M. (eds.) EUROCRYPT 2015. LNCS, vol. 9057, pp. 281–310. Springer, Heidelberg (2015). https://doi.org/10.1007/978-3-662-46803-6_10
16. Pass, R., et al.: Analysis of the Blockchain Protocol in Asynchronous Networks. In: Coron, J.-S., Nielsen, J.B. (eds.) EUROCRYPT 2017. LNCS, vol. 10211, pp. 643–673. Springer, Cham (2017). https://doi.org/10.1007/978-3-319-56614-6_22
17. Bentov, I., et al.: Cryptocurrencies without proof of work. In: Clark, J., Meiklejohn, S., Ryan, P.Y.A., Wallach, D., Brenner, M., Rohloff, K. (eds.) FC 2016. LNCS, vol. 9604, pp. 142–157. Springer, Heidelberg (2016). https://doi.org/10.1007/978-3-662-53357-4_10
18. Lewenberg, Y., et al.: Bitcoin mining pools: a cooperative game theoretic analysis. In: AAMAS 2015, pp. 919–927. Istanbul, Turkey: ACM (2015)
19. Birnbaum, A.: Statistical methods for poisson processes and exponential populations. J. Am. Stat. Assoc. **49**, 254–266 (1954)
20. Scheur, E.M.: Reliability of an m-out of-n system when component failure induces higher failure rates in survivors. In: IEEE Transactions on Reliability 37 (1988)
21. Amari, S.V., Misra, R.B.: Closed-form expressions for distribution of sum of exponential random variables. In: IEEE Transactions on Reliability 46 (1997)
22. https://en.wikipedia.org/wiki/Gamma_function#Main_definition
23. D'Arco, P., Mogavero, F.: On (multi-stage) Proof-of-Work blockchain protocols. Cryptology ePrint Archive, Report 2020/1262 (2020). https://eprint.iacr.org/2020/1262
24. https://reference.wolfram.com/language/ref/HypoexponentialDistribution.html
25. Bano, S., et al.: SoK: consensus in the age of blockchains. In: AFT 2019, pp. 183–198. Zurich, Switzerland: ACM (2019)

Towards Informational Self-determination: Data Portability Requests Based on GDPR by Providing Public Platforms for Authorised Minimal Invasive Privacy Protection

Dominik Schmelz(✉), Karl Pinter, Phillip Niemeier, and Thomas Grechenig

Industrial Software (INSO), Vienna University of Technology, 1040 Vienna, Austria
{dominik.schmelz,karl.pinter,phillip.niemeier,
thomas.grechenig}@inso.tuwien.ac.at

Abstract. The Universal Declaration of Human Rights (UDHR) defines that no human being should be subjected to arbitrary interference with his privacy. Yet last decade's digital platform progress has been legally widely unframed and untamed. Therefore, both collection and commercial use of personal data has become a widespread and profitable business model in which individuals currently practically have very little power. European Union's General Data Protection Regulation (GDPR) rebalances rights and obligations of data controllers processing personal data and data subjects whose data are being processed. Well-tailored and targeted use of blockchain technologies enables system transactions that strengthen individual regain of control over personal data and securely transfer it. The proposed system (PPAMIPP, public platform for authorised minimal invasive privacy protection) allows data subjects to claim the personal data processed by them and request their transfer in accordance with the GDPR by defining a respective novel process and supporting technical architecture. The proposed system is validated using a prototype implementation. In addition to demonstrating the feasibility of the system while maintaining confidentiality and integrity, the trade-offs between privacy and usability, as well as general problems of the defined process from legal and technical viewpoints, are highlighted.

1 Introduction

Technological advances allow for more data being recorded, transferred and processed in a shorter time [24]. Technologies such as Internet of Things (IoT) record profileable information about people, whereas Big Data applications allow data analysis in near real-time. These advances call for regulation as developments in information technology have raised concerns about information privacy, and its implications [2].

The European Union's regulation 2016/679, better know as General Data Protection Regulation (GDPR) has enabled on the one hand citizens of the

J. Prieto et al. (Eds.): BLOCKCHAIN 2021, LNNS 320, pp. 106–116, 2022.
https://doi.org/10.1007/978-3-030-86162-9_11

European Union to exercise given rights regarding their personal data [1, Art. 12–23], and on the other hand, forces companies, among other things, to secure data processing activities and enforce a certain level of transparency [1, Art. 24–43].

The GDPR grants a variety of rights to the data subject, including, among other things, the right of data subjects to request information about what data is being processed, for what purposes and by whom [1, Art. 15]. The GDPR also allows the data subject to apply for data portability [1, Art. 20]. These requests must be processed by the controller within one month of receipt [1, Art. 12 (3)]. Only in complex cases, the controller can extend the period to a maximum of three months [1, Art. 12 (3)].

Digital requests for data portability and answers to these request are not further defined in the GDPR on a technical level. Currently, most companies either implement a privacy web form to be filled out online or provide an email address for customers to contact.

This bares these three major tensions and problems:

- Easy, but secure transfer of data
- Secure authentication of the data subject, without further data disclosure
- Reproducible requests, without public disclosure of personal information

These problems lead to time-consuming processes, both for the controller and for the data subject. Most often, they also result in insecure processes that transfer personal data in an insecure manner and additionally lack proof of submission. The currently most implemented process for data portability is to write an email to a data protection officer who then sends the data via unencrypted email [13].

Blockchain technology, originally used in cryptocurrencies, is used in many fields such as logistics, identity management and insurances. The technology makes it possible to create a transparent, nearly immutable, decentralised record database appendable by multiple parties. In general, one advantage of blockchain can be seen when the parties involved share a distrust of each other but need to work together so that each party can rely on the data and applications stored on the blockchain rather than trusting each other [12].

The contribution of this publication is a structured analysis of the issue at hand and a solution proposal for the discussed problems. It introduces a technical concept and prototype implementation of a new solution that withstands the high expectations regarding data protection and privacy.

The paper is organised as follows. First, we present and analyse current research, solutions and background information (Sect. 2). Afterwards, we describe processes and roles relevant to the proposed solution (Sect. 3) and discuss the current state of the prototype implementation and architecture (Sect. 4). The data protection aspects of the prototype and decisions taken to implement privacy by design are discussed in Sect. 5. Next, we shed light on implications of the solution (Sect. 6). Finally, we reflect on the findings, draw the conclusion and discuss possible future work (Sect. 6).

2 Related Work

The research fields of this topic cover procedural, legal and technical areas, particularly e-government, blockchain technology, and data protection laws in the considered jurisdictions.

The prototype presented in this paper builds on previous work done by the authors and represents a further development in the direction of supporting individuals [13,14,18].

Identity Management is the process of ensuring that the right individuals have the proper access to resources. Governments and unions have been working on local and cross-country identification and authentication schemes for years. In 2014 the European Union started the process of cross-country identification, authentication, and trust services for electronic transactions (eIDAS) [11]. Sullivan et al. [20] have been working on blockchain and digital identity in the E-Government (E-Gov) environment. Ishmaev [5] researched Self-Sovereign Identity (SSI) solutions based on blockchain concerning sovereignty, privacy, and ethics. SSI developments on the basis of blockchain technology with compatibility to eIDAS are currently developed by the European Self-Sovereign Identity Framework (ESSIF) Lab[1] on the European Blockchain Services Infrastructure (EBSI).

In the field of E-Gov, there is a general trend toward digitising and optimising processes, focusing on data economy and data sovereignty [3,4]. The escalating collection of user and usage data has led to problems in terms of data protection and ethics, and laws such as GDPR are intended to counteract this. [19] therefore proposes a governance framework for data sovereignty. One problem is that governments around the world often use cloud service providers from other countries. This raises questions about data sovereignty, as [10] has explored.

Yuming et al. [25] dealt with the theory of data sovereignty. Fang et al. [4] worked on the topic of cyberspace sovereignty. Janssen et al. [6] clearly states that there is an urgent need for systems that supports data subjects in terms of their rights.

Data transfer via central data storages has been challenging. The ideal of end-to-end encrypted communication already existed in email communication (e.g. PGP) [9, p. 147ff] and also in the context of Blockchain [23]. File sharing platforms and cloud storages applied the concept in products such as Firefox Send. They either use a simple symmetric key transferred via a second channel or a private-public key pair. Tresorit[2] that added group access features with zero knowledge user authentication.

Blockchain has controversially presented solutions and problems regarding data protection at the same time. On the one hand, personal data processing operations on the blockchain have raised open legal questions with regard to GDPR [17] and on the other hand, consent management solutions, such as [15],

[1] https://essif-lab.eu.
[2] https://tresorit.com.

and [22], support the implementation and enforcement of the GDPR in the blockchain context.

The paper at hand uses the research mentioned above and combines it with a new approach to support individuals creating blockchain-based requests for data portability.

3 Privacy Enhanced Portability Process

A request for data portability is currently provided by means of an unstructured email or, at most, a proprietary portal. The authentication process is mainly performed manually. For lack of options, this is usually done by photographing or scanning the ID card or passport. This process has apparent shortcomings, such as the recipient being able to reuse the ID, and having no way to verify the ID. From a data protection perspective, the transmission of an ID card via a non-encrypted and non-integrity-protected channel is not recommended. Also, the requested data itself is oftentimes transferred via the same channels.

These factors lead to a process that is devastating from a data protection perspective. A system must be in place to help data subjects efficiently exercise their rights without putting their data at further risk.

The analysis of the GDPR Article 15 lead to the identification of the following roles, that were consequently used for the prototype:

- **Data Subject:** An identified or identifiable natural person to whom personal data relates.
- **Controller:** Is responsible for the means and purposes of the processing activities. Provides information in accordance with the GDPR.
- **Processor:** Processes personal data on behalf of the controller.
- **Data Protection Authority (DPA):** The DPA is a national supervisory authority for data protection in each country responsible for processing complaints.

According to the GDPR, data subjects have several rights concerning their personal data. These are, among other things, the right to information, access, rectification, withdrawal of consent, object, and the right to be forgotten. The authors developed a data request platform in the form of a prototype according to [1, Art. 20], namely the right of data portability. The data request platform uses a simple, standardised process for the data subject to create an inquiry faced to the controller or later in the process to the DPA if needed.

The standard process is illustrated in Fig. 1 and can proceed as follows: The data subject visits the data request platform, chooses a controller and submits info to request data including a customer ID or an equal identifier known to the controller. The data request platform generates a private and public key pair in the frontend and sends *only the public key* to the server (1). The data request platform client prints the private key without sending it to the server. The data request platform generates a Portable Document Format (PDF) with the request for data portability and link to the upload platform, including the public key of the data subject, the mail address of the data subject and a signature.

The data request platform asks the user to identify to and sign the PDF with an electronic IDentification, Authentication and trust Services (eIDAS) Identity Provider service (2). The data request platform puts the PDF in an email file and sends the email to the controller and data subject (4), and stores Simple Mail Transfer Protocol (SMTP) proof and a hash of the PDF on the blockchain using openTimeStamp (3). As a result, it can be proven later that the email with the provided content was sent to the data protection officer. This information may be provided later to the DPA in case of a dispute.

The email also contains a calendar entry file as a reminder to the data subject that a complaint can be lodged with the DPA after the expiry of the statutory deadline. The controller receives the request, clicks the link and uploads data (5). The data is automatically hybrid encrypted with the public key of the data subject. Hybrid encryption is used to increase the efficiency of large documents because in this method, the file is encrypted with a random symmetric session key, and this is asymmetrically encrypted with the recipient's public key. The data is only contained in clear text in the Controller User Interface (5a) and only sent encrypted to the Data Request Platform (5b). After the upload is finished, the controller receives proof that he or she uploaded the file using the openTimeStamp stored on the blockchain. The data request platform takes the email address out of the request, checks the signature and send a notification to the data subject.

The data subject receives an email with a link and opens it. The data subject enters the private key to retrieve the encrypted file (6a), which is decrypted in the Subject Interface (6b).

In case of a delay of more than three months (see Sect. 1), a complaint can be filed with the DPA, in which the proof is stored, and the DPA can carry out a review of the request (7). Included are the signature of the data subject, proof of the application including the time and proof of the delivery of the application to the data protection officer.

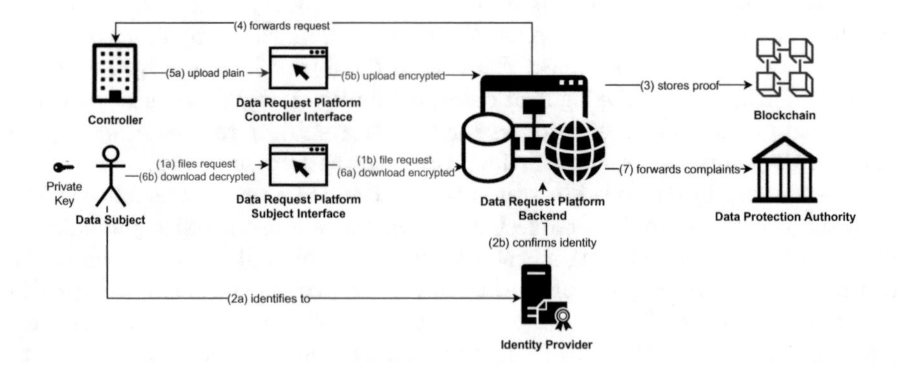

Fig. 1. Actors and processes in the context of the system.

With this mechanism, the following aims are achieved:

- The data subject is identified without disclosing much information to the data request platform or data controller
- The data is transferred with guaranteed confidentiality and integrity
- The data request platform has no access to the transferred files
- The request can be proven to authorities when filing a complaint with the authorities

Disclosure of less information is achieved through an identification system that does not reveal more information than needed for identification (eIDAS). eIDAS also provides for a legally binding authentication and signature on the request documents, and therefore prevents that a user can request data from a stranger. Confidentiality and integrity are obtained through the use of hybrid asymmetric encryption and blockchain stamping. This also provides the opportunity to prove the request to the authorities in case of escalation. The browser-based end-to-end encryption (E2EE) results in the property that the operator has no access to the transferred files.

4 Architecture and Prototype

The data request platform itself can be provided and managed by anyone (e.g. DPA, private companies, governments). It is not decentralised, but uses the decentralised blockchain to secure the data on it. It does not necessarily need a direct interface to the DPA nor to the controller. All national data protection agencies are contactable via email[3] while the contact details of each Data Protection Officer (DPO) must be published by the controller according to the GDPR, in many cases including an email address [1]. Therefore, the interfacing of the platform with these entities is executed via email.

Since the data request platform itself is processing personal data, it must comply with data protection regulations and be a role model regarding data protection implementation. It implements privacy by design and default and minimises the processing and storage of data. The platform does not store any information entered by the data subject but rather hashes the inputs as a filled-out PDF form or an email and deletes the original information after sending it to the controller and data subject via email. The filled-out form then is only saved by the data subject, not on the platform. For convenience reasons, a link containing the entered information is sent to the user, so he/she does not have to enter it again in case of a complaint (see Sect. 3).

The prototype that was created to demonstrate the potential of such a platform consists of a server-side application to provide basic functionalities regarding front-end content delivery, PDF generation, SMTP communication and interfacing with an Identity Provider, Notary Service and Object Storage (see Fig. 2).

[3] A list of all national data protection agencies of EU member states can be found on https://edpb.europa.eu/about-edpb/board/members_en.

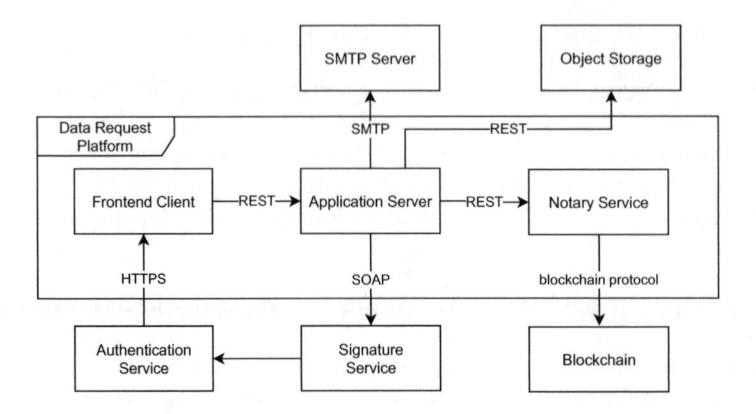

Fig. 2. Architectural overview of the system.

The Identity Provider provides an Authentication Service and a Signature Service. The prototype's front-end is realised in plain web technologies providing the user-facing forms and sending requests to the application server. In order to publish hashes on the Bitcoin Blockchain, the OpenTimestamps [21] protocol is used by a separate notary service. The OpenTimestamps protocol achieves scalability through aggregation. Hashes that are to be stored on the blockchain are collected on the OpenCalendar servers and committed in bulk as a Merkle tree root. This is done to minimise the transaction costs. This approach uses a public blockchain without additional transaction costs for more documents. Therefore a private blockchain is not used, nor recommended in this case.

As visualised in Fig. 2, the three components of the platform communicate via restful HTTP requests amongst themselves and with the authentication service, i.e. the respective provider of the eID. In the prototype the local eIDAS provider of Austria[4] was used. A further development could be the integration of an SSI such as eSSIF (see Sect. 2). The signature service provided by the authentication service is called via a SOAP interface. Requests to the data controllers are sent using the SMTP protocol. The communication with the blockchain is executed via the corresponding Blockchain protocol.

5 Data Protection

The data request platform processes personal data and therefore needs to be analysed regarding its processing activities. The platform was designed to implement 'Privacy by Default' and 'Privacy by Design' [8]. The data subject transmits the personal data to the platform, being a controller in the means of GDPR itself [1, Art. 4 (7)]. The data collection and processing happen on the basis of consent [1, Art. 6 (1)] of the data subject. The data is processed only for the purpose of creating the request and is deleted immediately after processing. No

[4] ID Austria https://eid.egiz.gv.at.

special categories of data are processed. The filled-out request is sent only to the controller. The responsible party uploads the data of the data subject to the data request platform in encrypted form. The operator or order processor of the data repository never has access to the data of the data subject because of the end-to-end encryption. The data at the order processor is deleted at the push of the "Delete Data" button by the data subject. The data subject has the right to view the processing operations at any time and to have his or her data restricted or deleted with immediate effect. The data stored on the blockchain only contains a hash and a timestamp of the requested and returned data. Neither does it contain any conclusions about the individual user. To exclude temporal conclusions, no IP addresses are stored in the logs of the server.

The blockchain service, therefore, does not store the document itself or any personal relatable information. It only receives a hash of the signed document and stores (as described in Sect. 4) the root of a Merkle tree of document hashes.

6 Implications and Conclusion

The current research shows blockchain technologies are used in several ways to achieve decentralised authority. Requests for data portability can be complicated, insecure and hard to prove to data protection authorities. The presented prototype shows that transparency and legal certainty can be created for all parties involved in a real-world environment. The process of data portability can be handled digitally and in a standardised way. Currently, there is no uniformity for such requests. The prototype presented solves this problem. With this prototype, the problems described in Sect. 1 can be solved. The request for data portability can be proven to authorities in the case of an escalation, and the data is transferred confidentially and with integrity while the data request platform has no access to the data. This mitigates a possible breach through an attack that results in a data disclosure.

Because of this, the data request platform can be hosted and run by governments, data protection authorities or private companies. The blockchain supports the data transfer in a way that the properties of the blockchain are partially inherited. The costs of the system are dependent on the usage and configuration. Data retention policies and automatic deletion were not implemented in the prototype but could lead to reduced hosting costs.

While the aforementioned features mainly concern the technical implications of the system, the research further revealed deficiencies in the current GDPR-relevant processes. The following issues were encountered during the process of the design and implementation of the prototype:

- Lack of definition of data protected authentication mechanisms;
- No defined communication processes, formats or interfaces;
- Lack of motivation for companies to make data easy to understand.

These can be overcome by a step by step migration to and introduction of a new platform but need legal backing in order to make them mandatory.

In designing the platform, design decisions had to be made regarding the privacy of the platform itself. In particular, there are trade-offs between privacy and usability that can make it challenging to introduce privacy features for widespread use of the system.

Specifically, the following trade-offs were found:

1. Local private key storage vs zero-knowledge platform;
2. Data transfer between process steps vs centralised storage of information;
3. Identification vs profiling of subjects.

These trade-offs were decided in favour of the most data-protected option during the design phase of the platform. In concrete terms, this meant that in the case of the first trade-off (1) it was decided that a locally stored private key is better from a data protection point of view than creating an account. It was also decided (2) that between the process steps, such as uploading the files by the controller, the central storage of the data subject's email address would be worse than transferring it via the corresponding link. The identification service (3) choice was the most challenging decision from a data protection point of view since attention had to be paid to the additional possibility of profiling by third parties. For this reason, the prototype was based on a connection to a centralised identity provider. A future extension in the direction of DID would be possible, should this prove to be sufficiently protected against profiling.

Compared to both the traditional process mentioned in Sect. 3 and proprietary, purely centralised systems, this method is able to transport data from the data processor to the data subject with authentication, confidentiality and integrity and therefore more data protected. Because of its centralised nature it does not have reliability features as blockchain-based decentralised communication applications [7,16].

The General Data Protection Regulation has been in force for several years now. Nevertheless, the exercise of rights is burdensome for individual citizens. The rights support platform presented in this paper is designed to help citizens access their rights and redress the data imbalance. It allows citizens to securely retrieve their data stored with a data controller whilst preserving data integrity and confidentiality. The prototype of this development showed problems in handling these processes and possible solutions with blockchain technology. Further developments regarding other legal issues that can be supported with blockchain technology could be elaborated in future research.

This supporting rights platform is intended to be a further step in the direction of restoring the balance between responsible companies and affected parties.

References

1. Council of European Union: Regulation (EU) 2016/679 of the European Parliament (General Data Protection Regulation). Official Journal of the European Union (2016). http://eur-lex.europa.eu/legal-content/EN/TXT/?uri=OJ:L:2016:119:TOC

2. Crossler, R.E.: Privacy in the digital age: a review of information privacy research in information systems. Manage. Inf. Syst. Q. 1017–1041 (2011)
3. Falk, S., Römmele, A., Silverman, M. (eds.): Digital Government. Springer, Cham (2017). https://doi.org/10.1007/978-3-319-38795-6
4. Fang, B.: Cyberspace Sovereignty: Reflections on Building a Community of Common Future in Cyberspace. Springer, Singapore (2018). https://doi.org/10.1007/978-981-13-0320-3
5. Ishmaev, G.: Sovereignty, privacy, and ethics in blockchain-based identity management systems. Ethics Inf. Technol. (2020). https://doi.org/10.1007/s10676-020-09563-x
6. Janßen, C.: Towards a system for data transparency to support data subjects. In: Abramowicz, W., Corchuelo, R. (eds.) Business Information Systems Workshops. BIS 2019. LNBIP, vol. 373. Springer, Cham (2019). https://doi.org/10.1007/978-3-030-36691-9_51
7. Khacef, K., Pujolle, G.: Secure peer-to-peer communication based on blockchain. In: Barolli, L., Takizawa, M., Xhafa, F., Enokido, T. (eds.) Web, Artificial Intelligence and Network Applications. WAINA 2019. Advances in Intelligent Systems and Computing, vol. 927, pp. 662–672. Springer, Cham (2019). https://doi.org/10.1007/978-3-030-15035-8_64
8. Klitou, D.: Privacy-Invading Technologies and Privacy by Design. T.M.C. Asser Press, The Hague (2014)
9. Kościelny, C., Kurkowski, M., Srebrny, M.: PGP systems and TrueCrypt. In: Modern Cryptography Primer, pp. 147–173. Springer, Heidelberg (2013). https://doi.org/10.1007/978-3-642-41386-5_6
10. Kushwaha, N., Roguski, P., Watson, B.W.: Up in the air: ensuring government data sovereignty in the cloud. In: 2020 12th International Conference on Cyber Conflict (CyCon), pp. 43–61 (2020)
11. Morgner, F., Bastian, P., Fischlin, M.: Securing transactions with the eIDAS protocols. In: Foresti, S., Lopez, J. (eds.) Information Security Theory and Practice. WISTP 2016. LNCS, vol. 9895, pp. 3–18. Springer, Cham (2016). https://doi.org/10.1007/978-3-319-45931-8_1
12. Peck, M.E.: Blockchain world - Do you need a blockchain? This chart will tell you if the technology can solve your problem. IEEE Spectrum 38–60 (2017)
13. Pinter, K., Schmelz, D., Grechenig, T.: Koordination der Informationspichten laut DSGVO mithilfe der Blockchain. In: IRIS 2020 Internationales Rechtsinformatik Symposion (2020)
14. Pinter, K., Schmelz, D., Lamber, R., Strobl, S., Grechenig, T.: Towards a multi-party, blockchain-based identity verification solution to implement clear name laws for online media platforms. In: Business Process Management: Blockchain and Central and Eastern Europe Forum, pp. 151–165. Springer, Cham (2019). https://doi.org/10.1007/978-3-030-30429-4_11
15. Rantos, K., Drosatos, G., Demertzis, K., Ilioudis, C., Papanikolaou, A.: Blockchain-based consents management for personal data processing in the IoT ecosystem. ICETE **2**, 738–743 (2018)
16. Sarıtekin, R.A., Karabacak, E., Durgay, Z., Karaarslan, E.: Blockchain based secure communication application proposal: cryptouch. In: 2018 6th International Symposium on Digital Forensic and Security (ISDFS), pp. 1–4. IEEE (2018)
17. Schmelz, D., Fischer, G., Niemeier, P., Zhu, L., Grechenig, T.: Towards using public blockchain in information-centric networks: challenges imposed by the European Union's general data protection regulation. In: 2018 1st IEEE International Conference on Hot Information-Centric Networking (HotICN), pp. 223–228 (2018)

18. Schmelz, D., Pinter, K., Brottrager, J., Niemeier, P., Lamber, R., Grechenig, T.: Securing the rights of data subjects with blockchain technology. In: 2020 3rd International Conference on Information and Computer Technologies (ICICT), pp. 284–288 (2020)
19. Singi, K., Choudhury, S.G., Kaulgud, V., Bose, R.J.C., Podder, S., Burden, A.P.: Data sovereignty governance framework. In: Proceedings of the IEEE/ACM 42nd International Conference on Software Engineering Workshops, pp. 303–306. Association for Computing Machinery (2020)
20. Sullivan, C., Burger, E.: Blockchain, digital identity, e-government. In: Treiblmaier, H., Beck, R. (eds.) Business Transformation Through Blockchain, pp. 233–258. Springer, Cham (2019). https://doi.org/10.1007/978-3-319-99058-3_9
21. Todd, P.: OpenTimestamps: scalable, trustless, distributed timestamping with bitcoin (2016). https://petertodd.org/2016/opentimestamps-announcement
22. Wirth, C., Kolain, M.: Privacy by blockchain design: a blockchain-enabled GDPR-compliant approach for handling personal data. In: Proceedings of 1st ERCIM Blockchain Workshop 2018 (2018)
23. Yakubov, A., Shbair, W., State, R.: BlockPGP: a blockchain-based framework for PGP key servers. In: 2018 Sixth International Symposium on Computing and Networking Workshops (CANDARW), pp. 316–322 (2018)
24. Yin, S., Kaynak, O.: Big data for modern industry: challenges and trends. In: Proceedings of the IEEE, pp. 143–146 (2015)
25. Yuming, L.: Data sovereignty theory. In: Sovereignty Blockchain 1.0, pp. 37–77. Springer, Singapore (2021). https://doi.org/10.1007/978-981-16-0757-8_2

Blockchain and the Riemann Zeta Function

Paulo Vieira(✉)

Research Unit for Inland Development (UDI), Polytechnic Institute of Guarda, Av. Dr. Francisco Sá Carneiro 50, 6300-559 Guarda, Portugal
pavieira@ipg.pt
http://www.ipg.pt

Abstract. Proof of Work (PoW) mechanisms used as part of the consensus mechanisms in block chains often have a major drawback namely that resources are only spent on doing the PoW and nothing else. In this paper we propose an adaption of hash based PoW's. This adaption consists in two aspects, firstly by embedding the space of the hash's, $\mathcal{H}$, in the space of complex numbers $\mathbb{C}$, more concretely in the subset $\mathcal{B} = \{z \in \mathbb{C} : 0 < Re(z) < 1\}$, $\Phi : \mathcal{H} \hookrightarrow \mathcal{B}$ and secondly in designing a new cryptographic puzzle in the hash space. The motivation behind these adaptions is that in this way the PoW can also be used to explore the Riemann Zeta Function (RZF). The RZF is associated with an open problem about the localization of its zeros. This problem was left as a conjecture by Riemann and is considered one of the most important open mathematical questions. The permanence of the problem is due to the mysterious behavior of the RZF in the region $\mathcal{B}$. This region will be by translation the search region used in the new cryptographic puzzle. The PoW will thus be able to contribute to a better understanding of the behaviour of the RZF in the region $\mathcal{B}$.

1 Introduction

One of the key components of blockchain technology is the consensus mechanism. This is the mechanism that allows nodes to register transactions on the chain without the need of a central entity. In several blockchains the consensus mechanism is the concept that the nodes, in order to register transactions, must do some sort of work that consumes resources and they must be able to give a proof that they did the work. This proof should be easy to verify whilst the proof construction should be difficult in terms of computational resources necessary. The difficult level should also be easily parameterizable. This is called the Blockchain's Proof of Work (PoW), as an example in BitCoin the PoW consists of solving a cryptographic Hash puzzle.

However the often considerable resources that are spent in order to create the PoW are only spent on doing the PoW and nothing else. Thus the motivation behind this paper is a new PoW scheme that can also do some useful work.

J. Prieto et al. (Eds.): BLOCKCHAIN 2021, LNNS 320, pp. 117–127, 2022.
https://doi.org/10.1007/978-3-030-86162-9_12

In this paper an adaption of hash based PoW's is proposed. As well as designing a new cryptographic puzzle in the hash space we embed the space of the hash's, $\mathcal{H}$, in the space of complex numbers $\mathbb{C}$, more concretely in the subset $\mathcal{B} = \{z \in \mathbb{C} : 0 < Re(z) < 1\}$, $\Phi : \mathcal{H} \hookrightarrow \mathcal{B}$.

We use the embedding function, Φ, to translate the hash space for the complex numbers that are in $\mathcal{B}$.

In particular, suppose that we have an hash function $H : X^* \longrightarrow \mathcal{H}$ where X is an alphabet, $\mathcal{H}$ is the hash space, and $X^* = \{a_0 a_1 ... a_n : n \in \mathbb{N}, a_i \in X, i = 0, 1, ..., n\}$. In consequence we have:

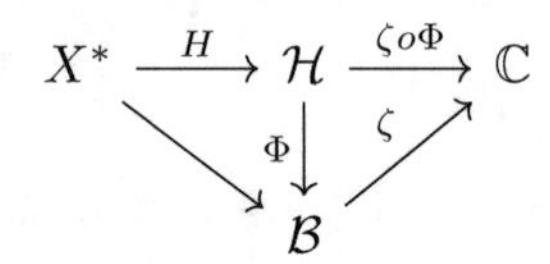

The adapted PoW is the following search algorithm, given in pseudo-code whose inputs are $h \in \mathcal{H}$ and $k \in \mathbb{N}$ and where $||$ denotes concatenation.

Algorithm 1. PoW algorithm

```
1: procedure PoW(h, k)
2:     Search
3:     generate a h' ∈ H
4:     calculate y = H(h||h')                        ▷ H is the hash function defined above
5:     if 0 < |ζ(Φ(h)) − ζ(Φ(y))| < 0.5 × 10^−k     ▷ Φ is the translation (embedding
   function) from H to B
6:     then return h'
7:     else search again
```

The PoW mechanism give us the computational power to be able to search complex numbers, $\Phi(y)$, from the hash space that are consistent with the transactions (the hash of the transactions h) by their proximity when evaluated by the RZF.

This establish a connection with the RZF Millenium problem listed by the Clay Mathematics Institute.

The motivation behind this is to introduce *usefullness* to the computation that is performed during the PoW computations made during the consensus mechanism.

This paper is divided in 5 sections. In the Sect. 2, called The PoW of a Blockchain, we define PoW, we identify its characteristics and we exemplify with the BitCoin's PoW. IN Sect. 3 the RZF is discussed. In the Sect. 4 we describe our proposal and we present an use case. In the last the section we present the conclusions and the future work.

2 The Proof of Work

The PoW is a way to show that someone performed a task. This task must be difficult to perform, easy to verify, and must have a configurable degree of difficulty.

One can summarize informally the characteristics of a good PoW in the following four items:

i. The PoW consists in a challenge to solve
ii. Under finite and computational conditions the challenge can be solved using computational brute force
iii. The challenge is hard to solve and this difficulty must be configurable
iv. It is easy to verify that an expression is a solution of the challenge

These characteristics were well synthesized by Ball et al. [1]. In this paper the authors created a framework for so-called "Proofs of Useful Work" and they define useful POW's in terms of three algorithms.

- Gen(1^n): a randomized algorithm that produces a challenge c
- Solve(c): an algorithm that solves the challenge c, producing a solution s
- Verify(c, s): a (possible randomized) algorithm that verifies the solution s to c

They also specify that Gen(.) and Verify(.) should run very quickly and that there should be a notion of hardness on the runtime of the Solve(.) algorithm. This means that:

i) for some pre-specified length of working time $t(n)$, Solve() should be able to produce solutions that Verify() accepts, and
ii) any attempted solution produced by an algorithm running in less time (e.g. $t(n)^{1-\epsilon}$ for any $\epsilon > 0$) should be rejected by Verify() with high probability.

In BitCoin the PoW consists in the following. Let H be an hash function $H : X \longrightarrow \mathcal{H}$ and $X = \{0, .., 9, a, ..., z, A, ..., Z\}^*$. The PoW consists in that for any piece of information x of X and a $value > 0$ to find an hash h in $\mathcal{H}$ such that $H(H(x)||h) < value$. The difficult of finding a solution to this problem increases as $value$ approaches zero. Also note that a solution is trivially verifiable.

In the definition referred above the BitCoin protocol should be written in the following way:

- $x \leftarrow G(1^n)$. For each x random generated the challenge, c, consist in to find an hash h' such that $H(H(x)||h') < value$ where $value = 0.5 \times 10^{-k}$ with $k \in \mathbb{N}$
- Solve(c) should find a solution $s = h' \in \mathcal{H}$ to the challenge c in time around 10 min. The value of k is used to parametrize the difficult of the challenge c. If the nodes are solving the puzzle in less 10 min k increase if are solving in more than 10 min k decrease.
- Verify(c, s) consists in verifying that $H(H(x)||h') < 0.5 \times 10^{-k}$. The time of verification in BitCoin is without significance.

3 The Riemann Zeta Function $\zeta(s)$

The best way to explain what is the Riemann zeta function is perhaps the definition given by Bombieri [2]. Citing him in italic

"The Riemann zeta function is the function of the complex variable s, defined in the half-plane $\mathcal{R}e(s) > 1$ *by the absolutely convergent serie* $\zeta(s) = \sum_{n=1}^{+\infty} \frac{1}{n^s}$, *and in the whole complex plane* $\mathbb{C}$ *by analytic continuation."*.

For historical reasons the complex variable is denoted by s.

A complex function is analytic in an open set if it is derivable in all the points of the set[1].

Analytic continuation is a technique to extend the domain of the function in a way that allows the new function to be analytic in all of the new domain. In the case of the RZF the extension happens to be $\mathbb{C}$. A full discussion of analytic continuation can be found in chapter 16 of the book [3].

Thus and continuing citing Bombieri [2] *"as shown by Riemann,* $\zeta(s)$ *extends to* $\mathbb{C}$ *as a meromorphic function with only a simple pole at* $s = 1$, *with residue 1, and satisfies the functional equation* $\pi^{-s/2}\Gamma(s/2)\zeta(s) = \pi^{-(1-s)/2}\Gamma(1-s/2)\zeta(1-s)$.*"*.

A meromorphic [3] function on an open subset D of the complex plane is a function that is analytic on all of the points in D except for a set of isolated points [3]. These points are called the poles of the function.

In Titchmarsh's book [4] a functional equation similar to the previous citation of Bombieri, appears written in the following way:

$$\zeta(s) = 2^s \pi^{s-1} sin(\frac{1}{2} s\pi)\Gamma(1-s)\zeta(1-s) \text{ that is valid for all } s \in \mathbb{C} - \{1\}.$$

Here $\gamma(.)$ is the Gamma function (GF) and can be explicitly defined as

$$\Gamma(s) = \int_0^{+\infty} x^{s-1} e^{-x} dx \text{ when } \mathcal{R}e(s) > 0. \tag{1}$$

Titchmarsh [4] presents several proofs, Chapter 2 Theorem 2.1, to the new functional equation.

As the RZF has an explicit formulation from the Dirichlet serie, $\sum \frac{1}{n^{1-s}}$, that is valid in $\mathcal{R}e(1-s) > 1$, the calculation of the ζ in the region $\mathcal{R}e(s) < 0$ is always possible.

There is only one region that is not able to be computed in the natural way from the Dirichlet serie. This region is the strip $0 < \mathcal{R}e(s) < 1$.

It is this mysterious strip that maintains the Riemann hypothesis as conjecture and has not let it arrive to a proven theorem.

The functional equation left by Titchmarsh allows is to show trivially that the odd integer negative numbers: $-2, -4, -6, ...$, are all zeros of the RZF, $\zeta(s)$. These zeros are called the trivial zeros of the RZF.

[1] The notion of derivability must be see in the sense of complex analysis, see [3].

The Riemann conjecture that was called the Riemann hypothesis is about the zeros of the ζ function. Riemann conjectured that all the other zeros are the complex numbers $s = 1/2 + i\alpha$ where α is a real number.

Riemann Hypothesis. The nontrivial zeros of $\zeta(s)$ have real part equal to $1/2$.

This conjecture is one of the seven problems classified by the Clay Mathematics Institute[2] as the problems of the Millenium: Today it is one of the six that are yet to be solved. The Clay Institute gives 1 million dollars for the resolution of each one of this problems. In the opinion of many mathematicians the Riemann hypothesis, and its extension to general classes of L-functions [5], is probably today the most important open problem in pure mathematics [2].

The Riemann conjecture and the RZF have a strong relation with other areas of interest such as the distribution of primes [6]. The knowledge of the values of the RZF has a direct use in understanding how the primes are distributed $\mathbb{N}$ in [7].

There are others applications of the RZF that we not refer here because it is out the scope of this paper.

4 Our Proposal

In blockchains an unavoidable step is the register of the transactions in the Ledger by the nodes. The set of transactions are registered by a node within a block in its ledger. After it is sent to the other nodes that update its ledgers. From this description it is clear that we have two important steps: the construction of the blocks and the register in the ledger.

To register a block one first needs to design an architecture of the blocks. Figure 1 shows the block architecture proposed by Nakamoto [8].

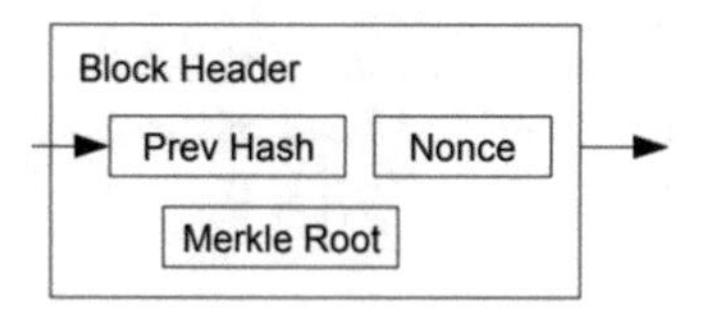

Fig. 1. Blocks of a blockchain

The nonce proposed by Nakamoto in [8] is a variable in cryptographic puzzle of hash's that need to be completed. The nonce is a variable, an hash h, in an equation $H(h_1||h) = h'$ where h_1 is the hash of the information that is contained in the block: the hash of the previous block and the hash of the Merkle tree (the Merkle root); H is an hash function; and h' is $h' = 0^k z$ for some $z \in \mathbb{N}$ and whose first k elements are zeros.

[2] https://www.claymath.org/millennium-problems.

This parameter k is a parameter that is *automatically readjusted* by the platform. In Bitcoin the readjustment is a regulation to maintain the construction of a block in around 10 min. When the blocks have been created in less (resp more) than 10 min the value of k increases (resp decrease) turning the puzzle more (resp less) difficult to solve. The computational effort made and the resources spent by a node to find the nonce is called a PoW (Proof of Work). This effort general is a computation completely wasted. This is the focus of our proposed. With our proposal we want to give utility to this computation creating utility for the PoW beyond the use to the Blockchain. We are not alone in this idea. The idea is shared by the work of Ball, Rosen, Sabin, Vasudevan in [1,9]; Miller, Juels, Shi, Parno, Katz in [10]; and by the existence of the blockchain PrimeCoin. PrimeCoin[3] is a blockchain that implements a PoW that consists in searching for chains of prime numbers (Cunningham chains and bi-twin chains)[4]. The Primecoin blockchain also has a mechanism to regulate the difficulty of the search for chains. This mechanism is the length of the chains, the number of the primes that appear in the chain. The need of calculating the chains of prime numbers referred to allow to register a block produced useful results leading to the discovery of unknown Cunningham chains and bi-twin chains[5].

4.1 The PoW Scheme

In our proposal the PoW consists in searches of values of the RZF in the mysterious strip of the Argand plan $0 < \mathcal{R}e(z) < 1$ with $z \in \mathbb{C}$.

Let $X = \{A, ..., Z, a, ..., z, 0, ..., 9\}$ $H : (X^*) \longrightarrow \mathcal{H}$ be and let Φ be a function $\Phi : \mathcal{H} \longrightarrow \{z \in \mathbb{C} : 0 < Re(z) < 1\})$ that allow to transform hash's in complex number. This function can be seen in the Subsect. 4.1.1.

The nonce in our proposal is also a variable in cryptographic puzzle of hash's that need to be completed. More concretely, the nonce is an hash variable, h', in an equation $\Phi(H(H(x)||h')) = y$ where H is an hash function, x is the information that result of the concatenation of the following data: the hash of the previous block, the timestamp and the hash of the Merkle tree (the Merkle root), $H(x) = h$, such that $0 < |\zeta(\Phi(h)) - \zeta(y)| < 0.5 \times 10^{-k}$ for a $k \in \mathbb{N}$.

4.1.1 Generating the Value of z

The value z is a complex number, $z \in \mathbb{C}$, and is a representation in $\mathbb{C}$ of the information, x, that consists in the concatenation of the data of three natural components of a block: the hash of the previous block (*prevHash*), the timestamp (*Timestamp*), and the Merkle tree root (*Merkle Root*)

$$x = prevHash||Timestamp||MerkleRoot, h = H(x).$$

[3] https://primecoin.io [11-06-2020].

[4] An easy way to know what this chains are is to searching in the site of wolfram https://www.wolframalpha.com [11-06-2020]. This is a website used by mathematicians for procedures similar to the one mentioned before.

[5] https://github.com/primecoin/primecoin/wiki/World-records [11-06-2020].

The way as z is built from h is described in the following paragraph. As h is an hexadecimal we represent by $\dot{h}$ its representation in decimals. The description uses explicitly an hash function sha256 but all the process can be done using a general hash function as can be easily seen. The idea of to use a concrete function is only to give a more practical approach.

Let $HEX = \{a, ..., f, 0, ..., 9\}$ be a set used to represent hexadecimal values.

The value h is an hash got from the hash function $sha256$.

Note that sha256 returns 32 byte hashes which can be written as 64 digits (one byte = 2 hex chars)

$$h = sha256(prevHash||Timestamp||MerkleRoot)$$

such that $h = h_0h_1h_2h_3h_4h_5$, $|h_0h_1h_2h_3h_4h_5| = 64$ with $h_0, h_1, h_2 \in HEX$, $h_3 \in HEX^{l_3}$, $h_4 \in HEX^{l_4}$, $h_5 \in HEX^{l_5}$ where $l_3, l_4, l_5 \in \mathbb{N}$, and $3+l_3+l_4+l_5 = |h_0| + |h_1| + |h_2| + |h_3| + |h_4| + |h_5| = 64$.

The expression $h = h_0h_1h_2h_3h_4h_5$ is the representation of a complex number, z, whose real part is in the interval [0, 1], $\mathcal{R}e(z) \in [0, 1]$, and the imaginary part $Im(z) \in \mathbb{R}$. The interpretation of h to be

$$z = (-1)^{\dot{h}_0}u + i\{(-1)^{\dot{h}_1}v10^{r(-1)^{\dot{h}_2}}\} \text{ with } u, v, r \in \mathbb{R}$$

(example, $z = -0.56 + i\{0.9534 \times 10^3\}$) follows the following description:

- $h_0 \in HEX$ allows to give the signal of the real part. It must be converted to its value decimal when it is used in z. Therefore, $h_0 = 0, ..., 9, a, b, c, d, e, f$ when used in z it is converted respectively by the decimal $\dot{h}_0 = 0, 1, ..., 9, 10, 11, 12, 13, 14, 15$.
- $h_1 \in HEX$ allows to give the signal of the of the imaginary part. When in the context of z all what was written to h_0 is valid also to h_1
- $h_2 \in HEX$ it is the signal of the exponent. When in the context of z all what is written to h_0 is valid also to h_2.
- $h_3 = h_{30}h_{31}...h_{3(l_3-1)}$ is the mantissa of the real part, $u = h_3 \times 10^{-1}$. It is the value $\dot{h}_3 = \dot{h}_{30}16^{-1} + \dot{h}_{31}16^{-2} + ... + \dot{h}_{3l_3-2}16^{-(l_3-1)}$ with $h_{3j} \in HEX$, $j = 0, 1, ..., l_3 - 1$,
- $h_4 = h_{40}h_{41}...h_{4(l_4-1)}$ is the value of the mantissa of the imaginary part, $v = \dot{h}_4$. It is the value $\dot{h}_4 = \dot{h}_{40}16^0 + \dot{h}_{41}16^1 + ... + \dot{h}_{4l_4-2}16^{(l_4-1)}$ with $h_{4j} \in HEX$, $j = 0, 1, ..., l_4 - 1$,
- $h_5 = h_{50}h_{51}...h_{5(l_5-1)}$ is the mantissa of the real part, $r = \dot{h}_5$. It is the value $\dot{h}_5 = \dot{h}_{50}16^{-1} + \dot{h}_{51}16^{-2} + ... + \dot{h}_{5l_5-2}16^{-(l_5-1)}$ with $h_{5j} \in HEX$, $j = 0, 1, ..., l_5 - 1$

Thus, the complex number z is $z = (-1)^{\dot{h}_0}u + i(-1)^{\dot{h}_1}v10^{r(-1)^{\dot{h}_3}}$ with $u, v, r \in \mathbb{R}$, $\mathcal{R}e(z) = (-1)^{\dot{h}_0}a$ and $Im(z) = (-1)^{\dot{h}_1}b10^{r(-1)^{\dot{h}_3}}$. Taking z the next step is calculating $\zeta(z)$ and $\Gamma(z)$. In this process the nodes can use any strategy.

In the Fig. 2 can be seen an application where an hash h is transformed to represent a complex number z. In the figure the hash h is

$$h = e8a44ee9e0cd1831817a9429dd3d9b88def4555575d676326a0d07875f29f05f,$$

$l_2 = 29$, $l_3 = 28$, $l_4 = 4$, and the complex number got z is

$$z = 0.2692660123 + i\{0.870915730 \times 10^{35}\}$$

l3	29									
l4	28		-1	4	0.25	d	0.8125	0	f	15
l5	4		-2	4	0.015625	e	0.0546875	1	0	0
w			-3	e	0.00341796875	f	0.003662109375	2	5	5
x=sha256(w)	e8a44ee9e0cd1831817a9429dd3d9b88def4555575d676326a0d07875f29f05f		-4	e	0.000213623046	4	0.000061035156	3	f	15
x0	e	+	-5	9	0.000008583068	5	0.000004768371			
x1	8	+	-6	e	0.000000834465	5	0.000000298023			
x2	a	+	-7	0	0	5	0.000000018626			
x3	44ee9e0cd1831817a9429dd3d9b88		-8	c	0.000000002793	5	0.000000001164			
x4	def4555575d676326a0d07875f29		-9	d	0.000000000189	7	0.000000000101			
x5	f05f		-10	1	0	5	0			
			-11	8	0	d	0			
			-12	3	0	6	0			
			-13	1	0	7	0			
			-14	8	0	6	0			
			-15	1	0	3	0			
			-16	7	0	2	0			
			-17	a	0	6	0			
			-18	9	0	a	0			
			-19	4	0	0	0			
			-20	2	0	d	0			
			-21	9	0	0	0			
			-22	d	0	7	0			
			-24	d	0	8	0			
			-24	3	0	7	0			
			-25	d	0	5	0			
			-26	9	0	f	0			
			-27	b	0	2	0			
			-28	8	0	9	0			
			SUM		0.2692660123		0.8709157308			35
			z = 0.2692660123 + i 0.870915730 X 10^(35)							

Fig. 2. Generating a complex number from an hash

4.1.2 Calculating the Values of ζ

The calculation of values of the ζ functions is done using values $z \in \mathbb{C}$ that are in the band $0 < \mathcal{R}e(z) < 1$. The value z is built as described above. To calculate the RZF in a point z the nodes can doing what understanding but there is a known functional equation (formula) that can be used. The formulas (2),

$$\zeta(z) = \frac{1}{1 - 2^{1-z}} \sum_{n=1}^{+\infty} \frac{(-1)^{n-1}}{n^z} : z \in \mathbb{C}, 0 < \mathcal{R}e(z) < 1 \tag{2}$$

The use of the formula 2 in general will give approximated values, because it is used in the context of numerical approximation. To calculate that values are used algorithms and techniques of numerical analysis to series and integration functions of complex values [11–13]. Thus, the values calculated for $\zeta(z)$ using this formulas or others are generally roughly calculated. There are visible uses of numerical calculus to calculate the values of ζ function. This can be seen in software as *Mathematica* and *Matlab*[6].

[6] https://www.mathworks.com/help/symbolic/sym.zeta.html.

4.2 The Register of a Transaction and the Verification

With the new proposed PoW the ledger can be seen as a repository of values of the RZF resulting from the translation of the transactions. This can help in knowing the local behavior of the RZF in the neighborhood of the transaction values translated to the region $\mathcal{B}$.

In the following we describe the behavior of the node when it is registering the transaction and we refer the verification process.

Registering a Block. The behavior of the node by steps (registering a block in the ledger):

1. There are the data of the transaction: x
2. It is done an hash of x: $h = H(x)$
3. The hash h is translated to a complex number z in the band $\mathcal{B}$: $\Phi(h) = z$
4. It is calculated the value z by the RZF: $\zeta(z) = w$, $w \in \mathbb{C}$
5. The node tries to find an hash h' such that: $0 < |w - \zeta(\Phi(H(h||h')))| < 0.5 \times 10^{-k}$ (it is assumed that it is a difficult problem find h')
6. The node publishes the 3-tuple: (h, k, h') to be used in the verification process as PoW)

Verification. $|\zeta(\Phi(h)) - \zeta(\Phi(H(h||h')))| < 0.5 \times 10^{-k}$ (It is easy to calculates the values numerically by the zeta function)

4.3 An Use Case

Following we show an use case. We use the repository of RZF values, the ledger, to calculate approximately RZF values of V that have not representations in machines. This complex numbers it is for example a number $V = a + ib$ where a is an irrational or b is an irrational number. The use case show how we can calculates $\zeta(V)$ using the ledger of the blockchain. What follows can be seen as a numerical way to calculates approximately RZF values of complex number similar to V (Algorithm 2).

In the Algorithm 2, the SET_V is a set of values of 3-tuples (h, k, h') published on the blockchain. Each one of these 3-tuples, (h, k, h'), is such that V is in the line segment formed by $\Phi(h)$ and $\Phi(h||h')$, $\overline{\Phi(h)\Phi(h||h')}$.

Algorithm 2. Calculation of V

1: **procedure** CALCULATE($\zeta(V)$)
2: build the set $SET_V = \{(h, k, h') : V \in \overline{\Phi(\text{h})\Phi(\text{h}||\text{h}')}\}$
3: If SET_V is empty **return** null
4: Take $(h_0, k_0, h_0') = arg\ min\{|\Phi(h) - \Phi(h||h')| : (h, k, h') \in SET_V\}$
5: $\zeta(V) = \zeta(\Phi(h_0)) + 0.5 \times 10^{-k_0-1}$
6: **return** $\zeta(V)$

4.4 Systematizing

In this subsection we release our concerns about the formalization of our PoW scheme and we indicate where the formulas present in the scheme can be found.

For an $n \in \mathbb{N}$ and an Hash function[7], H, $H : X^* \longrightarrow HEX^n$ we define our PoW scheme in the Ball's PoW definition as:

- $(x, k) \leftarrow G(1^n)$. Taking $H(x) = h$, to work in the hash space
  The challenge is denoted by c:
  c: For each $h \in \mathcal{H}$ random generated the challenge, c, consists in to find an hash h' such that $0 < |\zeta(\Phi(h)) - \zeta(\Phi(H(H(x)||h'))))| < 0.5 \times 10^{-k}$
- Solve(c) should find a solution $s = h \in \mathcal{H}$ to the challenge c in time around 10 min (adaption to the BitCoin). The value of k parametrizes the difficult of the challenge c. If the nodes are solving the puzzle in less 10 min k increase if are solving in more than 10 min k decrease (adaption to the BitCoin).
- Verify(c, s) consists in (through of c using (x, k), and through of s using h') to verify that $0 < |\zeta(\Phi(H(h||h'))) - \zeta(\Phi(h))| < 0.5 \times 10^{-k}$. The time of verification it is without significance.

Formulas:

1. (formula 1) the GF, $\Gamma(z) = \int_0^\infty x^{z-1}e^{-x}dx$, $\mathcal{R}e(z) > 0$ ([14], page 1–10).
2. (formula 2) the Riemann Zeta Function (RZF), $\zeta(z) = \frac{1}{1-2^{1-z}} \sum_{n=1}^{+\infty} \frac{(-1)^{n-1}}{n^z}$ with $0 < \mathcal{R}e(z) < 1$ ([4], page 16 and page 17).

5 Conclusion and Future Work

The merit of this paper is with the development of a mechanism that makes it possible to use the PoW associated with one of the biggest problems of the millennium allowing to have a better understanding of the RZF. In our PoW, the ledger of the blockchain it is a repository of collection of values of the RZF. This values are building having in attention the knowledge of the local behaviour of the RZF. In all the transactions are calculated two values evaluated by the RZF, the RZF values that is got from the traduction to the complex numbers and more one distinct value in its neighborhood. This will allow with time to have a repository of values of the RZF in neighborhood allowing local knowledge of the RZF.

This repository will allow to have local knowledge of the RZF, for example allow to calculates approximately, values by the RZF, that are not possible to represent in computers. This is was showed in the use case. Another application obvious result from the fact that the repository allow local knowledge of the RZF, this will allow to induce a topology in the hash space from the euclidean metric topology of the RZF. This only to refer some obvious uses form the repository.

[7] Note that $HEX^n = HEX \times ... \times HEX$ (n-times) is $\mathcal{H}$, where $HEX = \{a, ..., z, 0, ..., 9\}$.

In this paper it is assumed by us, without prove, that for each $(h, k) \in \mathcal{H} \times \mathbb{N}$ the search of an hash h' such that: $0 < |w - \zeta(\Phi(H(h||h')))| < 0.5 \times 10^{-k}$ it is a difficult problem in complexity theory sense. The proof of this it is out of the scope of this paper and it is left for a new paper that it is following. We are going also to building the underlying technology.

References

1. Ball, M., Rosen, A., Sabin, M., Vasudevan, P.N.: Proofs of useful work. IACR Cryptology ePrint Archive 2017/203 (2017)
2. Bombieri, E.: Problems of the millennium. The Riemann hypothesis (2000)
3. Rudin, W.: Real and Complex Analysis. Tata McGraw-Hill Education, New York (2006)
4. Titchmarsh, E.C., Titchmarsh, E.C.T., Heath-Brown, D.R., et al.: The Theory of the Riemann Zeta-Function. Oxford University Press, Oxford (1986)
5. Rudnick, Z., Sarnak, P., et al.: Zeros of principal L-functions and random matrix theory. Duke Math. J. **81**(2), 269–322 (1996)
6. Hardy, G.H., Littlewood, J.E., et al.: Contributions to the theory of the Riemann zeta-function and the theory of the distribution of primes. Acta Math. **41**, 119–196 (1916). https://doi.org/10.1007/BF02422942
7. Zagier, D.: The first 50 million prime numbers. Math. Intell. **1**(2), 7–19 (1977). https://doi.org/10.1007/BF03039306
8. Nakamoto, S.: Bitcoin: a peer-to-peer electronic cash system. Bitcoin (2008). https://bitcoin.org/bitcoin.pdf
9. Ball, M., Rosen, A., Sabin, M., Vasudevan, P.N.: Proofs of work from worst-case assumptions. In: Shacham, H., Boldyreva, A. (eds.) CRYPTO 2018. LNCS, vol. 10991, pp. 789–819. Springer, Cham (2018). https://doi.org/10.1007/978-3-319-96884-1_26
10. Miller, A., Juels, A., Shi, E., Parno, B., Katz, J.: Permacoin: repurposing bitcoin work for data preservation. In: 2014 IEEE Symposium on Security and Privacy, pp. 475–490. IEEE (2014)
11. Hoffman, J.D., Frankel, S.: Numerical Methods for Engineers and Scientists. CRC Press, Boca Raton (2018)
12. Sauer, T.: Numerical Analysis. Pearson Addison Wesley, Boston (2006)
13. Chapra, S.C., Canale, R.P., et al.: Numerical Methods for Engineers. McGraw-Hill Higher Education, Boston (2010)
14. Poularikas, A.D.: Transforms and Applications Handbook. CRC Press, Boca Raton (2018)

Case Study: The Automation of an over the Counter Financial Derivatives Transaction Using the CORDA Blockchain

Andrei Carare[1], Michela Ciampoli[1], Giovanni De Gasperis[2(✉)], and Sante Dino Facchini[2]

[1] Armundia Group, Rome, Italy
{a.carare,m.ciampoli}@armundia.com
[2] DISIM, Universitá degli Studi dell'Aquila, Via Vetoio, L'Aquila, Italy
giovanni.degasperis@univaq.it

Abstract. We implemented the automation of a real-world Bilateral Derivatives Over-The-Counter Post Trade Confirmation financial transaction model adopting the Corda permissioned blockchain technology; this kind of transaction is currently performed manually by specialized clerks in most private banking organizations. The proposed solution design is an innovative certified sequence of the workflow states of the transaction, based on the asynchronized interaction among two parties through a smart contract coordinated and validated by the notary node. We show the business logic model of the application and the flow diagrams compatible with the current Corda R3 distribution.

Keywords: Distributed ledger technology · Financial transactions · Distributed ledger · Blockchain

1 Introduction

Many Portfolio Managers want to stipulate Over-the-Counter Derivative (OTC) contracts because of their bespoke nature. Differently from the securities or derivatives traded on exchange platforms, OTC offers full flexibility for each transaction, avoiding intermediation or regulation. Components such as Price, Quantity, Maturity or Upfront fee can be freely defined between the parties to fit the specific needs of investors. Conventionally, OTC trades have been concluded over the phone, but - thanks to the improvements of technology and the birth of the Fintech sector - electronic networks are now taking the lead and allowing the parties involved to execute trades in an easier and faster way, without errors. Contracts are often executed with counterparties with limited public information; avoiding Counterparty bankruptcy, investors ask a cash Collateral, according to the Market Value of the contract.

Once a trade is executed between the two parties, the Counterparty will have to send the Term sheet to the Investor, who is going to sign it. At the same

J. Prieto et al. (Eds.): BLOCKCHAIN 2021, LNNS 320, pp. 128–137, 2022.
https://doi.org/10.1007/978-3-030-86162-9_13

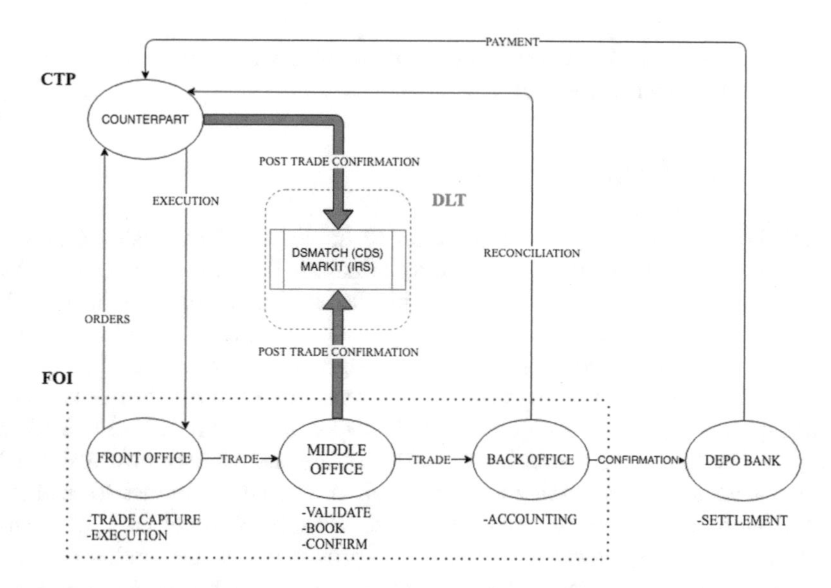

Fig. 1. The Over-the-Counter (OTC) trade scheme between a Counterparty (CTP) and a Front Office Investor (FOI). DSMATCH (Credit Default Swap) and MARKIT (Interest Rate Swap) functions of the Post Trade Confirmation (PTC) process is designed through a Distributed Ledger Technology (DLT) proposal.

time, both will book the trade in their *Front Office* system to allow the necessary verification checks and validate the trade by signing the Term Sheet. Checks are necessary not only for the signing but also because the buyer and seller need to ensure that the trade details are correctly injected in the accounting system, and that the correct instructions are sent for settlement and asset custody, were most of the time is not an automated procedure that can lead to significant operative or cash errors. On the market there are some platforms that allow to receive trades from the counterparty and allow the investor to check and validate - if correct - the trade details, but they always require the investor's manual intervention through a human operator checking numerical values at a computer screen, because there isn't any automatic validation on the same trades received from both parties. Our solution, exploiting the Blockchain technology, provides the financial clients with an innovative public platform that connects all the single pieces together and allows to the parties to retrieve in real time each trade executed on OTC contacts, thus avoiding fat-finger and other human related errors. Clerks must verify and confirm the trade details by matching them with a public unalterable reference, and to certify the validation thanks to a unique code assigned to the specific trade and acknowledged by the market - creditors and supervisory commissions included. Furthermore, OTC transactions are mostly not public, and their prices are privately agreed between two parties, they require very fast deal closing in order to keep agreed price fixed. Each transaction's activity, if the outcome is successful, is thus invisible to anybody else but the parties involved [6].

2 Related Work

An OECD 2020 report [1] on blockchain technologies in Italy shows how the national economy, diversified and export-oriented, is suitable for the development, experimentation and adoption of blockchain-based solutions in a variety of sectors. International trade is one of the most interesting areas for the development of solutions based on Distributed Ledger Technologies (DLT). The sectors for which the origin of the products represents an important market value, can benefit strongly from the qualities of transparency, security and traceability offered by the blockchain. Innovative SMEs across the country are testing DLT solutions with a view to putting them to work in a variety of industries and some are starting to commercialize them. Although the application of DLT/blockchain applications are best known in the field of cryptocurrencies, new industrial applications are constantly emerging. For example, innovative projects are underway by the SIA[1] (the "SIAChain") or the Italian Banking Association (ABI), which is developing specific applications for the Italian market, based on the blockchain infrastructure proposed by the players in the international market ("Spunta Project") [1]. As for the utilities sector, since 2016 Enel has been experimenting with various systems that leverage the blockchain [1].

2.1 Analysis on the Spunta Project

Spunta Banca DLT is a project[2] coordinated and promoted by ABI Lab[3] aimed at applying the new DLT technology to the interbank process of associated partners. The Italian interbank check-off procedure is linked to an operation traditionally carried out by the back office, aimed at reconciling the flows of transactions that generate accounting records on the reciprocal accounts and managing the suspended ones. ABI Lab has signed a collaboration agreement with software company R3 to develop innovative applications and solutions on the Corda blockchain platform. In 2017, 'Spunta Project' was born to operationalize the potential and benefits of blockchain in banking, through the application of a digital ledger distributed among the banks participating in the project, NTT Data for application development and SIA as the infrastructure provider. As reported in the article of the newspaper IlSole24Ore of 28 April 2020: *"the process has been operational for the first grouping of banks since the beginning of October 2020 with 97 banks participating in the project."*

[1] https://www.sia.eu Società Interbancaria per l'Automazione.
[2] https://www.abilab.it/aree-ricerca/blockchain-dlt/spunta-banca-dlt Spunta Banca DLT.
[3] https://www.abi.it Associazione Bancaria Italiana.

3 Distributed Ledger Technology Design on CORDA

The aim of this paper is to design a possible DSMATCH and MARKIT software module running on a blockchain platform in order to transfer the benefits of a DLT solution - in terms of security, privacy and reliable single source - to the whole process of Post Trade Confirmation. The requirements to be matched, in order to guarantee parties and service providers involved in the confirmation process, are the following:

- Transactions must be unique and not modifiable once confirmed by parties.
- Occurring time and date of transactions must be marked with unique and read-only timestamps.
- The system must reconciliate all transactions, matching orders from parties and confirming the transactions once all parties involved have approved the correctness from their side.
- Transactions must be visible only to the parties involved and to the service provider.
- Provider of service must save all operations and disclose them to the police and regulatory authorities when needed, according to the UE-GDPR regulation.

The DLT platform choose to implement such a modeling - among the many solutions nowadays available - is **Corda**[4]. This is an open-source distributed ledger software system, and it is meant for financial institutions in need of designing, processing and storing documents and contracts involved in financial transactions. Corda [2] is an open-source project born in 2016 as a permissioned DLT in order to allow the conservation and certification of contracts, agreements and obligations between parties trusting each other. It's based on traditional blockchain architecture with some important differences [4]. The funding block of the chain is the **"State"** or a contract instance that represent, in each instant, the evolution of the contract. States can be consumed and replaced by news ones as the contract workflow goes on, forming a ledger of non-mutable **"contract state objects"**. Transactions are only visible to parties involved, legitimate "validators" (notaries) and entities which will need to refer in a future time to their contract validity for new transactions.

To reach consensus among parties Corda defines three instruments. A special group of nodes – called **Notary Parties** - offering notarization services in terms of uniqueness of transactions, disputes resolution and timestamping. Secondly **Smart Contracts** to define the constraints on data and implementing the business logic of the agreement. Lastly a **Flow framework** to define the protocols of multi-step actions between the parties without a central controller. The overall set of this consensus instruments with data - represented as state objects - and all ancillary interface modules (APIs, communication plugins, GUIs) constitutes a ledger application called Corda Distributed Application

[4] https://www.corda.net The Corda permissioned distributed blockchain.

or CorDapp. We propose a solution aimed to implement a CorDapp representing DSMATCH and MARKIT functions and designed to be deployed on Corda Network: the official platform of R3 Consortium[5] using a private permissioned configuration, where the central authority validates user/node identity. Corda Network is natively designed to manage commercial and financial transactions and is directly owned and managed by the R3 Consortium, has its pillar is the pre-existing trust among the nodes of the network as well known and identified by the central authentication authority.

The main features are the following:

- Network protocol allows messages exchange regarding possible state transitions, once it is agreed older states are consumed.
- Each node verifies if the transaction is acceptable on its side on the business perspective of his organization.
- Consensus is reached only among nodes involved in the transaction.
- Double-spending problem is avoided through special nodes, called Notary, that certify the erasing of old states and the creation of new ones.

The structure of such a network is based thus on nodes which identity is well known and verified by R3 Consortium, on each node multiple CorDapps that can be deployed and used for the business purposes of the owner.

3.1 CorDapp Design Language

A useful tool to help designing and modelling CorDapps is represented by **CorDapp Design Language** (CDL)[6]. It is a set of 3 instruments, or diagrams, that allow to model a high-level representation of the CorDapp itself and how it interacts with the business environment the CorDapp is supposed to integrate with. The three logical views of CDL consist of the **Business Process Modelling Notation** a pool and swim-lane diagram to contextualize the CorDapp logic and flows within the Organization's business logic and applications, the **Smart Contract view** an acyclic graph where nodes are represented by states of the smart contract and connections represent the flows of the CorDapp and the **Ledger view** that based on the Smart Contract diagram shows the time evolution of the states defined through the possible flows.

3.2 Armundia DSMATCH (CDS) and MARKIT (IRS) CorDapp Solution

The scenario considered is where two parties – the **Front Office Investor** (FOI) and the **Counterpart** (CTP) - interacts directly to confirm independently the validity of the transaction's agreement. The preliminary step is to define the

[5] https://www.r3.com R3 Consortium.

[6] https://docs.corda.net/docs/cdl/cdl/cdl-overview.html CorDapp Design Language (CDL) overview.

business environment: the idea here is to have a double confirmation by CTP and FOI, both could be the initiators of the deal – as both could act as buyer or seller - and both could reject the deal in case of non-compliance with the expected result of the contract. The process of agreement starts with the initiator proposing the deal, the receiver then decide to approve or disapprove the contract through a simple logic: the Post Trade Confirmation data packet matches the one in his hand, as shown in Fig. 2. If it's the case the receiver pre-agrees the deal and sends his PTC packet to the sender; in similar way the sender decide if agreeing or not the deal matching the packet with the one in his hand. Upon the second confirmation the deal is completed, and the transaction registered and validated by a Notary node. The following Corda flows are defined:

- **Propose Deal:** initiated by seller with data related to his PTC packet
- **PreAgree Deal:** initiated by buyer with data related to his PTC packet
- **Agree Deal:** initiated by seller
- **Complete Deal:** initiated by buyer
- **Cancel Deal:** can be initiated by both parties in case of non-compliance

This business logic and interactions between Corda application flows and other non-DLT business processes and applications are modelled in the BPMN diagram (Fig. 2), in this scenario the FOI act as the seller and the CTP act as a buyer.

The composition of the **Post Confirmation Packet** – that is the core data managed by the CorDapp - must reflect and resume properly the nature of the deal between the parties, the following fields are the one selected for our proposal:

- Seller, Buyer: name of the party acting as a seller or buyer respectively;
- SellerID, BuyerID, TradeID: unique identifiers;
- InstrumentIDc, InstrumentIDp: unique identifier of the underlying guarantee title expected by consenter and used by proposer;
- AmountP, AmountC: value of the transaction set by proposer and expected by consenter;
- Proposer, Consenter: party proposing or consenting the flow;
- rejectionReason, rejectedBy: party and reason of rejecting the deal;

The states of the smart contract are defined as per flows determined in the BPMN and are the following: **Proposed, Preagreed, Agreed** and **Rejected**. In order to shape a business logic adherent and consistent with the Post Trade Confirmation process the Corda paradigm requires the definition of constraints at various logical levels and for all modules and components, the ones defined for this model are the following:

- Buyer and Seller, Consenter and Proposer must be different parties;
- Proposer must be either Buyer or Seller, Consenter must be either Buyer or Seller;
- InstrumentID must be a valid ISIN (International Security Identification Number) Code [3];

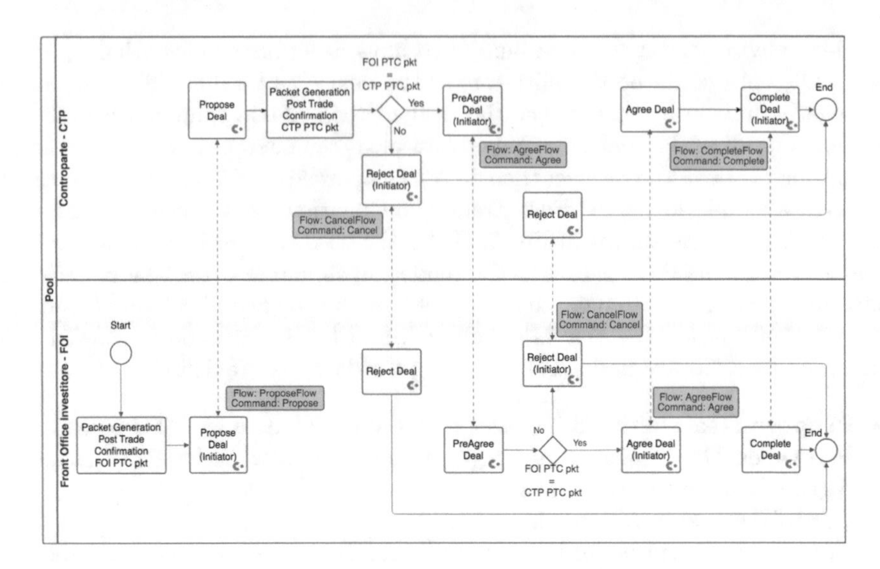

Fig. 2. Business process modeling notation view of DSMATCH (CDS) and MARKIT (IRS) CorDapp

- BuyerID and SellerID must be valid LEI (Legal Entity Identifier)[7] Codes;
- TradeID must be a valid UTI (Unique Transaction Identifier) Code [5];

Flows and Status constraints are also defined in order to establish how flows are initiated, agreed and rejected and what status changes rules are.

Once the Smart contract view is defined (Fig. 3), the business logic governing the corDapp is set. To describe all possible evolutions the smart contract may have, there is a final instrument represented by the Ledger evolution view.

In this example (Fig. 4) we describe the smart contract evolution in the scenario where both parties agree the deal. The first transaction $T \times 1$ represent the proposal from the FOI party that brings the status of the agreement to "Proposed", all variables are set to their values except rejectionReason and rejectedBy that can be set to NULL as per constraint rules of this state. In $T \times 2$ the CTP checks the variables AmountP and InstrumentIDp finding them compliant to what expected and thus the smart contract state is updated to "Preagreed", accordingly to the PreAgree arrow in Smart Contract view, and the AmountC and InstrumentIDc are written with relevant consenter values. At this point the FOI double-checks the Amount and Instrument values and finding them compliant with their business logic expectations activate the $T \times 3$ bringing the status of the contract to "Agreed". At this point both the parties close the process through the $T \times 4$ of completion of the deal. To complete the back-end design of the Post Trade Confirmation module we have to define the way the Service provider can access the transactions in case of authority request for information. As stated before, we designed the CorDapp in order to be deployed on Corda

[7] https://www.gleif.org Global Legal Entity Identifier Foundation.

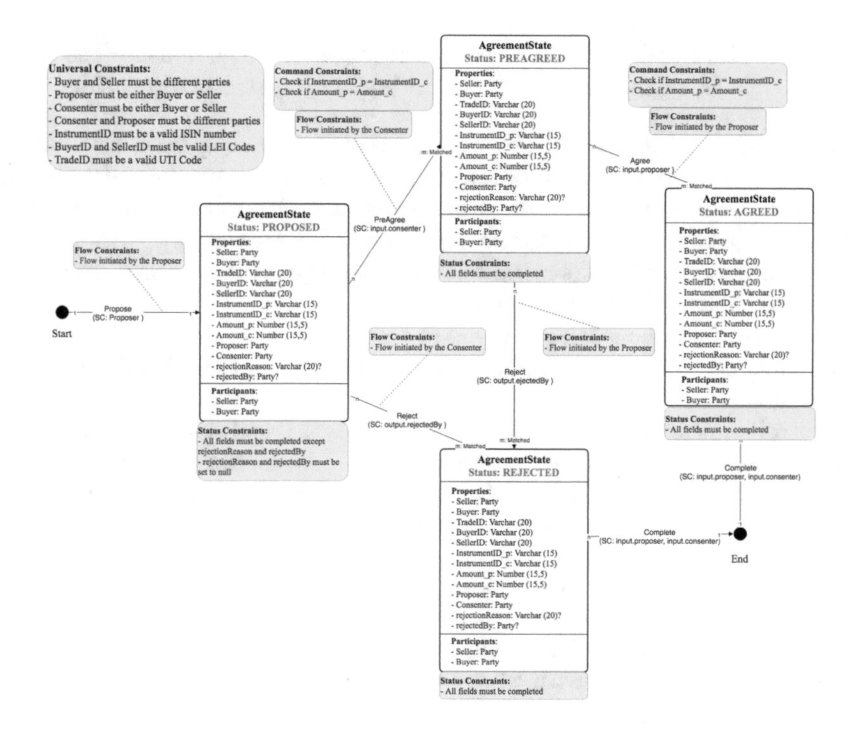

Fig. 3. Smart contract view of DSMATCH (CDS) and MARKIT (IRS) CorDapp

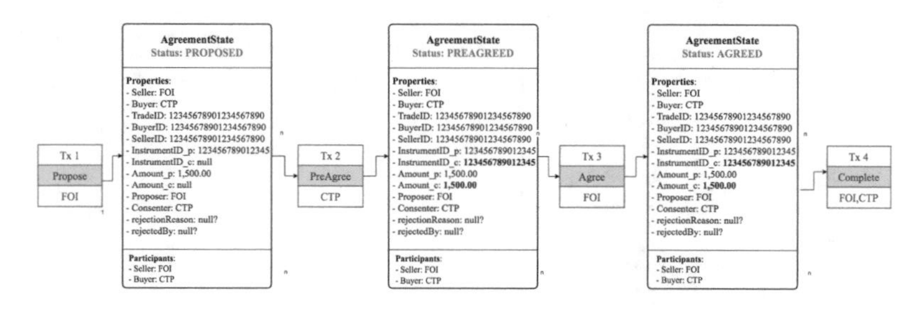

Fig. 4. Ledger evolution view of DSMATCH (CDS) and MARKIT (IRS) CorDapp

Network. This cloud service run by R3 gives basically remote access to nodes on dedicated virtual machines to deploy clients' applications; the access to each single node is possible through password protected accounts.

4 Front Office Application

A front-end application, based on the DLT described in Sect. 3.2, may be proposed for Armundia's clients in need to perform Over-the-Counter operations as per Fig. 1. We refer to the module allowing the Post Trade Confirmation activities. A first preliminary analysis of the problem brings to define the Actors involved in such a process both on the internal and on the external side:

- Middle Office Operator of the FOI: personnel operating on client's side in order to book, validate and confirm securities exchanges on the platform They interact both directly with the counterpart and with other actors internal to the FOI (Front Office Operators and Back Office Operators);
- Counterpart Operator: personnel of the counterpart office entitled to manage the Post Trade Confirmation duties.
- Armundia Compliance: supervisors of the system that manage the technical and financial functions of both the DLT back-end and Front office front-end modules.
- Control Authority Officer: Police, Institutional and Commercial entities meant to check and retrieve information from Armundia's system
- Front Office Operator of the FOI: secondary actor representing the internal part of the FOI interacting with the Middle Office operator.
- Back Office Operator of the FOI: secondary actor representing the internal part of the FOI interacting with the Middle Office operator.

A simple use case to define the functionalities think for the front office application could be the one described in Fig. 5:

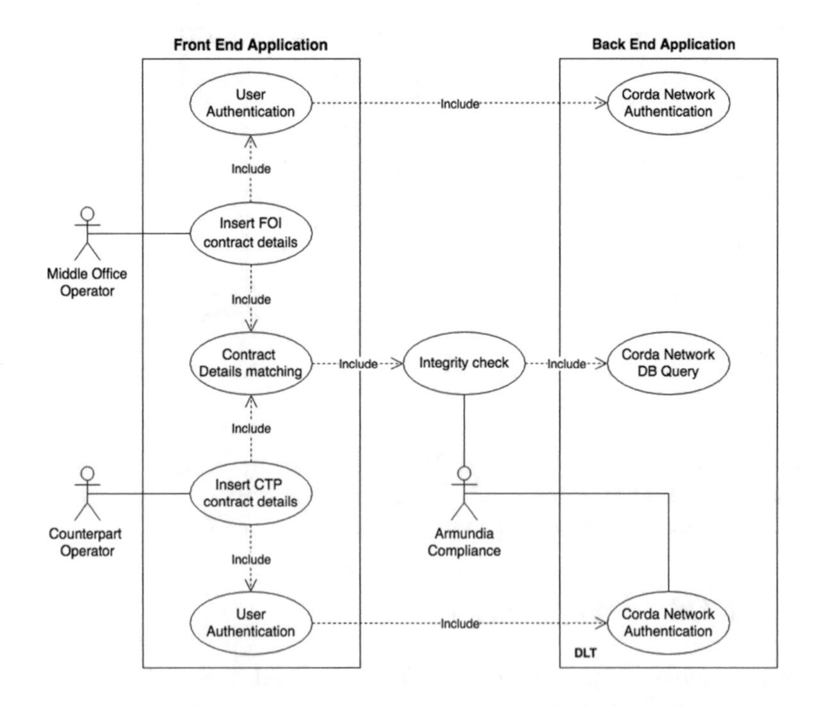

Fig. 5. Use case for Post trade confirmation operations

The use case described explains the interactions among actors and the system's modules, during the post trade confirmation process. The Counterpart and Middle Office operators place their respective orders details, after authentication on the front-end, and ask Armundia's system to execute the MARKIT an DSMATCH functions along with an integrity check of all relevant information.

The Front-end software application could be imagined as a web-based solution installed on Armundia facilities – for its technical and administrative personnel - and on its client's premises; in both cases interacting through a VPN encrypted connection with the Corda Network set of nodes. Armundia's instances could have access to all nodes of Corda Network with administrative privileges in order to check and retrieve all the relevant information may be requested by regulation bodies in case of fraud and unlawful behaviour. Each instance of the application thus exchanges information regarding the transactions only with its own dedicated Corda node; a messaging function with Armundia and between counterparts, outside the Corda Network, may be considered in order to facilitate communications and synchronizations of back-office's processes.

5 Conclusions

We have shown a scheme of a Derivative OTC Post Trade Confirmation financial transaction implemented over the Corda blockchain, ready to be integrated with conventional fintech software tools, that can automate with all the necessary warranties a procedure so far carried out manually by human clerk between banks. The system is implemented on a Corda test net, to validate how a real DLT scenario can reduce the errors and save time compared to a classic deployment configuration with manual operations.

Acknowledgements. This work was funded as a feasibility research by Armundia Group.

References

1. Bianchini, M., Kwon, I.: Blockchain for SMEs and entrepreneurs in Italy (2020)
2. Brown, R.G.: The corda platform: an introduction (2018). Accessed 27 May 2018
3. Josyula, V.: Coordinated Portfolio Investment Survey Guide. International Monetary Fund, Washington, D.C. (2018)
4. Sandner, P., Valenta, M.: Comparison of ethereum, hyperledger fabric and corda (2018). Accessed 6 July 2018
5. Smith, M.: A privatized approach to derivatives regulation: the CPMI-IOSCO's proposed unique transaction identifier scheme and its practical effects on transparency and regulatory arbitrage. Ga. J. Int'l Comp. L. **45**, 411 (2016)
6. Zhu, H.: Finding a good price in opaque over-the-counter markets. Rev. Financ. Stud. **25**(4), 1255–1285 (2012)

Application of Blockchain to Peer to Peer Energy Trading in Microgrids

Manuel Sivianes[1(✉)] and Carlos Bordons[1,2]

[1] Systems Engineering and Automatic Control Department, Universidad de Sevilla, Seville, Spain
{mscastano,bordons}@us.es

[2] Laboratory of Engineering for Energy and Environmental Sustainability, Escuela Técnica Superior de Ingeniería, Universidad de Sevilla, Sevilla, Spain

Abstract. A distributed energy management algorithm taking full advantage of the blockchain technology is proposed. It serves as a one day-ahead energy schedule that allows each networked entity to make peer to peer (P2P) safe power trades with the rest of the microgrid agents.

Blockchain serves as a global aggregator that verifies power trades and evaluate convergence across iterative steps. The proposed scheme has been implemented within Ethereum blockchain and its benefits are compared simulating different scenarios.

Keywords: Blockchain · Peer to peer · Microgrid · Smart contract · Energy trading

1 Introduction

The energy industry is being reshaped by a huge expansion during the past years within the field of renewable energy sources (RES) [1], which leads into energy production being more reliant on the weather, and consequently more unpredictable, as well as producing potential energy imbalances [2,3]. Furthermore, the energy system is becoming more decentralized due to the recent inclusion of distributed energy resources (DER), such as photovoltaic solar panels, battery storage systems or electric vehicles (EV) [4,5]. All these incorporations make central management increasingly daring. These challenges must be addressed by enhancing the electricity system flexibility, granting it the ability of handling this paradigm while maintaining the system integrity and stability [6].

In deregulated power systems, agents using DERs can potentially trade energy by handling their own generation, storage capacity and consumption profile.

For example, in [7], a P2P energy trading software platform enables buyers, sellers, suppliers and system operators to trade energy in local power networks; and, in [8], bill sharing, mid-market rate and an auction based pricing strategy

J. Prieto et al. (Eds.): BLOCKCHAIN 2021, LNNS 320, pp. 138–148, 2022.
https://doi.org/10.1007/978-3-030-86162-9_14

are studied to allow flexible P2P trades within local customers. In this context, multidirectional flows of energy are enabled in contrast with centralized energy trading, granting users the possibility of establishing energy matching preferences. This leads into more data and control information been exchanged between end points, which can involve security or privacy issues.

To deal with these challenges, and allow decentralization, we propose to use blockchain. With the goal of removing intermediaries and making peer to peer (P2P) safe interactions possible, blockchain was created alongside with Bitcoin by Satoshi Nakamoto [9]. This technology reduces drastically the possibility of a breach and alteration of data by using cryptography and decentralization. A blockchain is conformed of data packages, called blocks, each of which contains multiple transactions. Every block is cryptographically linked with the previous one, except the first block which is known as the genesis block [10]. By adding the concept of smart contracts to the inherent properties of blockchain is possible to erase completely the human interface and, thus, third parties. In this scenario, smart contracts contain the pre-established rules for enabling direct energy transactions between endpoints based on local consumer preferences, with no intermediaries [11].

This work proposes an energy management platform that focuses on maximizing the global wellness of their agents by solving a distributed optimization problem through iterative steps. The algorithm makes use of a smart contract deployed in the Ethereum blockchain which serves as a global aggregator that erases the need of a third party controlling and distributing the data. The concept of global wellness refers to minimizing the financial costs from the electricity retail market. Other articles also following a decentralized approach for energy management using a blockchain are [19,20]. In [19], a blockchain-based scheme for energy trading between electric vehicles and critical loads in a logical network to meet temporary energy demands is presented; and, in [20], the bidding process and dynamic pricing based on supply and demand for energy are automated through blockchain smart contracts.

The rest of the paper is organized as follows. Section 2 describes the mathematical formulation for the optimization problem. Section 3 and 4 presents the proposed trade mechanism and distributed algorithm, respectively. Section 5 describes the blockchain implementation. Section 6 presents simulations of the proposed algorithm, and a comparison of its performance in different scenarios. Finally, conclusions are given in Sect. 7.

2 Problem Formulation

The following formulation is developed to operate within a microgrid where households may have access to EV, photovoltaic panels and battery systems. It serves as a one day-ahead energy program. Let us define a microgrid of $n > 1$ nodes, indexed by $i = 1, \ldots, n$, by the complete graph K_n modeled over T timesteps, indexed by $t = 1, \ldots, T$. Every agent is connected to the microgrid and the utility grid. Power imported from the latter is defined as $p^g_{i,t}$. The cost function for each node i in timestep t is modeled as:

$$C_{i,t}^{g}(p_{i,t}^{g}) = c_t p_{i,t}^{g} \Delta t \quad \forall i,t, \tag{1}$$

where c_t is the monetary cost of purchasing each kWh per hour and $\Delta t = 24/T$ determines the amount of hours contained in each timestep.

Each household presents a fixed load demand $p_{i,t}^{l}$ that is uncontrollable. Power generation through solar panels $p_{i,t}^{pv}$ is divided into the photovoltaic energy used $p_{i,t}^{pv_u}$, and the photovoltaic energy surplus $p_{i,t}^{pv_s}$, as can be seen in (2). $p_{i,t}^{pv}$, $p_{i,t}^{pv_u}$ and $p_{i,t}^{pv_s}$ are constrained as can be seen from (3) to (5).

$$p_{i,t}^{pv} = p_{i,t}^{pv_u} + p_{i,t}^{pv_s}, \quad \forall i,t, \tag{2}$$

$$0 \leq p_{i,t}^{pv} \leq \overline{p_{i,t}^{pv}}, \quad \forall i,t, \tag{3}$$

$$0 \leq p_{i,t}^{pv_u} \leq \overline{p_{i,t}^{pv}}, \quad \forall i,t, \tag{4}$$

$$0 \leq p_{i,t}^{pv_s} \leq \overline{p_{i,t}^{pv}}, \quad \forall i,t. \tag{5}$$

The availability of batteries and electric vehicles incorporate additional constraints. Electric vehicles are modeled as flexible controllable charges that allow their users to choose the charging power for each timestep t, $p_{i,t}^{ev}$. This charging power is constrained as follows:

$$0 \leq p_{i,t}^{ev} \leq \delta_{i,t} \overline{p_{i,t}^{ev}}, \quad \forall i,t, \tag{6}$$

where $\delta_{i,t}$ is a binary variable that represents the possibility of charging the EV at timestep t. The EV charging efficiency is defined by η^{ev} and the EV daily total charge must accomplish the charging energy demand E_i^{ev} as follows:

$$\sum_{t=1}^{T} \eta^{ev} p_{i,t}^{ev} \Delta t = E_i^{ev}, \quad \forall i. \tag{7}$$

The net battery power $p_{i,t}^{b}$ represents the difference between the discharging power $p_{i,t}^{bd}$ and the charging power $p_{i,t}^{bc}$ of agent i at timestep t as seen below:

$$p_{i,t}^{b} = p_{i,t}^{bd} - p_{i,t}^{bc}, \quad \forall i,t. \tag{8}$$

Energy stored within batteries is defined as $E_{i,t}^{b}$, and the efficiency of both charging and discharging are defined as η_c^b and η_d^b, respectively. The mathematical relation among them can be seen in (9).

$$E_{i,t}^{b} = E_{i,t-1}^{b} + (\eta_c^b p_{i,t}^{bc} - p_{i,t}^{bd}/\eta_d^b)\Delta t, \quad \forall i,t. \tag{9}$$

Upper and lower bounds for battery variables are defined below:

$$\underline{p_{i,t}^{bd}} \leq p_{i,t}^{bd} \leq \overline{p_{i,t}^{bd}}, \quad \forall i,t, \tag{10}$$

$$\underline{p_{i,t}^{bc}} \leq p_{i,t}^{bc} \leq \overline{p_{i,t}^{bc}}, \quad \forall i,t, \tag{11}$$

$$\underline{E_{i,t}^{b}} \leq E_{i,t}^{b} \leq \overline{E_{i,t}^{b}}, \quad \forall i,t. \tag{12}$$

The power balance for every agent is defined in (13), where offer must accomplish demand $\forall t$, and $p^g_{i,t}$ is calculated.

$$p^g_{i,t} = p^l_{i,t} + p^{ev}_{i,t} - p^{pv_u}_{i,t} - p^b_{i,t}, \quad \forall i, t. \tag{13}$$

Taking into consideration these constraints the optimization problem to be solved for each household is as follows:

$$\min_{\forall p^g_{i,t}} \sum_{t=1}^{T} C^g_{i,t}(p^g_{i,t}) \tag{14}$$

$$s.\ t. \quad (2)\text{–}(13) \tag{15}$$

3 Proposed Trade Mechanism

For the trading strategy it is proposed a model where every household can trade energy with any agent in order to minimize the global economic effort of the community:

$$\min_{\forall p^g_{i,t}, p^t_{ij,t}} \sum_{t=1}^{T} \left(C^g_{i,t}(p^g_{i,t}) + \sum_{j \neq i}^{n} c_t(p^g_{j,t} - p^t_{ij,t}) \right) \quad \forall i, \tag{16}$$

where $p^t_{ij,t}$ is the power sent by i to j at timestep t. Note that the sign of the $p^t_{ij,t}$ trade is negative since it is preventing agent j from purchasing this power from the utility grid, and the optimization problem (14)–(15) needs to be solved first in order to include $p^g_{j,t} \forall j, t$ in (16). This new objective function implies adding the following features into the model presented in Sect. 2:

- $p^{bd}_{i,t}$ is divided into $p^{b_u}_{i,t}$ and $\sum_{j \neq i}^{n} p^{b_t}_{ij,t}$, as shown in (17). $p^{b_u}_{i,t}$ represents the energy that is extracted from the battery and is consumed without being transferred to any agent. On the other hand, $p^{b_t}_{ij,t}$ is the energy sent from i to j in t. Note that (18) forces $p^{b_t}_{ij,t}$ to be lower than the energy that j needs to purchase from the external grid in t, $p^g_{j,t}$.

$$p^{bd}_{i,t} = p^{b_u}_{i,t} + \sum_{j \neq i}^{n} p^{b_t}_{ij,t}, \quad \forall i, t, \tag{17}$$

$$0 \leq p^{b_t}_{ij,t} \leq p^g_{j,t}, \quad \forall i, t. \tag{18}$$

- The solar panel energy balance from (2) is modified to (19) by adding $\sum_{j \neq i}^{n} p^{pv_t}_{ij,t}$, which represents the photovoltaic energy sent from i to the rest of agents in every timestep t. Again, every $p^{pv_t}_{ij,t}$ trade is bounded in (20) to ensure that they are not greater than every individual power deficit for $\forall j, t$.

$$p^{pv}_{i,t} = p^{pv_u}_{i,t} + \sum_{j \neq i}^{n} p^{pv_t}_{ij,t} + p^{pv_s}_{i,t}, \quad \forall i, t, \tag{19}$$

$$0 \leq p^{pv_t}_{i,t} \leq p^g_{j,t}, \quad \forall i, t. \tag{20}$$

– The power balance (13) is reformulated as follows:

$$p^g_{i,t} = p^l_{i,t} + p^{ev}_{i,t} + p^{bc}_{i,t} - p^{pv_u}_{i,t} - p^{b_u}_{i,t}, \quad \forall i, t. \tag{21}$$

– $p^t_{ij,t}$ from (16) is defined in (22) and constrained in (23) and (24):

$$p^t_{ij,t} = p^{pv_t}_{ij,t} + p^{b_t}_{ij,t} \quad \forall j, t, \tag{22}$$

$$0 \leq p^t_{ij,t} \leq p^g_{j,t} \quad \forall j \neq i, t. \tag{23}$$

$$p^t_{ii,t} = 0 \quad \forall t. \tag{24}$$

Taking into consideration the updated electric and trading model, the following optimization problem is obtained:

$$\min_{\forall p^g_{i,t}, p^t_{ij,t}} \sum_{t=1}^{T} \left(C^g_{i,t}(p^g_{i,t}) + \sum_{j \neq i}^{n} c_t(p^g_{j,t} - p^t_{ij,t}) \right) \quad \forall i. \tag{25}$$

$$s.\ t. \qquad (3)\text{–}(12), (17)\text{–}(24) \tag{26}$$

4 Proposed Distributed Algorithm

The distributed algorithm to minimize the optimization problem (25–26) is composed of 5 steps:

1. First iteration is started with every household solving locally its own optimization problem (14)–(15) which does not take into consideration the rest of the network, and $P^g_i \in \mathbb{N}^{1 \times T}$ is obtained, which contains the energy that i needs to purchase $\forall t$ for the next day, $p^g_{i,t}$.
2. The global demand matrix for the next day $P^d \in \mathbb{N}^{n \times T}$ is built with every P^g_i from the previous step and must be known for all households.
3. P^d is used to configure the upper bounds (18), (20) and (23) from the optimization problem (25)–(26), which is solved locally by each agent and $P^t_i \in \mathbb{N}^{n \times T}$ is calculated, which contains the power $p^t_{ij,t}$ that agent i aims to send to each agent at each timestep.
4. The global trade matrix $\Phi \in \mathbb{N}^{n \times T \times n}$ is built with every P^t_i from last step and a consensus process is started to guarantee that the total energy received $\forall i, t$ is not higher than the respective $p^g_{i,t}$ from P^g_i. It is known that every individual $p^t_{ij,t}$ complains with constrain (23), but the sum of every $p^t_{ij,t}$ can violate it. Thus, after this consensus phase is completed, the compliant trades $p^{t_c}_{ji,t}$ are calculated and it is ensured that $\sum^n_{j \neq i} p^{t_c}_{ji,t} \leq p^g_{i,t} \forall i, t$, leading to the consensus trade matrix Φ^c.
5. **If** $\Phi^c - \Phi < \varepsilon$, where ε is the permitted tolerance, all proposed trades are feasible according to the established threshold; **or** *iteration* $> \Psi$, where Ψ is the maximum number of iterations, **the algorithm finishes**.
   Else, all agents recalculate locally the optimization problem from step 1

adding the energy trades from Φ^c. Energy received and sent $\forall i, t$ are calculated as $\sum_j^n \Phi_{i,t,j}^c$ and $\sum_j^n \Phi_{j,t,i}^c$, respectively, and included in (13):

$$p_{i,t}^g = p_{i,t}^l + p_{i,t}^{ev} + \sum_j^n \Phi_{j,t,i}^c - p_{i,t}^{pv_u} - p_{i,t}^b - \sum_j^n \Phi_{i,t,j}^c, \quad \forall i, t.$$

Once the optimization problem is solved, a new P_i^g is obtained that includes the compliant energy trades from Φ^c. **Go to step 2.**

5 Blockchain Implementation

The distributed algorithm from Sect. 4 is executed along with a blockchain network to provide full traceability and the ability to audit the process. This is achieved using Ethereum, which is a public and permissionless blockchain that provides a feature called smart contracts [12]. Ethereum enables users to create smart contracts by using a built-in fully fledged Turing-complete programming language. The blockchain network is run by executing a local Ethereum node with Ganache CLI [13]. Agents can interact easily with the smart contract using a React-based [14] graphic user interface that makes use of web3.js [15] libraries. This collection of libraries allow agents to connect with a local or remote ethereum node using HTTP, IPC or WebSocket. It must be pointed out that the proposed blockchain implementation only focuses on realizing the functionalities needed for the control strategy: gas consumption efficiency or execution speed are not considered.

The smart contract coded with Solidity programming language and implemented in the Ethereum blockchain fulfills the following functions [16]:

1. Control flow of the distributed algorithm.
2. Exchange of information between all participants.
3. Execute the consensus algorithm for Φ^c.

Adding the smart contract functionalities to the proposed algorithm leads to the following modification within the distributed algorithm:

- The optimization variables P_i^g obtained in steps 1 and 6 are uploaded to the smart contract, where the global demand matrix P^d is built and available to be checked by every household.
- Once P^d is built, agents access the smart contract and call for different methods that provide them the data required for step 3. This means that every agent knows the energy deficit of their neighbors.
- When every P_i^t from step 3 is submitted to the smart contract, the Φ matrix is built, and the consensus algorithm is executed, starting the consensus process that leads to Φ^c.
- Termination condition $\Phi = \Phi^c$ is verified.

6 Simulations and Discussion

In this section, the benefits of using the proposed distributed algorithm through blockchain are illustrated through numerical examples. Two scenarios are recreated to observe the impact of using a classic constant tariff (CT) or an hourly discrimination tariff (HDT). The topology of the considered microgrid is depicted in Fig. 1, where 11 households take part within the network, and 4 of them have power generation and storing facilities. For the uncontrollable load of households, various demand profiles are used with a total daily consumption of 9.55 kWh, which is the average daily electric energy consumption for a Spanish household according to IDAE [17]. The cost of withdrawing power from the utility grid c_t(€ /kWh) are gathered from [18] and depicted in Fig. 2.

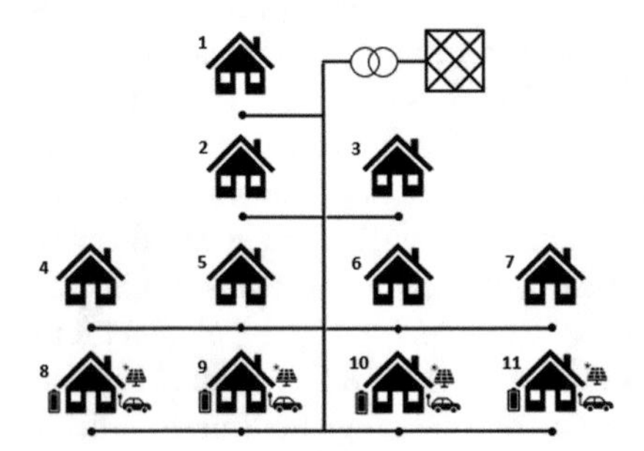

Fig. 1. Simulated microgrid topology.

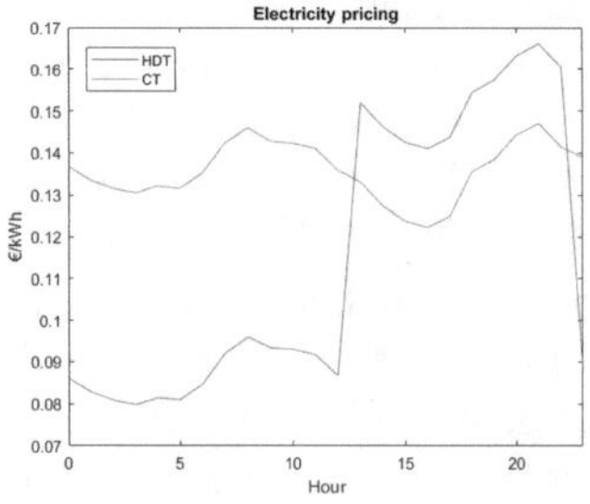

Fig. 2. Electricity tariffs.

A daily energy consumption of 7 kWh and a charging efficiency of 89% are set for EVs, and every household freely chooses the charging hours where their EVs may be charged. For the photovoltaic generation, a single profile is used for those households that present this facility. Batteries in this study are able to store 1250 kWh, and their charge and discharge efficiencies are set at 94.5%.

6.1 Test 1: Distributed Algorithm Using CT

In this test the microgrid showed in Fig. 1 is simulated with the proposed distributed algorithm from Sect. 4 using the CT from Fig. 2.

From step 1 it is obtained P_i^g for all households for the next day by solving locally the isolated optimization problem (14–15). Results from all households are sent to the smart contract. Results from agent 8 are illustrated in Fig. 3. P^d is built in the smart contract and depicted in Fig. 4. Optimization problem from step 3 is solved locally taking into consideration the P^d matrix, which is retrieved from blockchain. Power trades P_i^t $\forall i, t$ are submitted to blockchain. Φ is built within the smart contract and the hourly power sum expected to be sent $\forall i, t$ before consensus is calculated as $\sum_j^n \Phi_{j,t,i}$, and depicted in Fig. 5. Then, the consensus method is executed to calculate all $p_{ij}^{t_c}$, and build Φ^c. The energy received after consensus $\sum_j^n \Phi^c_{j,t,i}$ is depicted in Fig. 6. Since $\Phi \neq \Phi^c$, second iteration is started including Φ^c within the isolated optimization problem.

The algorithm finished after 4 iterations. The monetary cost evolution across all iterations is gathered, and shown in Fig. 7.

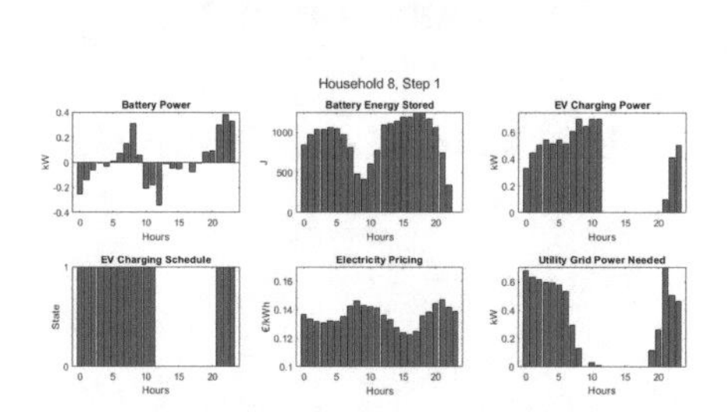

Fig. 3. Agent 8, step 1, test 1.

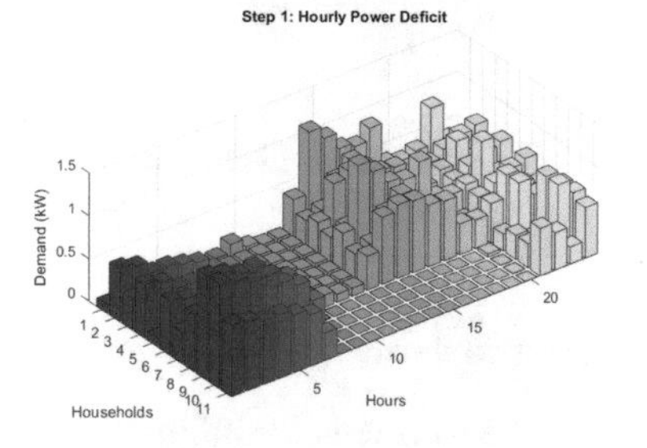

Fig. 4. P^d, first iteration, test 1.

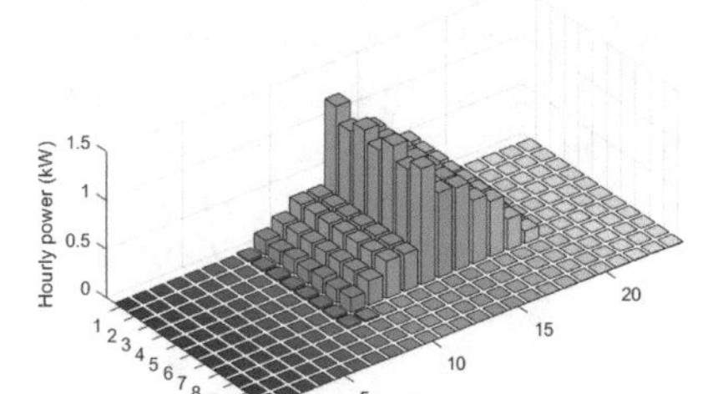

Fig. 5. $\sum_j^n \Phi_{j,t,i}$, first iteration, test 1.

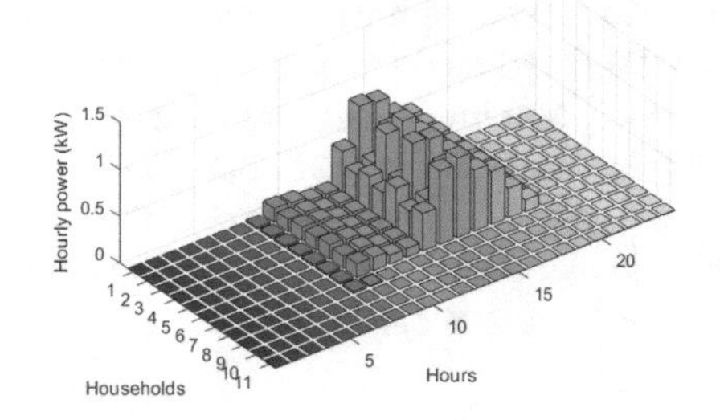

Fig. 6. $\sum_j^n \Phi^c_{j,t,i}$, first iteration, test 1.

6.2 Test 2: Distributed Algorithm Using HDT

In this test the electricity pricing tariff used is the HDT from Fig. 2 instead of the CT. It is only depicted the monetary cost evolution across all iterations since the steps to be fulfilled are identical from those covered in Test 1.

The algorithm finished after 6 iterations. The monetary cost across evolution for all iterations are shown in Fig. 8.

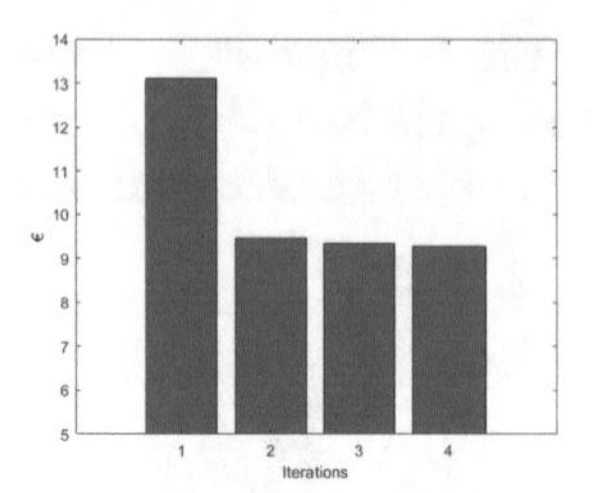

Fig. 7. Microgrid daily costs, test 1.

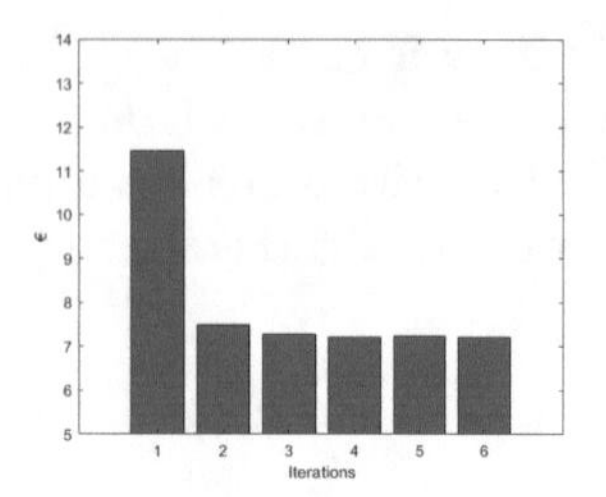

Fig. 8. Microgrid daily costs, test 2.

6.3 Comparison

In this subsection it is highlighted the positive effect of using the proposed distributed P2P blockchain enabled sharing platform comparing the final objective function, where P2P energy trading is enabled, with its initial state, where households solve an isolated optimization problem.

In Test 1, where a CT tariff is used, the daily monetary cost for the microgrid from solving the isolated optimization problem (14–15) is 13, 10 € , and, after the algorithm reaches finalization, is reduced 29, 08% to 9, 29 € , which corresponds to 1390 € reduced per year. In Test 2, where a HDT tariff replaces the CT used in Test 1, the initial daily monetary cost for the microgrid is 11, 45 € , and, after 6 iterations, it is reduced 36, 94% to 7, 22 € , and 1544 € annually. As expected, the usage of HDT leads to lower monetary costs since prosumers can adapt their flexible loads and battery power to take full advantage of those timesteps when electricity prices are lower.

7 Conclusions

This article proposes a distributed energy management platform within a microgrid that serves as a day-ahead energy scheduling program that minimizes the utility bill of a neighborhood by enabling power trades among households. These features are built in a blockchain network to avoid relying on third parties, have full traceability of shared data, and enable safe P2P interactions, among other benefits.

Thanks to the distributed algorithm proposed, the utility grid bill is reduced, and consensus around power trades is always reached. In particular, when an HDT is used, DERs can fully unleash their potential since not only they are used to reduce its user bill but also its neighbors'. Furthermore, the usage of blockchain erases the possibility of any agent manipulating the algorithm for their own benefit since the smart contract, and every transaction are immutable, proving blockchain a reliable tool to enable decentralization in a transparent and safe manner.

Regarding future research, we plan to add an intra-daily demand response to address possible imbalance between the day-ahead prediction and the live power demand. Also, it could be of interest using a private/hybrid blockchain network to study performance and viability.

Acknowledgments. The authors would like to acknowledge the project GESVIP funded by Junta de Andalucía (ref. US-1265917) and SAFEMPC funded by Ministerio de Ciencia e Innovación (PID2019-104149RB-I00).

References

1. Panwar, N.L., Kaushik, S.C., Kothari, S.: Role of renewable energy sources in environmental protection: a review. Renew. Sustain. Energy Rev. **15**(3), 1513–1524 (2011)
2. Banshwar, A., Sharma, N.K., Sood, Y.R., Shrivastava, R.: Real time procurement of energy and operating reserve from Renewable Energy Sources in deregulated environment considering imbalance penalties. Renew. Energy **113**, 855–866 (2017)
3. Chiu, W.Y., Sun, H., Poor, H.V.: Energy imbalance management using a robust pricing scheme. IEEE Trans. Smart Grid **4**(2), 896–904 (2012)
4. Jiayi, H., Chuanwen, J., Rong, X.: A review on distributed energy resources and MicroGrid. Renew. Sustain. Energy Rev. **12**(9), 2472–2483 (2008)
5. Akorede, M.F., Hizam, H., Pouresmaeil, E.: Distributed energy resources and benefits to the environment. Renew. Sustain. Energy Rev. **14**(2), 724–734 (2010). https://doi.org/10.1016/j.rser.2009.10.025
6. Eid, C., Codani, P., Perez, Y., Reneses, J., Hakvoort, R.: Managing electric flexibility from Distributed Energy Resources: a review of incentives for market design. Renew. Sustain. Energy Rev. **64**, 237–247 (2016)
7. Zhang, C., Wu, J., Zhou, Y., Cheng, M., Long, C.: Peer-to-Peer energy trading in a Microgrid. Appl. Energy **220**, 1–12 (2018)
8. Long, C., Wu, J., Zhang, C., Thomas, L., Cheng, M., Jenkins, N.: Peer-to-peer energy trading in a community microgrid. In: 2017 IEEE Power & Energy Society General Meeting. IEEE (2017)
9. Nakamoto, S.: Bitcoin: a peer-to-peer electronic cash system. Manubot (2019)
10. Nofer, M., Gomber, P., Hinz, O., Schiereck, D.: Blockchain. Bus. Inf. Syst. Eng. **59**(3), 183–187 (2017). https://doi.org/10.1007/s12599-017-0467-3
11. Wohrer, M., Zdun, U.: Smart contracts: security patterns in the ethereum ecosystem and solidity. In: 2018 International Workshop on Blockchain Oriented Software Engineering (IWBOSE). IEEE (2018)
12. Ethereum organization. https://ethereum.org/en/whitepaper/. Accessed 06 May 2021
13. Ganache CLI. https://github.com/trufflesuite/ganache-cli. Accessed 28 June 2021
14. React. https://es.reactjs.org/. Accessed 28 June 2021
15. web3.js - Ethereum JavaScript API. https://web3js.readthedocs.io/en/v1.3.4/. Accessed 28 June 2021
16. Solidity. https://docs.soliditylang.org/en/develop/index.html. Accessed 06 May 2021
17. IDAE. https://www.idae.es/uploads/documentos/documentos_Documentacion_Basica_Residencial_Unido_c93da537.pdf. Accessed 06 May 2021
18. System Operator Information System. https://www.esios.ree.es/es/pvpc. Accessed 06 May 2021

19. Umoren, I.A., Jaffary, S.S.A., Shakir, M.Z., Katzis, K., Ahmadi, H.: Blockchain-based energy trading in electric-vehicle-enabled microgrids. IEEE Consum. Electron. Mag. **9**(6), 66–71 (2020)
20. Khattak, H.A., Tehreem, K., Almogren, A., Ameer, Z., Din, I.U., Adnan, M.: Dynamic pricing in industrial Internet of Things: blockchain application for energy management in smart cities. J. Inf. Secur. Appl. **55**, 102615 (2020)

The Value and Applications of Blockchain Technology in Business: A Systematic Review of Real Use Cases

Oscar Lage[1,2](✉), María Saiz-Santos[2], and José Manuel Zarzuelo[2]

[1] TECNALIA, Basque Research and Technology Alliance (BRTA), Parque Científico y Tecnológico de Bizkaia #700, 48160 Derio, Spain
oscar.lage@tecnalia.com

[2] Faculty of Economics and Business Administration, The University of the Basque Country, Av. Lehendakari Aguirre, 83, 48015 Bilbao, Spain

Abstract. This work provides an empirical study to identify the specific objective pursued by companies that are currently investing to develop blockchain technologies to improve their processes. Unlike other studies based on the theoretical potential application of blockchain technology in different sectors, the main objective of this paper is to analyse real projects and investment of companies in blockchain technology to identify the main value or use that managers prioritize in their technology development efforts. More than one hundred blockchain projects from different sectors have been examined with the aim of extracting the perceived value of blockchain technology by managers, customers, and partners. Identifying the business value that is most demanded in each sector and company size, as well as the relationship between the values that are jointly demanded. This article assesses the main values attributed to blockchain, highlighting those really appreciated by companies to invest in them and identifying new applications of blockchain technology in different sectors, and generating organisational change.

Keywords: Blockchain · IT business value · Technology innovation · Technology strategy · Organizational transformation

1 Introduction

The birth of bitcoin [1] at the end of 2008 was much more than the beginning of the first decentralized cryptocurrency. The technology designed to support bitcoin cryptocurrency would later be called blockchain and give rise to a new family of decentralized technologies. Blockchain is a distributed peer-to-peer architecture that introduces major disruptions to traditional business by decentralising governance through the creation of a secure design that does not require trusted third parties to establish transactional relationships between two parties.

According to 83% of C-suite executives [2], their companies will lose competitive advantage if they do not adopt blockchain. Leaders are increasingly investing in blockchain and digital assets as one of the top five strategic priorities.

J. Prieto et al. (Eds.): BLOCKCHAIN 2021, LNNS 320, pp. 149–160, 2022.
https://doi.org/10.1007/978-3-030-86162-9_15

There are several overall reviews regarding potential blockchain-based applications [3–6]. However, what is the real value provided to the companies? what specific business needs can be addressed by blockchain?

Beyond theoretical studies about potential applications, this work contributes to the understanding of the business value of blockchain technology for companies, and identifies which features are most required. Everything based on a systematic review of all the blockchain innovation projects that Tecnalia has carried out for private companies and public administration.

The work is organized as follows. Section 2 will review previous work on the state of the art and its contributions. Section 3 presents the methodological approach used to carry out the research. Sections 4, 5 and 6 describe the results and the contribution, explaining the benefits of blockchain for companies and the statistical impact of each one, which is expected to help researchers to identify new projects and applications of this technology. Finally, Sect. 7, ends with the presentation of the most relevant conclusions, trends, and further research lines.

2 Background and Related Work

Seebacher et al. [3] conducted one of the first systematic reviews of the literature analysing the common characteristics identified in 32 articles that examined potential uses of the technology. The two main features identified in the work are trust and the decentralized nature of the technology.

Tama et al. [4] identify in their analysis four main areas of potential applicability of blockchain technology, which are financial services, healthcare, business markets, and others, such as digital right management system or reputation system. Hawlitschek et al. [5] evaluate blockchain from the perspective of its possible applicability in the shared economy.

More recently, Casino et al. [6] elaborated a taxonomy of blockchain-based applications after the study of 260 articles and 54 reports of the grey literature. This taxonomy identifies the following potential application domains: financial, integrity verification, governance, Internet of Things, health, education, privacy and security, business/industry and finally data management.

These authors also point out that although the literature attempts to propose blockchain as a panacea it should not be used in every case. In particular, Seebacher et al. [3] remarked that for further research would be interesting to explore the contribution of blockchain in non-theorical projects by performing an empirical analysis of real cases. They suggested that significant deviations are expected between theoretical applications and those finally adopted by industry.

This research is in the same spirit of the research conducted by the previous authors, that is, our aim is to analyse real projects in which companies have invested, to understand and measure the perceived value of blockchain technology in business.

3 Methodology

A mixed research method combining a qualitative and quantitative analysis was carried out to answer our research question: "What is the real business value of blockchain for companies regardless of the sector and application?". Therefore, a qualitative analysis of actual project documentation was combined with the statistical analysis of the database created from the qualitative analysis.

The starting point was the information associated with blockchain projects carried out by Tecnalia's cybersecurity and blockchain research group. To endorse the significance of this information it should be noted that Tecnalia is a leading research and technological development centre in Europe. Actually, it is the first private Spanish organisation in contracting, participation, and leadership in the European Horizon 2020 and ranked second in European patent applications.

To obtain the information, 104 blockchain-related contracts in the period from 2017 to 2020 were selected from the corporate ERP (Enterprise Resource Planning) software. The information on these contractual agreements was extracted in a spreadsheet and those contracts belonging to the same project in which several entities were involved were grouped together. There were 55 individual projects and 12 consortium projects with a maximum of seven partners and an average of 4.08 research participants per consortium.

The projects widely reflect the main economic sectors, so we consider it a broad enough sample to obtain significant results. Figure 1 shows the representation and investment by sector in the analysed sample (in percentages).

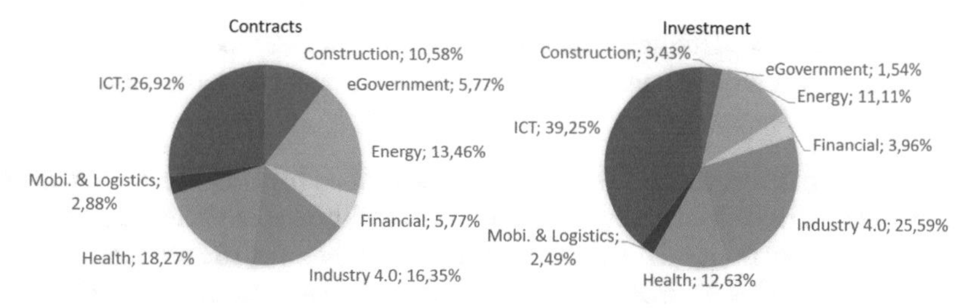

Fig. 1. Blockchain investment distribution by sector.

Regarding the relative impact on the investments, it can be observed how the ICT (Information and Communication Technologies) projects are positioned as the ones with the largest budget, surely because this sector has understood, and demanded, the blockchain technology before others.

Concerning the size of the company, Fig. 2 shows the distribution of investments according to company size. As can be seen, medium-sized companies are the ones that prevail both in terms of the number of contracts and their investment.

The next step after the creation of the database was to compile the documentation of contracts, project reports and consortium agreements. In all the documents there was a section that expresses, under different titles and denominations, the needs of the customer or consortium, the opportunity, and the challenge faced by the project using blockchain in combination with different technologies. A qualitative content analysis of the documentation was carried out to identify the keywords that define the needs that motivated the investment.

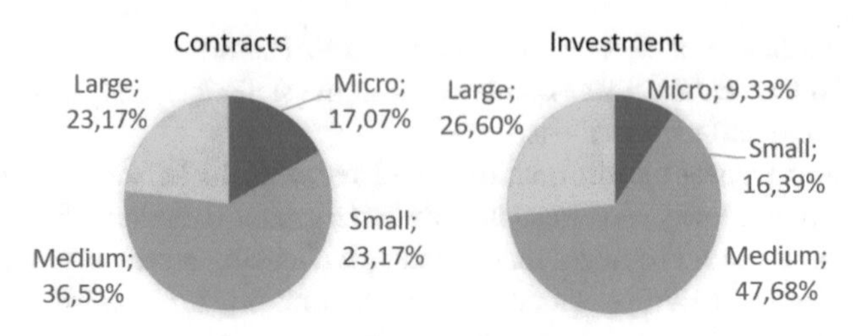

Fig. 2. Blockchain investment distribution by company size.

Thus, to obtain the list of benefits of the technology, a thorough reading of the project documentation was carried out and all the keywords were noted down in the project documentation. Once the keywords expressing the needs covered by blockchain technology had been isolated, they were standardised and grouped together, resulting in 12 business values for which companies are investing in blockchain, which we analyse in the following section. Finally, we performed an analysis of the relationships between the 12 benefits using hierarchical clustering, the results of which are presented in Sect. 6.

4 Business Value of Blockchain Technology

The different benefits identified in the research are discussed below, starting with a definition of each value, and continuing in the following sections with an analysis of the relationships between them.

Decentralization of Processes: one of the most obvious benefits of blockchain technology because of its distributed architecture and its ability to create reliable peer-to-peer relationships without the need for a trusted third party to intermediate between the parties [7]. Blockchain enables the management of business-to-business processes in a decentralised way, allowing all parties to directly participate in the management of business processes. Blockchain makes it possible to audit each process, as well as its compliance with the business rules established by the community in the blockchain network [8].

Decentralization of Business Models: the academic literature is now starting to reflect this value [9–11]. The research has found how companies want to disrupt current business models, especially platform business models [12]. Internet has enabled the creation of new business models based on digital platforms, like Airbnb and Uber, that allows them to create ecosystems in which some participants offer products/services. Precisely the business model that companies are trying to adopt in the last years, is the one that companies in the study try to destroy. An open and free blockchain application can offer to the suppliers and demanders of products/services a very similar service than the current digital platforms.

Traceability and Provenance of Assets: blockchain technology provides an unalterable record of the asset's history, in which both its origin and components can be analysed, as well as all changes, updates, maintenance, and operations throughout its history. Every change made to the resource that is relevant for the ecosystem can be recorded on the blockchain, guaranteeing both the integrity of the data and the date on which it occurred.

The provenance issue arises precisely in several projects of traceability of both tangible and intangible goods [13]. Several projects focused on the origin of the materials used in the industrial or construction process were also analysed, all of them linked to the circular economy [14].

Traceability/Certification of Processes and Regulatory Compliance: this is closely linked to the previous section, however, in this case the traceability is done over a process instead of an asset. In general, it refers to those internal or external processes to organizations that must be monitored, and in some cases certified because in some domains this is necessary for regulatory compliance. Processes linked to different domains have been identified in the literature [15–18].

Transparency of Processes: closely linked to traceability because transparency is based on giving visibility on the traceability of a process. In addition, blockchain based records give more confidence to the transparency process because there is no doubt that the records have not been modified at a later point in time.

Shared Asset/Process Management: Adams et al. [19] in 2003 analysed the problem of shared asset management, and blockchain can solve the challenges identified in the study by providing all stakeholders with a single synchronised view of the shared resources and their states. This perfect synchronisation can reduce frictions between the different actors, enabling complex ecosystems.

Trustworthiness and Integrity: the decentralization of governance, as well as the integrity and immutability of blockchain transactions, make blockchain networks a "source of truth" [20]. Additionally, each participant must digitally sign every transaction they make on the blockchain, which is important for potential claims.

Automation of Processes: currently companies do not rely on automating certain decisions based on information that may be published by a third party on a web service because the third party may modify such information, which could result in the company making incorrect decisions, and also not having evidence that the third party has acted in bad faith [21]. Thanks to blockchain, companies can automate decisions based on third party information, which can substantially increase the business value of the companies.

Smart Contracts: in 1994 Szabo [22] devised the term Smart Contract and sometime later, specified the concept of Smart Contracts in more detail [23], but Smart Contracts could not be implemented until blockchain technology had been conceived. When in this study we refer to the term Smart Contract we refer to Szabo's original concept in which there is a custody of value (tokens) in the intermediation between two or more parties.

Digital Identity: many ecosystems and users demand a new model of interoperable digital identity, focused and managed by the user himself, and ensuring privacy (Privacy by Design). Allen [24] defined the basis of about a new decentralised identity model called "Self-Sovereign Identity" (SSI). SSI ensures that users maintains control of their identity and do not have to rely on any central entity. Thus, users will be able to present the attributes of their identity (age, nationality, academic qualifications, etc.) to third parties minimizing the presented information using zero knowledge proofs [25] to maximize their privacy.

Sovereignty of Data and Data-Driven Services: it is a growing need in business [26]. Today, most business data are not being exploited, nor is artificial intelligence being able to play a greater role, precisely because of the desire to have control over the data. Companies are afraid of sharing data with third parties because once they do it they no longer have control over it, data can be replicated and distributed without their consent, losing its economic value [27].

Machine-to-Machine (M2M) Transactions and Machine Economy: this is a new paradigm that emerges by transferring the sharing economy to the IoT [28]. Thanks to the capability of decentralization and tokenization of assets and services of blockchain, the Machine Economy allows the creation of a new sharing economy among the machines themselves; putting in value their data/services in this ecosystem, and operating in an autonomous way with tokens [29]. The token economy [30] in the IoT field is still very novel but in the set of analysed projects it can be discovered tokenized transactions between machines in Industrial Internet of Things (IIoT) (manufacturing and energy) as well as other proofs of concept in the fields of autonomous vehicles.

5 Blockchain Business Value Impact Analysis

The demand for each business value has been studied according to different impact criteria. This will help to understand the needs of the companies, as well as to identify patterns. The first analysis determines the investment in each of these values by company size. Table 1 shows how trustworthiness and integrity is the most demanded need, it represents between 16,96% and 21,51% of the investment, depending on the size of the company. On the other hand, automation of processes is the only aspect in which some of the company sizes do not invest, specifically in the case of micro and small enterprises. Probably because large and medium-sized companies have more complex processes where blockchain can provide greater value and justify the return on investment.

Micro enterprises are most active investing in the decentralisation of business processes and Machine Economy, although with less difference compared to other company sizes. Micro and large companies are investing in the decentralisation of processes, in contrast to small and medium-sized companies that mainly invest in traceability of processes. We can observe how medium-sized companies are the most active investors in shared asset management; and large companies stand out for their investment in process automation and limited investment in Smart Contracts.

Table 2 presents the percentage of investment made by sector in each of the identified business values. The construction sector is mainly investing in process traceability, shared asset management, and trustworthiness. No investments have been identified in more

Table 1. Business value by company size (percentages)

Size	Decentralis. of processes	Decentralis. of Busines. Mod.	Traceab. .& Prov. Assets	Traceab./Cert. Processes	Transp. Processes	Shared asset Management	Trust & Integrity	Automation of Processes	Smart Contracts	Digital Identity	Sovereignty of Data.	Machine Economy
Micro	14,96	9,16	4,60	9,92	2,12	4,28	16,96	-	5,81	12,11	9,50	10,58
Small	8,11	2,44	8,03	13,00	4,11	8,04	21,53	-	5,34	13,11	8,11	8,17
Med.	8,50	1,49	6,36	15,33	3,23	12,04	17,32	5,91	5,53	9,61	6,42	8,25
Large	15,14	2,45	6,45	7,23	1,89	5,22	21,51	11,03	1,13	14,32	4,20	9,42

Table 2. Business value by sector (percentages)

Sector	Decentralis. of processes	Decentralis. of Busines. Mod.	Traceab. .& Prov. Assets	Traceab./Cert. Processes	Transp. Processes	Shared asset Management	Trust & Integrity	Automation of Processes	Smart Contracts	Digital Identity	Sovereignty of Data.	Machine Economy
Cons.	-	-	10,78	22,21	10,78	22,21	22,21	6,51	-	5,31	-	-
eGov	9,93	-	-	24,34	5,26	24,34	13,98	1,75	-	16,04	4,35	-
Energ.	3,74	4,23	15,42	15,38	1,68	13,77	15,79	10,77	3,82	-	-	15,42
Finan.	0,69	1,38	0,69	22,14	8,46	22,83	10,49	22,14	11,18	-	-	-
Ind4.0	4,08	0,90	4,31	4,66	0,79	10,79	20,90	10,57	1,83	6,44	15,84	18,89
Health	13,53	5,89	5,90	12,39	-	17,88	17,88	5,19	-	12,69	8,64	-
Mobil.	11,20	6,81	7,99	7,99	3,60	14,80	14,80	14,80	-	11,20	-	6,81
ICT	9,03	1,94	2,25	7,35	0,33	4,34	17,64	10,29	3,48	19,23	11,75	12,36

decentralised characteristics such as process decentralisation, business models, smart contracts, or Machine Economy.

Governments are mainly investing in process traceability and shared asset management, followed by digital identity for their citizens and decentralisation of processes. Furthermore, is one of the few sectors that is not investing in asset traceability.

Energy sector is heavily investing in process and asset traceability, as well as process automation to improve its current processes. In addition, due to the decentralisation of power generation and emerging prosumers [31], the energy sector is investing in trust and integrity, shared asset management, and Machine Economy.

Financial industry is focused on improving its current processes, with most of its investment focused on shared asset management, process automation, and traceability. Precisely because the disintermediation of financial relations could jeopardise its medium/long-term business viability, there is no investment in the decentralisation of processes.

The fourth industrial revolution is reflected in the industrial investments that have trust and integrity of relationships at their core, as well as industrial data sovereignty and Machine Economy. In a second tier, shared asset management and process automation could be highlighted.

Health sector invests mainly in shared resource management, digital identity, and trust, with traceability and decentralisation of processes in a secondary plane. All this fits with a more patient-centric vision in which health services aim to interoperate and empower the user over their health data.

In the case of mobility, we should note that they mainly invest in shared asset management, digital identity, and trust, but in this case their commitment is linked to process automation rather than process decentralisation.

ICT sector stands out for its clear commitment to digital identity and trust, which aims to change the identity model towards a Self-Sovereign Identity (SSI), followed by data sovereignty and Machine Economy linked to the data economy and IoT ecosystem.

Finally, Fig. 3 shows the investment made in each of the business values identified in the study. Evaluated projects indicate a higher investment in trustworthiness of data and relationships between different stakeholders. This characteristic represents the 18% of global investment.

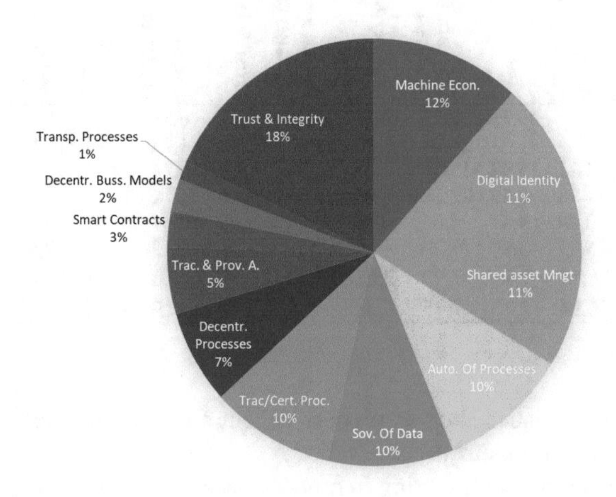

Fig. 3. Blockchain investment distribution by business value.

Machine Economy is the next value with the highest investment. This distinctive IoT trend, both for consumer and industrial applications, represents 12% of global investment; probably driven by the high investment in ICT and Industry 4.0 projects, as shown in Fig. 1. These four business values together account for 56% of the overall analysed investment.

Surprisingly, the three business benefits that have received the least investment are process transparency, decentralisation of business models and smart contracts. Together, they represent only 6% of the analysed investment, followed by asset traceability which represents just 5% of the investment. These features are some of the most frequently mentioned in the studies reviewed in Sect. 2, but they are not the ones in which companies are making their main investments.

6 Relational Model of Blockchain Provided Value

After discussing the business value and impact that blockchain technology offers to companies, the next step was to analyse if there is any relationship in the demand of these values. For this purpose, a clustering analysis using a hierarchical cluster was carried out [32]. To perform the analysis, we have associated with each busines benefit a 104-tuple vector, corresponding to the 104 projects examined in this paper. Each component has been assigned the worth 1 if the business value was present in the purposes of the project, and 0 otherwise. Then we used the squared Euclidean distance as a measure of the relationship between the business values. In this way we have obtained a dendrogram (Fig. 4) where we can appreciate the relationship between the different business benefits of the blockchain technology in the analysed projects [33].

The dendrogram shows clear relationships between the decentralization of business models and smart contracts. The result is natural since one of the most common ways of creating decentralized businesses is through a Decentralized Autonomous Organization (DAO) which is based on a smart contract that governs the organization [34]. In the case of M2M transactions and Machine Economy, it is clearly related to the above as it is a decentralization of the current IoT business models through smart contracts. The fourth value in the first cluster is process decentralization, which is also found when business models are decentralized.

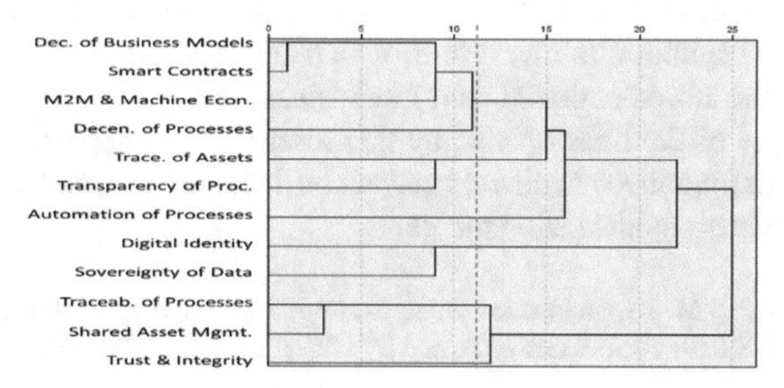

Fig. 4. Hierarchical Clustering Dendrogram using an average linkage between groups (rescaled distance).

Traceability and provenance of assets is a highly demanded value in transparency projects, that is why these two variables are closely related in many of the analysed projects.

Something similar occurs with the traceability of processes and the management of shared assets, which are closely related and form a cluster, as many shared asset management projects also require traceability of the processes carried out on shared assets.

Digital identity and data sovereignty are also closely linked, since in order to create data sovereignty systems it is common to use a decentralized identity on the basis of which to implement policies related to data consumption.

Finally, it should be noted that there are two factors that are not so closely related with others and, therefore, each one forms a cluster on its own. These characteristics are Trustworthiness & Integrity, and Process Automation.

7 Conclusions and Future Work

More than one hundred real blockchain contracts were analysed to answer the research question of what blockchain business values companies are investing in, regardless of the sector and application. The research has been done based on a systematic analysis of actual company investments rather than theoretical benefits and applications of the technology. The real needs of the companies regarding blockchain were identified, classified, and correlated.

Our findings allow us to obtain the business value of blockchain for companies in real projects, since until then the literature only had specific use cases or reviews of potential blockchain applications, but in no case a review and analysis based on a set of real projects in which companies have invested. The study will allow to identify new applications of the blockchain technology in scenarios that have not yet been identified, increasing the adoption of blockchain technology.

On the other hand, the study shows that although two of the most promising values of blockchain technology are the ability to decentralise processes and business models, companies are mainly demanding trustworthiness of their information, digital identity, and shared asset management. This might indicate that companies are making pragmatic decisions and are in a process of learning.

Despite the above, the work has identified that although decentralisation of business models currently only represents 2% of investment, all sectors, except construction and governments, are beginning to invest in it; which can represent a great risk to current models, such as the aforementioned platform business model.

Therefore, one of the lines opened by this research is to explore how blockchain technology will decentralize and disrupt current business models, and what kind of new decentralized business models will emerge.

Acknowledgments. J. M. Zarzuelo acknowledges support from the Spanish Ministry of Science and Innovation (PID2019- 105291GB-I00), and by UPV/EHU (GIU20/019).

References

1. Nakamoto, S.: Bitcoin: a peer-to-peer electronic cash system (2008)
2. Deloitte's 2020 Goblal Blockchain Survey (online). https://www2.deloitte.com/content/dam/insights/us/articles/6608_2020-global-blockchain-survey/DI_CIR%202020%20global%20blockchain%20survey.pdf. Accessed on 07 May 2021
3. Seebacher, S., et al.: Blockchain technology as an enabler of service systems: a structured literature review. In: International Conference on Exploring Services Science, pp. 12–23 (2017)
4. Tama, B.A., et al.: A critical review of blockchain and its current applications. In: 2017 International Conference on Electrical Engineering and Computer Science (ICECOS), pp. 109–113 (2017)
5. Hawlitschek, F., et al.: The limits of trust-free systems: a literature review on blockchain technology and trust in the sharing economy. Electron. Commer. Res. Appl. **29**, 50–63 (2018)
6. Casino, F., et al.: A systematic literature review of blockchain-based applications: current status, classification and open issues. Telematics Inform. **36**, 55–81 (2019)

7. Zheng, Z., et al.: An overview of blockchain technology: architecture, consensus, and future trends. In: 2017 IEEE International Congress on Big Data, pp. 557–564 (2017)
8. Mendling, J., et al.: Blockchains for business process management-challenges and opportunities. ACM Trans. Manag. Inf. Syst. **9**(1), 1–16 (2018)
9. Nowiński, W., et al.: How can blockchain technology disrupt the existing business models? Entrep. Bus. Econ. Rev. **5**(3), 173–188 (2017)
10. Lage, O., et al.: Blockchain and the decentralisation of the cybersecurity Industry. DYNA **96**(3), 239 (2021). https://doi.org/10.6036/10188
11. Morkunas, V.J., et al.: How blockchain technologies impact your business model. Bus. Horiz. **62**(3), 295–306 (2019)
12. Täuscher, K., et al.: Understanding platform business models: a mixed methods study of marketplaces. Eur. Manag. J. **36**(3), 319–329 (2018)
13. Montecchi, M., et al.: It's real, trust me! Establishing supply chain provenance using blockchain. Bus. Horizons **62**(3), 283–293 (2019)
14. Kouhizadeh, M., et al.: At the nexus of blockchain technology, the circular economy, and product deletion. App. Sci. **9**(8), 1712 (2019)
15. Shih, C.S., et al.: Design and implementation of distributed traceability system for smart factories based on blockchain technology. In: Proceedings of the Conference on Research in Adaptive and Convergent Systems, pp. 181–188 (2019)
16. Caro, M.P., et al.: Blockchain-based traceability in Agri-Food supply chain management: a practical implementation. In 2018 IoT Vertical and Topical Summit on Agriculture-Tuscany (IOT Tuscany), pp. 1–4. IEEE (2018)
17. Compert, C., et al.: Blockchain and GDPR: how blockchain could address five areas associated with GDPR compliance. IBM Security White Paper (2018)
18. Ølnes, S., et al.: Blockchain in government: benefits and implications of distributed ledger technology for information sharing. Gov. Inf. Q. **34**, 355–364 (2017)
19. Adams, W.M., et al.: Managing tragedies: understanding conflict over common pool resources. Science **302**(5652), 1915–1916 (2003)
20. Böhme, R., et al.: Bitcoin: economics, technology, and governance. J. Econ. Perspect. **29**(2), 213–238 (2015)
21. Ter Hofstede, A.H., et al., (eds.): Modern Business Process Automation: YAWL and Its Support Environment. Springer Science & Business Media, p. 492 (2009)
22. Szabo, N.: Smart contracts. Unpublished manuscript (1994)
23. Szabo, N.: Smart contracts: building blocks for digital markets. EXTROPY: J. Transhumanist Thought **18**(2) (16) (1996)
24. Allen, C.: The path to self-sovereign identity. Life with Alacrity (2016)
25. Goldreich, O., Oren, Y.: Definitions and properties of zero-knowledge proof systems. J. Cryptol. **7**(1), 1–32 (1994). https://doi.org/10.1007/BF00195207
26. Jarke, M.: Data Sovereignty and Data Space Ecosystems. Business and Information Systems Engineering (2019)
27. Harris, D., et al.: Standards for secure data sharing across organizations. Comput. Stand. Interf. **29**(1), 86–96 (2007)
28. Lage, O.: Blockchain: from industry 4.0 to the machine economy. In: Computer Security Threats. IntechOpen (2019)
29. Chen, Y.: Blockchain tokens and the potential democratization of entrepreneurship and innovation. Bus. Horiz. **61**(4), 567–575 (2018)
30. Lee, J.Y.: A decentralized token economy: How blockchain and cryptocurrency can revolutionize business. Bus. Horiz. **62**(6), 773–784 (2019)
31. Morstyn, T., et al.: Using peer-to-peer energy-trading platforms to incentivize prosumers to form federated power plants. Nat. Energy **3**(2), 94–101 (2018)

32. Ziberna, A., et al.: A comparison of different approaches to hierarchical clustering of ordinal data. Metodoloski zvezki **1**(1), 57 (2004)
33. Forina, M., et al.: Clustering with dendrograms on interpretation variables. Anal. Chim. Acta **454**(1), 13–19 (2002)
34. Norta, A.: Designing a smart-contract application layer for transacting decentralized autonomous organizations. In: International Conference on Advances in Computing and Data Sciences, pp. 595–604 (2016)

A Preliminary Review on Complementary Applications of Databases and Blockchain Technology

Renata Kramberger[1(✉)], Tatjana Welzer[2], and Aida Kamišalić[2]

[1] Zagreb University of Applied Sciences, Vrbik 8, 10000 Zagreb, Croatia
renata.kramberger@tvz.hr

[2] Faculty of Electrical Engineering and Computer Science, University of Maribor, Koroška cesta 46, 2000 Maribor, Slovenia
{tatjana.welzer,aida.kamisalic}@um.si

Abstract. Although blockchain was originally designed to support cryptocurrency transactions, its use has greatly expanded. Due to its security mechanisms and immutability feature, blockchain is now also used for a variety of applications. Nevertheless, blockchain is facing several challenges. On the other hand, databases have been on the market for decades, offering a wide range of affordable solutions. In this paper, we explore the possibility of combining blockchain and database technologies, taking into account the specifics of their corresponding characteristics and issues. The case study was conducted by analyzing and classifying papers into three categories: blockchain as support for databases, databases as support for blockchain, and hybrid systems. Our research focuses on the improvement of technologies features and not on the performance improvement. Obtained results show the potential in developing combined solutions in all three categories taking into account the main advantages and disadvantages of both technologies.

Keywords: Blockchain · Database · Hybrid systems

1 Introduction

With years of data digitization, the need for secure and reliable methods of data storage is constantly increasing. The development of data storage systems has been taking place since the 1960s [1] and, although it has come a long way, the need for improvement compels further research and development. In addition to the development of existing technologies, new and innovative technologies have also emerged. These technologies include NoSQL, NewSQL, and blockchain. Although relational SQL, NoSQL, and NewSQL differ significantly from blockchain technologies in terms of data storage methods, they all serve a similar purpose - to provide reliable and secure data storage. Blockchain was originally designed as a system for cryptocurrencies and transactions, but over time it has been adapted to store different types of data (e.g. patient records,

J. Prieto et al. (Eds.): BLOCKCHAIN 2021, LNNS 320, pp. 161–170, 2022.
https://doi.org/10.1007/978-3-030-86162-9_16

election voting data, IoT related data, student academic achievements, etc.). While each of the aforementioned technologies has its own advantages, they also have their limitations. Moreover, each of them was designed and developed in response to a different problem. Databases were designed to store, update, and query data. Blockchain was initially designed to store transactions and developed into a solution for secure and immutable data storage.

Although databases and blockchain are seen as two completely separate technologies, some blockchain systems (such as Ethereum and Hyperledger) use databases to store the global state of the network. Podgorelec et al. [20] show in their research that most blockchain technologies use key-value stores, which are essentially NoSQL databases. Less commonly, they also use relational and document-oriented NoSQL databases. The popularity of key-value stores lies in their ability to perform read and write operations quickly. There has also been research on whether it is beneficial to store data off the chain, and to what extent does this method improve the limitation of blockchain technologies. Eberhardt et al. [9] present five different patterns of off-chain data storage and the benefits and limitations of each of them. Furthermore, Marinho et al. [16] present a case study of three different scenarios: when data is kept on the blockchain, in a relational database, and in their hybrid system. Our research explores the possible applications of the combination of the two technologies, and whether these combinations could offer new features to either blockchain or database technologies.

The rest of the paper is organized as follows. Section 2 provides an overview of databases and blockchain development. The methodology used to conduct the research relevant to this paper is introduced in Sect. 3. Section 4 presents our analysis of the data collected, including the opportunities for their combinations. Final discussion and conclusions are presented in Sect. 5.

2 Background Concepts

Data storage technologies have evolved and been developed in various forms over many years. The development of relational databases began in the early 1960s and continued to develop over the years. NoSQL databases were introduced with the presentation of the CAP theorem and are still being improved today. Blockchain evolution started a cryptocurrency transaction platform and is still being perfected for multiple purposes (e.g. electronic voting, access control, etc.).

2.1 Databases

Databases play an important role in data management and are one of the key elements in the development of IT solutions. The development of database technology began in the early 1960s with the creation of the first database management system (DBMS) [1]. In 1972 E. F. Codd provided a set of rules (ACID) which guarantees reliable data processing. ACID is the foundation of relational databases, which are still the most used DBMS today [7,8,10]. Relational databases were originally intended to run on a single server and could only scale vertically. However, the growth of the data has increased the need for horizontal scalability and the development of distributed databases. Due to their original design, scaling traditional relational DBMSs has not shown to be an easy task [11]. In 2000, Brewer introduced the CAP theorem which is the foundation of NoSQL databases [5]. While relational databases must comply with all four ACID rules, NoSQL databases can only support two out of three rules proposed by the CAP theorem in case of failure. While relational databases are all based on the same data storage type, there are several types of NoSQL data stores: key value, document, wide-column, and graph [11]. Each of these types is intended to solve a different problem, and one should carefully choose which NoSQL database to use when considering what specific problem the system is intended to solve.

2.2 Blockchain

The development of blockchain technology began when S. Nakamoto proposed a new type of monetary system that relied on a peer-to-peer network without centralization. Bitcoin's goal was to provide a system that would allow for a transparent, immutable, and secure method of performing payments that did not depend on a single company or country [18]. Although Bitcoin was a revolution and opened the door to cryptocurrency transfers, Ethereum not only provided the ability to perform cryptocurrency transactions, but also the ability to create custom code that would run on top of the blockchain. This was done through the use of smart contracts. In Ethereum, smart contracts are written in Solidity and provide a method for creating business applications [24]. Ethereum paved the way for several similar solutions for building blockchain applications such as Ripple [22], Cardano [6], etc. Certain blockchain technologies further evolved and can now support a wide range of applications that are not related to cryptocurrencies. These applications leverage the strengths of blockchain technology to provide data immutability, transparency, and overall security. These applications include electronic voting, health care, identity management systems, access control systems, notaries, and supply chain management [15].

3 Methodology

The research methodology of this paper has been adapted to gather articles that would provide answers to the defined research questions:

1. RQ1: When is this combination of technologies used?
2. RQ2: What types of databases are used?
3. RQ3: What role does blockchain technology play in such systems?

The search string used to collect articles is the following, with some specific database related modifications:

```
(blockchain OR "block chain") AND
(database OR "data base" OR nosql) OR
"blockchain database"
```

The search was performed in January 2021 on some of the most popular databases: Web of Science, IEEE Explore, SpringerLink, ScienceDirect, and ACM Digital Library.

For articles to be considered, they had to have been published in the last 4 years and written in English. The elimination process also included the removal of all articles that were not related to the given topic, regardless of the search string. For articles to be taken into account, they had to contain precisely defined implementations - situations when they are used, what technologies are used, and why. Use cases that did not contain the needed information were not included in the final set. Since multiple search engines may display the same article in the results, duplicates also had to be removed. In addition, a snowballing process was performed to add specific articles and White papers that were essential to the list. The final list contained 13 articles that were used in this paper.

4 Analysis

The aforementioned articles were collected and evaluated. Conference papers, white papers and journal articles were taken into consideration. As can be seen in Fig. 1, the number of articles has increased over the years.

During the analysis of the obtained data, we found that some use databases as support and extension of blockchain technologies, while others use blockchain features in order to extend and enhance database technologies. Moreover, some combine certain features of each technology in a new hybrid system with the intention of solving a specific problem. Given these observations, we have divided the collected works into three categories:

1. **Blockchain as support to database** where blockchain are used to enhance the capabilities of databases,
2. **Database as support to blockchain** where databases provide support for blockchain technology, and
3. **Hybrid system** that combine both database and blokchain into a new custom system.

4.1 Blockchain as Support to Database

As can be seen in Table 1, there are a number of cases where blockchain is used as support to enhance the capabilities of databases. Although the development of relational databases has been ongoing for many years, blockchain enables the development of additional security and immutability features.

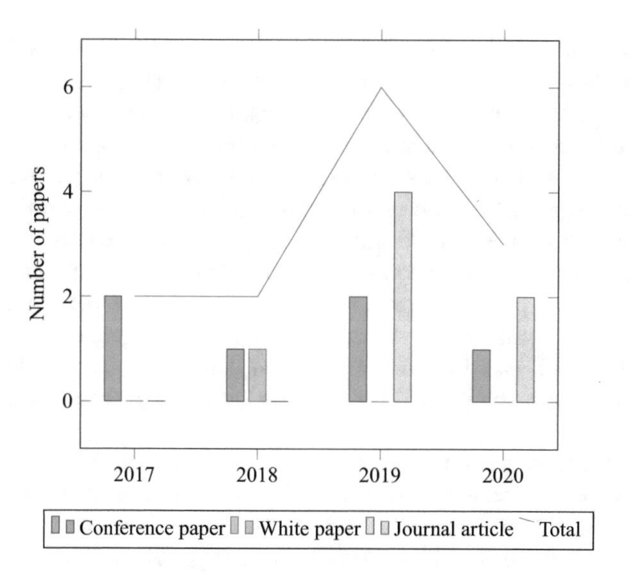

Fig. 1. Number of papers over the past four years

Table 1. Overview of articles that use blockchain as support to databases

Ref.	Purpose (RQ1)	Database (RQ2)	DB type (RQ2)	Blockchain (RQ3)	Blockchain role (RQ3)
[21]	Database management system	N/A	N/A	Ethereum	Access information, operation log
[23]	Data transparency and integrity	MySQL	Relational	Hyperledger Fabric	Providing that the data is tamper-proof and traceable
[27]	Photo forensics	N/A	N/A	Ethereum	File verification, immutability and history
[12]	Data sharing	N/A	N/A	N/A	Security
[3]	Database management system	N/A	Relational	N/A	Transaction verification and immutability

In their paper, Tharani et al. [21] describe a database management system based on blockchain technology. This DBMS differs from the classical DBMS as it helps to monitor operations that are performed on the database. These operations include CRUD (create, read, update, and delete). The blockchain is used to store data about the database (number of tables, table and column names, roles, permissions, etc.) and also data about each operation performed on the database. Another important role of blockchain in this DBMS is to verify the permissions that users can use to access and manipulate data within the database. Blockchain is known for its high data immutability, so this type of DBMS provides data integrity and a distributed control of the data in the database.

Beirami et al. [3] also focus on developing better database management systems in their work. They present a design for a trusted database management system. Using blockchain technology, they extend the associated tables to support immutable transactions and transaction verification.

Vo et al. [23] focus on the problem of uninsured vehicles and insurance fraud. Their main goal is to create a transparent system for micro-insurance handling. Micro-insurances are insurance options that allow clients to pay for a lower premium than the annual one (an example is the "pay as you drive" option). The authors designed a mobile application that would record all the driver's movements, driving speed, and similar information. The data that they believe needs to be tamper-proof is stored on the Hyperledger Fabric blockchain. This data includes the details of the premium purchased, the geographical movements of the driver, speed, etc. In their application, they enhance the databases with blockchain to reduce the possibility of fraud by making certain important data tamper-proof.

The use of images without the permission of the copyright holder has become an increasing issue of concern. Zou et al. [27] use the combination of databases, image authentication technologies, and blockchain to develop a blockchain-based photo forensics system to address this copyright problem. The system makes it possible to verify the copyright and authenticity of a given image. In addition to copyright preservation, the system can also be used to see the transformation of a particular photo or image. Within the system, it is the miners' job to verify each image. Due to their size, the photos are not stored directly on the blockchain itself, but within a database.

DSSPS (Data Sharing Security Protection Scheme) [12] also deals with the security of data exchange on the Internet. Han et al. proposed using the core competencies of blockchain to create a system that solves this problem. They focus on consortium blockchain's and ciphertext policy attribute-based encryption to enhance security.

4.2 Database as Support to Blockchain

There are a number of cases where databases are used to support and enhance the capabilities of blockchain technologies (see Table 2).

One way to harness the potential of databases to enhance blockchain is to create a framework for analytics. Bartoletti et al. [2] have created a framework to store and display analytics for blockchains. Using a database to store analytics greatly speeds up data queries and analytics report generation. This framework is designed to be interchangeable. Due to its interchangeability, it can be used

Table 2. Overview of articles that use databases as support to Blockchain technology

Ref.	Purpose (RQ1)	Database (RQ2)	DB type (RQ2)	Blockchain (RQ3)	Blockchain role (RQ3)
[2]	Blockchain analytics	MySQL & MongoDB	Relational & NoSQL	Bitcoin and Ethereum	Data source
[13]	Handling UTXO	LevelDB 2.0	Key-value	Bitcoin Core 0.16.1	Transaction storage
[14]	Consensus algorithm for transaction coordination	IPFS	Distributed file system	Custom	Normal

with both Ethereum and Bitcoin blockchains, and one can choose between using the NoSQL (MongoDB) or relational (MySQL) database.

Another way to improve the blockchain's capabilities is to use databases for UTXO (Unspent transaction output) processing. UTXO is used to speed up verification time on a blockchain network. UTXO-based blockchains do not use the entire transaction history, but only the minimum set of transactions needed to verify the transaction. BZIP [13] uses a key-value NoSQL database (LevelDB 2.0) to store the UTXO data. The data is stored as key-value pairs, where the key is the transaction ID combined with an index indicating the location of the target output transactions, and the value stores all the data about the transaction (account address, coin value, block height, etc.). Since key-value databases are not able to merge redundant copies of user addresses, but can create a hash table with null values, they introduce a two-level hash table to reduce redundancy.

Khan et al. focused their work on creating a new consensus algorithm [14]. They created FAST, a consensus algorithm based on the MapReduce paradigm used for block transaction processing. Unlike other systems discussed in this paper, FAST does not use a traditional database, but rather IPFS (InterPlanetary File System), a data storage system. The advantages of their solution include faster handling of block transactions and sustainability for higher workloads.

4.3 New Hybrid System

Another way to leverage the strengths of databases and blockchains is to combine them in a new hybrid system. Unlike the first two categories, the most useful properties of both technologies are used to form a unique solution that doesn't emphasize on upgrading of either database or blockchain. Table 3 shows the combinations and purposes in creating such systems.

BigchainDB [4] is a blockchain database that emerged from the idea of creating a system consisting of the best features of blockchain and database technologies. The main features of blockchain that improve the database are decentralization, immutability, owner-controlled assets, and Byzantine Fault Tolerance. To this end, they use the Tendermint framework to create a custom blockchain. Using a MongoDB NoSQL database allows inheritance of features such as low latency, high transaction rate, indexing, and the ability to query structured data.

Table 3. Overview of articles that propose new hybrid systems

Ref.	Purpose (RQ1)	Database (RQ2)	DB type (RQ2)	Blockchain (RQ3)	Blockchain role (RQ3)
[4]	Blockchain based database	MongoDB	NoSQL	Tendermint framework for custom blockchain	Data storage
[25]	Data transparency and integrity	N/A	Relational	N/A	Data transparency and integrity
[19]	Blockchain based database	MySQL & IntegriDB	Relational & Relational	Tendermint framework for custom blockchain	Security
[17]	Middleware	N/A	N/A	Modified Ripple	Middleware between the user application and the database
[26]	Blockchain based database	MySQL	Relational	Tendermint framework for custom blockchain	Data storage

Wu et al. [25] have developed a blockchain recall management system used in the pharmaceutical industry. Their idea is to create a system that shortens the time it takes to initiate drug recalls. The system would allow for faster detection of signs that a particular product has contraindications and would allow for more transparency throughout the pharmaceutical product recall process.

FalconDB is a blockchain-based collaborative database [19]. The motive was to create a system where individual users can use a database that simultaneously has the security level of a blockchain, low storage costs, and high efficiency. They use the Tendermint framework to create a custom blockchain, and both MySQL and IntegriDB to store a data model that allows logging of changes to the database so that even in the event of a malicious accident, the correct database state can be restored. In addition to the queries used in relational databases, it provides the ability to write both history (that return the wanted data throughout the history of the database) and delta queries (that return the historical changes in a specific time range). This provides users with a detailed view of the changes made to the data over time and reduces the possibility of malicious changes to the database [19].

ChainSQL is an open source system that combines blockchain and distributed databases in a new system [17]. The modified Ripple blockchain is used as middleware between the user application and the database. In their paper [17], they present 3 use cases where their hybrid system is useful: as middleware between an application and a database, as a way to create database recovery, and to create database audits (tracing and access control). ChainSQL also provides the ability to write complex queries. They have developed an API that allows queries to be written in both SQL and JSON [17].

Zhu et al. [26] have addressed the problem of modeling complex data storage tasks on blockchains in their work. They propose a blockchain database called SEBDB (Semantics Empowered BlockChain DataBase) that leverages functionalities of both blockchains and relational databases. For the implementation the Tendermint framework and KAFKA were used for a custom blockchain, and MySQL was used to store off-chain data.

5 Discussion and Conclusions

The analysis provided in this paper to answer introduced research questions shows that databases and blockchain can be combined in three ways. Databases can be used to support and extend blockchain features, and vice versa. It is also possible to create custom solutions that combine both technologies and work together as a single fused entity.

Databases provide a way to store, process, and query data. This is also one of the main advantages that databases have over blockchain, where data cannot be queried in a simple and effective way. Both relational and NoSQL databases are used to extend blockchain features, depending on the specific requirements of the system.

Blockchain provides additional immutability, verification, and traceability which are its main advantages. Although relational databases have been around for years, new technologies can and will additionally improve the security and immutability of data.

Our research shows that a combination of databases and blockchain is possible and that this area has potential. The fact that only a few papers have been published in recent years considering this topic shows the need for further research and work. As part of our future work, we plan to expand our research to make a systematic literature review that will also include Distributed Ledger Technologies and other distributed storage solutions.

Acknowledgements. The authors acknowledge financial support from the Slovenian Research Agency (Research Core Funding No. P2-0057).

References

1. Bachman, C.W.: The origin of the integrated data store (IDS): the first direct-access DBMS. IEEE Ann. Hist. Comput. **31**, 42–54 (2009). https://doi.org/10.1109/MAHC.2009.110
2. Bartoletti, M., Lande, S., Pompianu, L., Bracciali, A.: A general framework for blockchain analytics. In: SERIAL 2017–1st Workshop on Scalable and Resilient Infrastructures for Distributed Ledgers, Colocated with ACM/IFIP/USENIX Middleware 2017 Conference, New York, New York, USA, pp. 1–6. Association for Computing Machinery Inc. (2017). https://doi.org/10.1145/3152824.3152831
3. Beirami, A., Zhu, Y., Pu, K.: Trusted relational databases with blockchain: design and optimization. Procedia Comput. Sci. **155**, 137–144 (2019). https://doi.org/10.1016/j.procs.2019.08.022
4. BigchainDB GmbH: BigchainDB: The blockchain database, pp. 1–14 (2018)
5. Brewer, E.A.: Towards robust distributed systems (abstract). In: Proceedings of the Nineteenth Annual ACM Symposium on Principles of Distributed Computing - PODC 2000, New York, NY, USA, p. 7. ACM Press (2000). https://doi.org/10.1145/343477.343502
6. Cardano. https://why.cardano.org/en/introduction/motivation/
7. Codd, E.F.: A relational model of data for large shared data banks. Commun. ACM **13**, 377–387 (1970). https://doi.org/10.1145/362384.362685
8. DB-Engines Ranking - popularity ranking of database management systems. https://db-engines.com/en/ranking
9. Eberhardt, J., Tai, S.: On or off the blockchain? Insights on off-chaining computation and data. In: De Paoli, F., Schulte, S., Broch Johnsen, E. (eds.) Service-Oriented and Cloud Computing, pp. 3–15. Springer, Cham (2017). https://doi.org/10.1007/978-3-319-67262-5_1
10. Haerder, T., Reuter, A.: Principles of transaction-oriented database recovery. ACM Comput. Surv. **15**, 287–317 (1983). https://doi.org/10.1145/289.291
11. Hagerty, P.: The rise of the distributed SQL database – CrateDB. https://crate.io/a/rise-distributed-sql-database/
12. Han, D., Chen, J., Zang, G., Wang, X., Gao, Y.: DSSPs: a data sharing security protection scheme based on consortium blockchain and ciphertext-policy attribute-based encryption. In: ACM International Conference Proceeding Series, New York, NY, USA, pp. 14–19. Association for Computing Machinery (2019). https://doi.org/10.1145/3376044.3376048

13. Jiang, S., et al.: BZIP: a compact data memory system for UTXO-based blockchains. J. Syst. Archit. **109**, 101809 (2020). https://doi.org/10.1016/j.sysarc.2020.101809
14. Khan, N.: FAST: a MapReduce consensus for high performance blockchains. In: BlockSys 2018 - Proceedings of the 1st Blockchain-Enabled Networked Sensor Systems, Part of SenSys 2018, New York, NY, USA, pp. 1–6. Association for Computing Machinery Inc. (2018). https://doi.org/10.1145/3282278.3282279
15. Di Francesco Maesa, D., Mori, P.: Blockchain 3.0 applications survey. J. Parallel Distrib. Comput. **138**, 99–114 (2020). https://doi.org/10.1016/j.jpdc.2019.12.019
16. Carlos Marinho, S.S., Filho, J.S.C., Moreira, L.O., MacHado, J.C.: Using a hybrid approach to data management in relational database and blockchain: a case study on the e-health domain. In: Proceedings - 2020 IEEE International Conference on Software Architecture Companion, ICSA-C 2020, pp. 114–121, 2020). https://doi.org/10.1109/ICSA-C50368.2020.00030
17. Muzammal, M., Qu, Q., Nasrulin, B.: Renovating blockchain with distributed databases: an open source system. Futur. Gener. Comput. Syst. **90**, 105–117 (2019). https://doi.org/10.1016/j.future.2018.07.042
18. Nakamoto, S.: Bitcoin: a peer-to-peer electronic cash system (2008)
19. Peng, Y., Du, M., Li, F., Cheng, R., Song, D.: FalconDB: blockchain-based collaborative database. In: Proceedings of the ACM SIGMOD International Conference on Management of Data, New York, NY, USA, pp. 637–652. Association for Computing Machinery (2020). https://doi.org/10.1145/3318464.3380594
20. Podgorelec, B., Turkanović, M., Šestak, M.: A brief review of database solutions used within blockchain platforms. In: Prieto, J., Pinto, A., Das, A., Ferretti, S. (eds.) Blockchain and Applications, vol. 1238, pp. 121–130. Springer, Cham (2020). https://doi.org/10.1007/978-3-030-52535-4_13
21. Samantha Tharani, J., Tharmakulasingam, M., Muthukkumarasamy, V.: A blockchain-based database management system. Knowl. Eng. Rev. **35** (2020). https://doi.org/10.1017/S0269888920000302
22. Schwartz, D., Youngs, N., Britto, A.: The ripple protocol consensus algorithm. https://ripple.com/files/ripple_consensus_whitepaper.pdf
23. Vo, H.T., Mehedy, L., Mohania, M., Abebe, E.: Blockchain-based data management and analytics for micro-insurance applications. In: Proceedings of the International Conference on Information and Knowledge Management, New York, NY, USA, pp. 2539–2542. Association for Computing Machinery (2017). https://doi.org/10.1145/3132847.3133172
24. Wood, G.: Ethereum: a secure decentralised generalised transaction ledge. Ethereum Project Yellow Paper, pp. 1–32 (2014)
25. Wu, X., Lin, Y.: Blockchain recall management in pharmaceutical industry. Procedia CIRP 590–595 (2019). https://doi.org/10.1016/j.procir.2019.04.094
26. Zhu, Y., Zhang, Z., Jin, C., Zhou, A., Yan, Y.: SEBDB: semantics empowered blockchain database. In: Proceedings - International Conference on Data Engineering, pp. 1820–1831. IEEE Computer Society (2019). https://doi.org/10.1109/ICDE.2019.00198
27. Zou, R., Lv, X., Wang, B.: Blockchain-based photo forensics with permissible transformations. Comput. Secur. **87**, 101567 (2019). https://doi.org/10.1016/j.cose.2019.101567

Blockchain Platform Selection for Security Token Offering (STO) Using Multi-criteria Decision Model

Richard[1,2(✉)], Harjanto Prabowo[2], Agung Trisetyarso[2], and Benfano Soewito[2]

[1] Information Systems Department, School of Information Systems, Bina Nusantara Univesity, Jakarta 11840, Indonesia

[2] Computer Science Department, BINUS Graduate Program - Doctor of Computer Science, Bina Nusantara Univesity, Jakarta 11840, Indonesia

{richard-slc,harprabowo,atrisetyarso,bsoewito}@binus.edu

Abstract. Security Token Offering (STO) is the new and reshaped form of the famous Initial Coin Offering (ICO). The thing that distinguishes STO from ICO is the inclusion of several rules from regulators and auditors. The control by regulators and auditors will generate a trade-off between control and decentralization of the STO. We did a simulation for selecting the most suitable blockchain platform for the STO process using the Multi-Criteria Decision Model simulator to cover this issue. The simulation result shows the relation between the selected features, software quality, and the most suitable solution for STO.

Keywords: Security Token Offering · Blockchain platform selection · Multi-Criteria Decision Model

1 Introduction

Security Token Offering (STO) is the evolution of the famous Initial Coin Offering (ICO), the most popular cryptocurrency use case during the 2017–2018 period. ICO raised around $140 million funds during its most popular era within more than 2.000 projects (https://www.icodata.io/stats/2018). However, ICO has an unpopular reputation with many projects created for *pumping and dumping* (scam). A report from *Satis Research* stated that 78% of ICO's were Identified Scams, ~4% Failed, ~3% had Gone Dead, and ~15% went on to trade on an exchange [1]. While ICO ended with scam projects, STO offers a new opportunity to provide a more secure and reliable token/coin offering process with the inclusion of regulators and auditors [2].

Since the ICO and STO happen in the blockchain, transaction transparency and auditability are enabled by default. The regulators and auditors are no need to worry about the completeness of the transaction data. Anyone can see all transactions on the blockchain from the search-engine-like web page in the public blockchain, usually called *explorer*. Nevertheless, the anonymity of a public blockchain will create another problem for the token offering related to the user's integrity and KYC. On the contrary, private blockchain could offer more reliability with limited access and more control to

J. Prieto et al. (Eds.): BLOCKCHAIN 2021, LNNS 320, pp. 171–178, 2022.
https://doi.org/10.1007/978-3-030-86162-9_17

the blockchain data. Still, the private blockchain will create drawbacks in decentralization, which is the main idea of blockchain. The selection of blockchain platforms will become an important thing to be discussed since several blockchain platforms offer different advantages from the platform. In this research, we are trying to find the most suitable blockchain platform for the STO process while generating more involvement from regulators and auditors without sacrificing decentralization.

This research paper will compare several options of blockchain platforms currently available in the blockchain industry. We will use Multi-Criteria Decision Model to select blockchain platforms previously published by Siamak Farshidi [3]. We will fill the indicators in the model with the requirements from Security Token Offering (STO) process. In the *Literature Review* section, we will discuss the literature review on a blockchain platform for STO and the feature requirements from STO. *The research Method* section will discuss the Multi-Criteria Decision Model used in this research. We will disclose the result of blockchain platform selection for STO in the *Discussion* section. Finally, the *Conclusion* section will conclude this research paper.

2 Literature Review

2.1 Literature Review on Blockchain Platform Selection for Security Token Offering (STO)

To design a security token, we need to define the token characteristics [4]. All the desired functions and rules of the token need to be written on the token definition [2]. The coin or token offering process needs to consider several factors from a blockchain platform, such as network effects and blockchain permission [5]. Currently, most blockchain platforms are not interconnected with each other [6], and this kind of issue is called an *interoperability* issue [7]. With the vast amount from the birth of various blockchain platforms, their interaction will be more challenging in the future [8].

The blockchain platform selection will be a complex task which is including several criteria such as *security*, *interoperability*, *consensus mechanisms*, and *transaction speed*. The proper platform selection will generate a long-term impact on the time and cost needed to develop a blockchain application [3]. The *scalability*, *sustainability*, and *transaction costs* also need to be considered in the blockchain platform selection [9, 10] The inclusion and application of *business rules* become necessary for blockchain platform selection [11]. This kind of *business rules* inclusion has been applied to the research for a permissioned digital asset trading mechanism platform [12].

In designing an STO process, we need to determine the *token characteristics* derived from the compliance requirement of the regulators [4]. Furthermore, we need to ensure that STO is held in a blockchain platform with vast *network effects* [5]. Other issues like *scalability* and *sustainability* also become essential in selecting a blockchain platform [10].

2.2 Requirements from STO for Blockchain Platform Selection

We summarize the essential requirements for the STO process based on the literature review in the previous section. The requirements that should be considered in selecting a blockchain platform for STO are defined as follows:

1. ***Blockchain Permission.*** The level of accessibility of a blockchain platform. It can be *public*, *private*, or *hybrid*. The blockchain permission needs to be considered related to the STO process that requires regulator and auditor's inclusion while not sacrificing the decentralization.
2. ***Network Effects.*** The network effect of a blockchain platform is essential consideration related to the market potential and token utility. The more substantial network effect will generate more advantages for the STO mechanism built on the platform.
3. ***Interoperability.*** As blockchain technology and application become mainstream, the blockchain platform will have its rapid growth in number as well. Several blockchain platforms are trying to solve this problem by creating interoperability functions (e.g., Polkadot - https://polkadot.network, Moonbeam - https://moonbeam.network, Polygon - https://polygon.technology).
4. ***Cross-chain interaction.*** Beyond interoperability, cross-chain interaction becomes a requirement as this capability will attract more investors, which has its asset on several blockchain platforms.
5. ***Security.*** This requirement includes the security of the blockchain network, smart contracts in the blockchain platform, and wallet & key management.
6. ***Transaction Speed.*** The faster transaction processing speed will generate better opportunities for the STO as every trade will happen in the perfect time.
7. ***Transaction Cost.*** The cost of every transaction becomes essential consideration in selecting a blockchain platform for STO. The lower cost will attract more investors to involve in the STO process.
8. ***Scalability.*** Blockchain as the technology platform will face the classic decentralization trilemma, which includes scalability, security, and decentralization.
9. ***Business Rules.*** The inclusion of business rules could be on the smart contract level or even on the blockchain consensus level.

3 Research Method

To address the considerations in selecting the blockchain platform for STO, we use a Multi-Criteria Decision Model published by Dr. Siamak Farshidi in his paper "*Decision Support for Blockchain Platform Selection*" [3]. The decision model includes all the decision criteria, alternative options, and the relation between them. The main blocks of the decision model are *Decision Meta-Model, Software Quality Model, Domain-Description, Feature Value,* and *Case Definition.*

1. ***Decision Meta-Model.*** This section gives an ontological description of the Multi-Criteria Decision Making (MCDM). There is two sub-section of this Decision Meta-Model, including *Quality* and *Features*. The *quality* represents the critical attributes of a blockchain, including interoperability, maturity, and blockchain performance. At the same time, Features include smart contract functions and on-chain transactions.
2. ***Software Quality Model***. Determine the software quality of a blockchain platform. This model is designed by software quality experts using ISO/IEC 25010 and ISO/IEC 9126 standards as the primary quality indicators. This model includes the quality requirements and their relationship with the other components.

3. ***Domain Description.*** The process to identify the criteria from several blockchain platforms, including features, documentation, and characteristics. The domain description can be used to generate an initial hypothesis for blockchain feature selection. The domain description is designed and validated by several blockchain experts.
4. ***Feature Value.*** A knowledge-based feature that is designed by mapping process between blockchain platform and features defined in the domain description. The identified blockchain platform is the result of a literature study and documentations from the current blockchain platforms.
5. ***Case Definition.*** The decision-maker defines the case or business process that wants to be mapped to the blockchain platform selection. In the case definition, the decision-maker will define the priority of each feature using MoSCoW [13].

4 Discussion

We use the simulator tool designed by Siamak Farshidi on the page: https://dss-mcdm.com/index.html. The simulator is created to match the *Multi-Criteria Decision Model* for blockchain platform selection defined in Sect. 3. The *Decision Meta-Model, Software Quality Model, Domain Description,* and *Feature Value* are pre-defined on the simulator. We made several modifications to the pre-defined value, including the platform options and features.

4.1 Case Definition and Feature Prioritization

We define the Case Definition by prioritizing the features based on the STO requirements we collected in Sect. 2 using MoSCoW [13].

1. ***Blockchain Permission.*** A blockchain network could be public or private. In the context of STO, the token offering process should be held in the public blockchain to reach a wider audience and interest in the offering. Therefore, public permission becomes a ***should-have***.
2. ***Network Effects.*** The end goal of STO is getting a company closer to the capital by the fund-raising process. Network effects are the level of usage and economic benefit created by a blockchain platform. The more substantial network effect will generate higher benefits to the STO project. Despite this, the network effect still does not determine the success rate of an STO project. Therefore, the network effect is categorized as a ***could-have***.
3. ***Interoperability.*** With interoperability, the token from the STO project will become more valuable for any investor. They could use the token generated from STO for any utility in another blockchain platform, for example, Loan, Staking, and so forth. To attract more investors and generate more value for the security token, interoperability categorized as a ***should-have*** feature in a blockchain platform.
4. ***Cross-chain interaction.*** Beyond interoperability, cross-chain interaction generates more value for every blockchain platform. With cross-chain interaction, we could create more valuable functions related to the utility of a token. The cross-chain interaction feature in a blockchain platform is categorized as a ***should-have***.

5. ***Security.*** Security becomes the ***must-have*** feature of a blockchain platform. We must ensure that a blockchain platform has security protection related to the network itself (consensus) and smart contract level.
6. ***Transaction Speed.*** Transaction speed matters because it will impact the liquidity and market price of a token. More reliable transaction speed will create more advantages for investors and token issuers. The transaction speed is categorized as a ***could-have*** feature.
7. ***Transaction Cost.*** Several blockchain platforms apply the transaction fee for executing a transaction. This fee mechanism is created to protect the network from suspicious activity that is executing the transaction repeatedly. On the contrary, this transaction fee will generate a trade-off as a disadvantage for the STO process. As the transaction cost will be charged to the user, transaction cost management is categorized as a ***must-have*** feature in a blockchain platform.
8. ***Scalability.*** This issue is a classic issue related to blockchain as the technology itself is in the early stage and very fast in development. The scalability is categorized as a ***could-have*** feature in a blockchain platform because it will create more trust from the investor to an STO project. The scalability means more assurance from the STO project in the future.
9. ***Business Rules.*** The one that distinguishes STO from ICO is the inclusion of regulators and auditors in the offering process. The STO also needs to translate required business rules to the token offering mechanism. Therefore, the inclusion of business rules becomes a ***must-have*** feature in a blockchain platform.

4.2 Decision-Making Simulation Result

We evaluate the required features and it's prioritization using the MCDM simulator. The pre-defined quality attributes and feature value are also being used in this simulation (Fig 1).

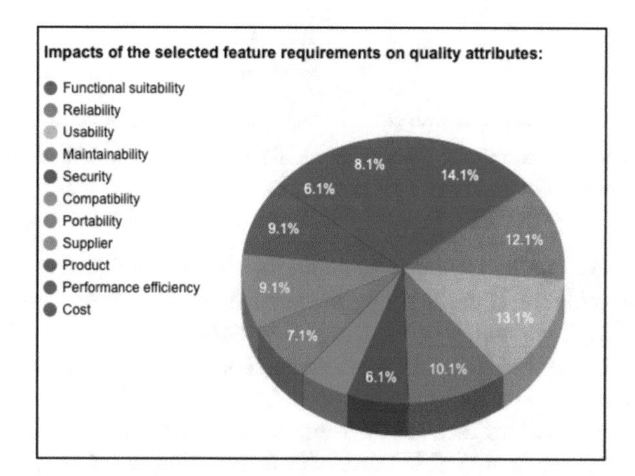

Fig. 1. The impact of the selected feature on the quality attributes based on the case definition

The required features from the STO process have a solid relation to functional suitability in quality attributes with a 14.1% impact. Besides, these features have a strong relation with reliability, usability, and maintainability. Each percentage is more than 10%. The other non-significant relations to quality attributes include portability, supplier, product, performance efficiency, cost, security, and maintainability. These relations to quality attributes show the view of requirements from the quality point of view (Fig. 2).

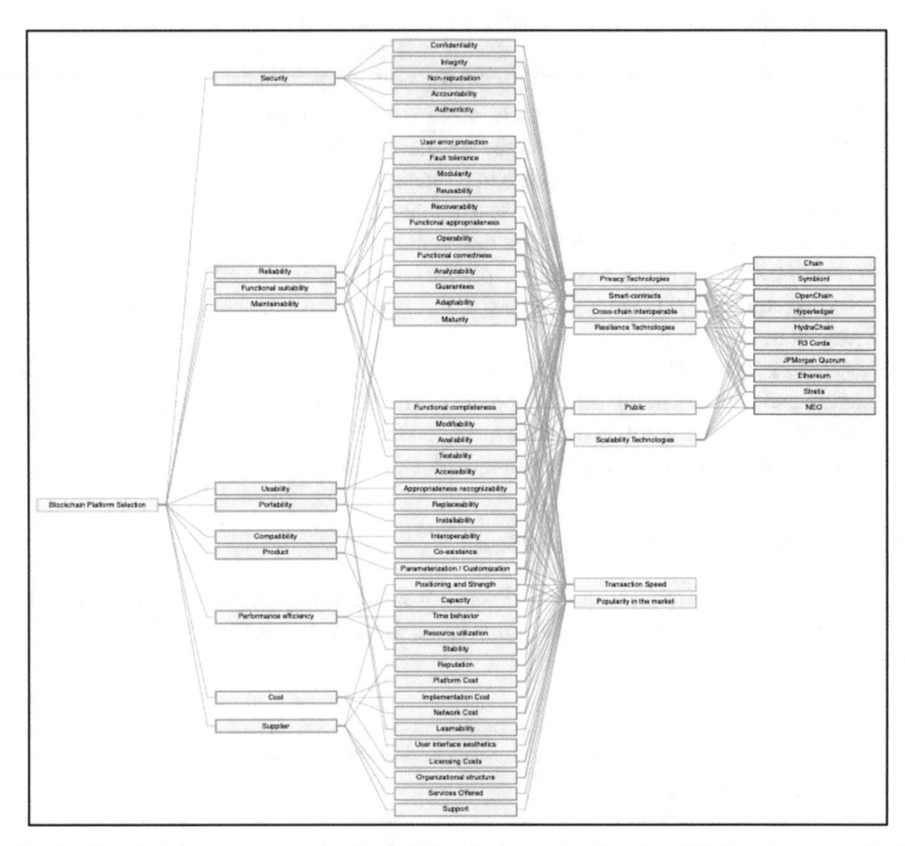

Fig. 2. The decision structure of a blockchain platform selection for STO based on identified features in the case definition

The relations of quality attributes (column 2), blockchain features (column 3), and case definitions (column 4) are deriving the decision to several options of blockchain platform including *Ethereum*, *Hyperledger*, *Openchain*, *NEO*, etc. (Fig. 3).

The most feasible solution is Ethereum, with a score of 99. Hyperledger, HydraChain, and Openchain become the following options with the score between 81 and 82. The other options are including Stratis and Neo, with a score of 65. The simulator also suggests other options such as R3, Corda, JPMorgan Quorum, Chain, and Symbiont, with a score of 47–48. The result of the simulation is choosing Ethereum as the first option of the blockchain platform for the STO process.

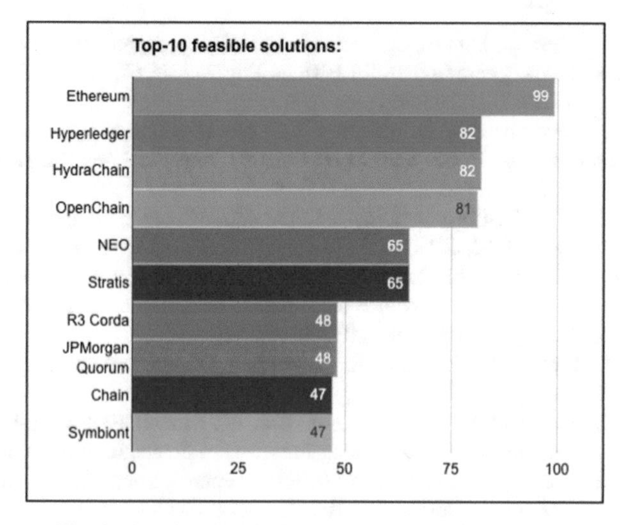

Fig. 3. Blockchain Platform Recommendation for STO

5 Conclusion

Blockchain platform selection becomes an important issue related to the incapability and incompatibility of the current solution in the blockchain industry. We use MCDM because there is multiple consideration in selecting the platform as STO requires a balanced trade-off between the control and decentralization.

From the feature definition, prioritization, and decision-making simulation we made in Sects. 3 and 4, the result of blockchain platform selection shows that Ethereum is the most suitable solution. The Ethereum platform is the most popular and industry-leading blockchain platform that supports smart contracts for creating decentralized apps. Several standards are related to a security token proposed and discussed by the Ethereum community, such as ERC-1400, ERC-1404, ERC-1462, ERC-884, and ERC-1450. Ethereum also has the most substantial network effect compared to the other options. Ethereum is the second-largest cryptocurrency as of May 2021, according to coinmarketcap.com, with dominance in the market cap of 19%.

The consideration and simulation in this research paper are based on current blockchain platform options in the blockchain industry and its current capabilities. The future is probable that the platform selection needs to be reconsidered and simulated as blockchain technology is growing very fast, and so many platforms rise in recent years. The new blockchain platform will set its starting point in the Ethereum and other existing platform positions to overcome the disadvantages of the current blockchain platform.

References

1. Dowlat, S.: Cryptoasset Market Coverage Initiation: Network Creation (2018)
2. Momtaz, P.P.: Entrepreneurial finance and moral hazard: evidence from token offerings. J. Bus. Ventur. 106001 (2020) https://doi.org/10.1016/j.jbusvent.2020.106001

3. Farshidi, S., Jansen, S., Espana, S., Verkleij, J.: Decision support for blockchain platform selection: three industry case studies. IEEE Trans. Eng. Manag. **67**, 1109–1128 (2020). https://doi.org/10.1109/TEM.2019.2956897
4. Lee, J.Y.: A decentralized token economy: how blockchain and cryptocurrency can revolutionize business. Bus. Horiz. **62**, 773–784 (2019). https://doi.org/10.1016/j.bushor.2019.08.003
5. Sharma, Z., Zhu, Y.: Platform building in initial coin offering market: Empirical evidence. Pacific-Basin Financ. J. **61**, 101318 (2020). https://doi.org/10.1016/j.pacfin.2020.101318
6. Xu, X., et al.: a taxonomy of blockchain-based systems for architecture design. In: Proceedings - 2017 IEEE International Conference on Software Architecture ICSA 2017, pp. 243–252 (2017). https://doi.org/10.1109/ICSA.2017.33
7. Hardjono, T., Lipton, A., Pentland, A.: Towards a Design Philosophy for Interoperable Blockchain System, pp. 1–27 (2018)
8. Qiu, H., Wu, X., Zhang, S., Leung, V.C.M., Cai, W.: ChainIDE: a cloud-based integrated development environment for cross-blockchain smart contracts. In: Proceedings of the International Conference on Cloud Computing Technology and Science, CloudCom. 2019-Decem, pp. 317–319 (2019). https://doi.org/10.1109/CloudCom.2019.00055
9. Pustišek, M., Umek, A., Kos, A.: Approaching the communication constraints of ethereum-based decentralized applications. Sensors (Switzerland) **19**, 2647 (2019). https://doi.org/10.3390/s19112647
10. Casino, F., Dasaklis, T.K., Patsakis, C.: A systematic literature review of blockchain-based applications: current status, classification and open issues. Telemat. Inf. **36**, 55–81 (2019). https://doi.org/10.1016/j.tele.2018.11.006
11. Aparecido Petroni, B.C., Gonçalves, R.F., Sérgio de Arruda Ignácio, P., Reis, J.Z., Dolce Uzum Martins, G.J.: Smart contracts applied to a functional architecture for storage and maintenance of digital chain of custody using blockchain. Forensic Sci. Int. Digit. Investig. **34**, 300985 (2020). https://doi.org/10.1016/j.fsidi.2020.300985
12. Wang, R., Tsai, W.-T., He, J., Liu, C., Deng, E.: A distributed digital asset-trading platform based on permissioned blockchains. In: Qiu, M. (ed.) SmartBlock 2018. LNCS, vol. 11373, pp. 55–65. Springer, Cham (2018). https://doi.org/10.1007/978-3-030-05764-0_6
13. Waters, K.: Prioritization using MoSCoW. Agil. Plan. **12**, 31 (2009)

Towards a Holistic DLT Architecture for IIoT: Improved DAG for Production Lines

Denis Stefanescu[1(✉)], Patxi Galán-García[1], Leticia Montalvillo[1], Juanjo Unzilla[2], and Aitor Urbieta[1]

[1] Ikerlan Technology Research Centre, 20500 Arrasate-Mondragon, Spain
{distefanescu,pgalan,lmontalvillo,aurbieta}@ikerlan.es
[2] University of the Basque Country (UPV/EHU), 48013 Bilbao, Spain
juanjo.unzilla@ehu.eus

Abstract. The Industrial Internet of Things (IIoT) aims to greatly improve the existing production procedures, enhancing customer experiences, reducing costs and increasing efficiency. IIoT will make a significant impact on existing business models in several areas. However, IIoT has several issues related to the security of the information and the excessive centralization which creates single point of failures, poor performance and scalability. Therefore, we introduce an Industry 4.0 case scenario, and, on top of it we propose a multi-layer distributed ledger architecture for IIoT that combines various types of Distributed Ledger Technologies (DLTs) in order to deliver a lightweight and efficient solution for Industry 4.0. However, in this paper we only focus on the design of the first layer of the architecture. We design a lightweight Directed Acyclic Graph (DAG) DLT architecture and propose several improvements regarding the lightweight devices participation, data storage, cryptography and consensus.

1 Introduction

The Industrial Internet of Things (IIoT), which can be defined as the use of the Internet of Things (IoT) in industrial sectors and applications, is one of the most important parts of Industry 4.0 [1].

However, the IIoT has several challenges that need to be addressed [2]. First, security is a major problem, as IIoT devices can be easily attacked, data can be tampered by malicious third parties, and are more vulnerable to denial of service attacks. Second, data privacy is also a major issue, since the data can be easily exposed. Finally, centralization is another problem, since traditional industrial systems rely on centralized servers, which not only are commonly threatened by weak connections and security vulnerabilities, but also become a significant bottleneck when the number of devices and data flow increases (limiting the performance of the network).

To solve aforementioned challenges, many researchers have proposed the use of blockchain in IIoT. Blockchain is a type of Distributed Ledger Technology

J. Prieto et al. (Eds.): BLOCKCHAIN 2021, LNNS 320, pp. 179–188, 2022.
https://doi.org/10.1007/978-3-030-86162-9_18

(DLT) in which all the transactions are stored in a chain of blocks that are cryptographically linked. A DLT consists of replicated, shared, and synchronized digital data spread across multiple sites, with no central authority. The participating nodes reach an agreement over the state of the ledger by following a consensus algorithm. This technology has many benefits such as decentralization, security, privacy, treaceability and interoperability, which makes it ideal for the deployment of the Industry 4.0.

Nonetheless, currently, blockchain is not well suited for resource-constrained environments such as the IIoT, since most of the existing consensus algorithms offer low throughput, require high resource consumption, have a lack of energetic efficiency, low scalability and a considerable delay in storing transactions. Furthermore, blockchain also requires huge storage capacity and network bandwidth, as the information has to be replicated in each one of the participating nodes. Indeed, in a recent systematic literature review we have conducted (currently under review for publication), we conclude that applying blockchain to IIoT is still a major challenge, and there is no DLT architecture that completely fulfills the requirements of IIoT scenarios in terms of computational power, storage, throughput and energetic efficiency.

Therefore, we contribute to this problem with the following solutions:

- We introduce a generic IIoT smart factory scenario, where we define different levels and analyze their challenges and needs.
- We propose a layered multi-DLT architecture suited for a generic smart factory scenario, that is highly efficient, secure, and fast. We design a lightweight and trustworthy DLT environment from when the data is generated in IIoT up until the data is processed at a business level.
- We delve deeper into the first layer of our proposed architecture (which is a Directed Acyclic Graph (DAG) type DLT), and propose several improvements over the existing state of-the-art of DAG type DLTs. Proposed improvements aim to improve lightweight IIoT devices participation in the DLT, storage, cryptography and consensus.

The remainder of the paper is organized as follows. Section 2 analyzes the existing lightweight blockchain (and DLT) proposals for IIoT and discusses their gaps and limitations. In Sect. 3 we describe a generic smart factory scenario, which is composed of 4 layers. In Sect. 4 we describe our multi-DLT architecture proposal and in Sect. 5 we describe in detail our improved lightweight DLT proposal for the production line layer. Finally, Sect. 6 includes the conclusion of the paper and the future work.

2 Related Work

In this section we summarize the main characteristics and contributions of other relevant DLT-based architectures for IIoT.

Seok *et al.* [3] propose a blockchain architecture for the IIoT that improves the performance of the cryptographic hash algorithm that is typically used in the mining process. This proposal reduces the computational burden and latency of blockchain, while reducing the enormous energy requirements of the mining process.

Liu *et al.* [4] introduce a lightweight, IIoT oriented blockchain system that improves many problematic aspects of blockchain such as the energy, storage and bandwidth consumption. First, they propose a more efficient Proof of Work (PoW) based consensus which employs a reputation system in order to reduce its energy consumption. Then, they introduce a more lightweight data structure for the blocks to streamline broadcast content, reducing the network bandwidth. Finally, they design a novel "unrelated Block offloading filter" to reduce the size of the ledger.

Yu *et al.* [5] use the Edge and Cloud technologies to create a layered architecture for the IIoT. The authors of this paper propose three layers: the IIoT devices layer, the Edge layer and the Cloud blockchain layer. The IIoT layer comprises heterogeneous IIoT devices that send their data to the Edge layer. The Edge layer processes and temporarily stores the data that comes from the IIoT layer. Finally, the Cloud blockchain layer stores the offloaded data that is sent from the Edge layer. They claim to improve the scalability and efficiency of blockchain by distributing the load among the layers.

Different from the typical blockchain-based proposals, a compacted Directed Acyclic Graph (DAG) architecture for the IIoT is presented in [6]. In this approach blocks are organized in a DAG structure that is ordered by levels and width. The CoDAG structure is part of a three layer architecture: the miners, the gateways and the nodes. The miners maintain the ledger using their computing power, similar to Bitcoin. The gateway nodes are powerful IIoT nodes that maintain part of the ledger. Finally, nodes are devices that participate in the network by submitting data or payments. This proposal claims to offer a much higher throughput than a typical blockchain.

3 Industry 4.0 Case Scenario

Smart manufacturing requires the eradication of offline operations. Factories and plants that are connected become much more efficient and productive [7]. However, inter-connecting all the components of a smart factory is not straightforward in terms of performance and security, especially when applying the DLT technology to this field [8].

Today, most factories are structured hierarchically from the shop floor level up to the enterprise resource planning level. The reference architecture diagram for Industry 4.0 is the pyramid of automation [9]. Our smart factory case scenario consists of four levels, as shown in Fig. 1, similar to the original automation pyramid scheme, but with slight changes that are adapted to the real world based on our industrial experience. In this practical scenario approach, we start by optimizing the smallest asset (e.g. machine, sensors, actuators, tools, etc.), and then progressively scale to bigger structures: from production lines, industrial plants, to a smart factory business consortium. This case scenario enables us to deliver a lightweight and complete DLT architecture where each aspect is carefully addressed.

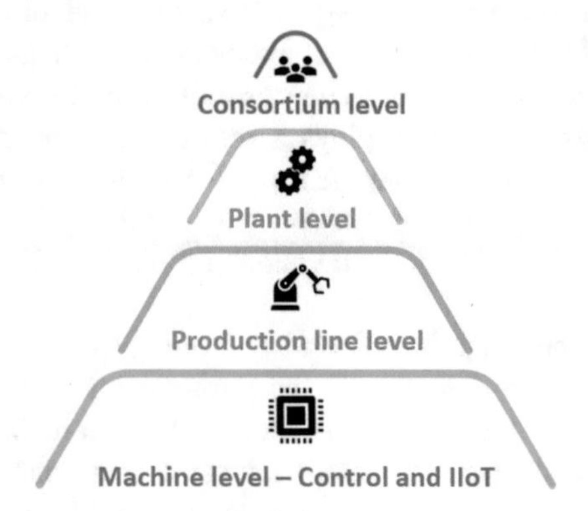

Fig. 1. Our smart factory case scenario

3.1 Machine Level

The machine level corresponds to Cyber-Physical Production Systems (CPPS). This layer is composed by IIoT devices that measure system parameters (sensors) and execute specific actions (actuators). The data is transmitted between IIoT devices and control systems, which, in our use case, are also part of this level. These are devices and systems that control the industrial processes based on information from IIoT devices. They include controllers (PLCs, RTUs), distributed control systems (DCS), operator panels (HMI) and supervision and control systems (SCADA). The control systems are meant to manage several lightweight devices. This layer usually generates a great amount of data, and requires fast and real-time management.

3.2 Production Line Level

The production line level is composed by various machines and robots that are inter-connected. Each production line forms a sub-network within the smart factory. This level, similar to the previous layer, is required to handle huge amounts of data in a fast and secure manner. Due to the great number of IIoT devices and data traffic (i.e. transactions), this layer must be highly scalable as well as energy and cost efficient.

3.3 Plant Level

The plant level is the linking of all production lines into a single global network of the whole smart factory. This layer has some physical drawbacks such as the size of the network and the distance between its component nodes when it comes to wireless connection. However, this network is not required to handle such massive and rapid exchanges of data as the previous layer. This layer must offer a highly optimized, secure and efficient distributed storage solution in order to feasibly manage the whole data of the factory.

3.4 Consortium Level

The consortium level integrates many smart factories into a consortium network in order to provide governance, interoperability, traceability, as well as a safe and private distributed storage solution for the entire smart factory ecosystem. This network is composed of a relatively low number of nodes that are physically separated by hundreds or thousands of kilometers. This layer must offer a moderate transaction speed, and great transparency and immutability of the data, as well as the possibility to execute smart contracts between multiple organizations.

4 Proposed Multi-DLT Architecture

In this section we explain the proposed lightweight DLT ecosystem architecture at a high level. We deploy it on top of the smart factory case scenario shown in Sect. 3.

From top to bottom, our DLT architecture consists of three layers:

1. Data source DLT layer
2. Bridge DLT layer
3. Business DLT layer

As shown in Fig. 2, each one of the layers of our smart factory architecture corresponds to one or two levels of the defined smart factory case scenario. We intend to use various DLTs in order to address all the challenges and needs of each level of our smart factory case scenario.

The first layer includes the first two levels of the case scenario: the machine level where IIoT and control devices are located, and the production line level, which is typically formed by various machines. This layer is in charge of processing the data from the IIoT devices. The middle layer of our architecture includes the plant level of our case scenario. This layer connects in a DLT network all the production lines of one or several plants. Furthermore, this intermediate network also acts like a data bridge between the other layers of the architecture. Finally, the consortium DLT network includes the top level of the case scenario, which is the consortium-business level. This network aims at establishing a lightweight secure DLT network between various smart factories, which would form a connected smart factory ecosystem in the context of Industry 4.0. However, as we mentioned before, in this paper we only focus on the design of the first layer of the architecture, which is further explained in Sect. 5.

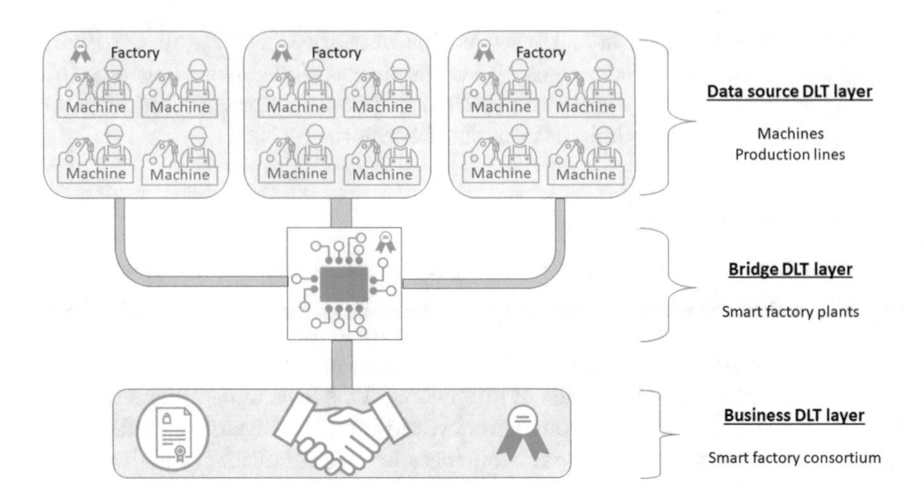

Fig. 2. Smart factory ecosystem multi-DLT architecture diagram

5 Data Source DLT Layer

For the first layer of our architecture, which corresponds to the machine and production line levels of our smart factory scenario, we propose a lightweight DAG-based architecture, as shown in Fig. 3. Our DAG architecture consists of clusters of lightweight devices that belong to an industrial machine (e.g. sensors, thermostats, actuators, etc.). These devices are managed by a more powerful Edge node, which corresponds to an industrial control system. Specifically, the lightweight devices send the data to the DAG, and the powerful Edge nodes manage it if necessary.

A production line network comprises several machines that are composed by numerous lightweight IIoT devices which are managed by an industrial control system that we identify as Cluster Head (CH). We chose a DAG-based DLT in order to promote scalability and fast transaction processing from the large scale IIoT. A DAG DLT does not use miners to validate transactions, instead, the nodes that issue a new transaction must approve two previous transactions and perform a small amount of PoW to avoid spam in the network. Transactions can therefore be issued without fees, facilitating micro-transactions. This mechanism gives DAG-based DLTs huge scalability and throughput (i.e. transactions per second) [6], as the more transactions are issued, the faster and more secure the network is. Furthermore, the lack of mining makes DAGs highly efficient. Thus, this type of DLT is much more suitable for resource-constrained environments that handle a huge number of transactions, which is the exact case of this type of industrial network.

However, DAG DLTs, despite being IoT-focused solutions, still have some room for improvement. Therefore, in the following sub-sections we propose several improvements for our DAG DLT layer.

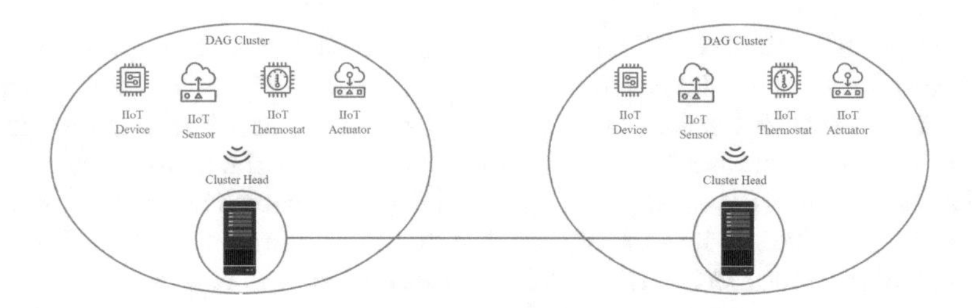

Fig. 3. DAG DLT architecture diagram

5.1 Lightweight Devices Participation

In this work, we try to encourage the participation of lightweight nodes in the DAG network, instead of following the usual approach where the lightweight devices only send their data to a more powerful node that does all the work that the DLT architecture demands. For this purpose we propose to make use of Solid-State Drive (SSD) swap memory. Promoting lightweight devices participation within DLTs, especially for sensing devices is a significant challenge, as shown in other works such as [10], where authors use a software architecture specifically designed for a trust-less water management system where IoT devices can directly transact sensed data on a blockchain network.

The authors in [11] propose the acceleration of a DAG-based ledger using Resistive Random-Access Memory (ReRAM), which is a type of non-volatile Random-Access Memory (RAM). However, this type of memory is currently rare and costly, as it has not been yet widely adopted. Therefore, the use of common RAM memory to improve the performance of the DAG DLT and to allow lightweight IIoT devices to participate in the network would be a more optimal choice. Nevertheless, RAM memory is volatile, which means that it requires power to maintain the stored information. It retains its contents while powered on, but when the power is interrupted, the stored data is quickly lost. Thus, for this purpose we propose the use of SSD swap memory. A SSD is much slower than the RAM memory, but, it has numerous relevant advantages such as the low cost, its high availability on the market, and the fact that is non-volatile. Using this approach we intend to achieve a better performance of our DAG DLT, as well as a greater participation of lightweight IIoT devices in the network.

5.2 Storage

According to [12], industrial data has reached a total volume of more than 1000 Exabytes annually, with a clear upward trend. However, storing such great amounts of IIoT data in the DAG DLT is costly and inefficient, as it requires the use of high storage capacity industrial control devices and high capacity networks. Furthermore, one of our top priorities is to encourage as much as possible the participation of the lightweight IIoT devices in the DAG network. Thus, the most optimal choice is the externalization of the data storage, while maintaining only references in the DLT. The externalization of the data storage to the cloud could also be studied, however, in this work we already propose an external consortium network that can be considered as "the cloud". Furthermore, in this section, we are discussing the data storage issue within the smart factory.

We use the InterPlanetary File System (IPFS) to solve the storage problem. With IPFS, the data is immutably secured and timestamped, without having to attach all of it to the DAG DLT. IPFS is a distributed peer-to-peer file system. In our architecture, IPFS storage could be handled by the powerful Edge nodes that we defined within our DAG DLT. When a file is added to IPFS, the file is divided into various blocks, and all of the blocks are given a unique cryptographic hash. Then, IPFS removes the duplications that are present across the network. Each network node in IPFS stores only the content of its interest, along with some indexing information. Thus, IPFS is more suitable than a centralized database for keeping the production lines databases linked in a secure and decentralized way. Hence, in our solution, the IIoT data that are generated by each production line DAG DLT are grouped and stored in IPFS, while the DAG stores the hash of the IPFS files containing the IIoT data. The cryptographic hashes of the files can then be used to find the actual location of the file.

However, data querying performance in IPFS could be a bottleneck, since for the read queries, IPFS requires to resolve remote nodes and download objects via the internet. Nonetheless, there are works such as [13] that address this problem and propose efficient solutions.

5.3 Cryptography

As stated in the introduction, cryptography is a key part of DLTs. Lightweight cryptography must take into account the trade-off between security, cost, and performance. A DLT usually includes two types of cryptographic functions:

- **Hash functions.** A hash function is a cryptographic function that is easy to check, but difficult to forge. In blockchain, hashes are used for block linking and for mining.
- **Public-key (asymmetric) encryption digital signatures.** The private key is used for signing messages and the corresponding public key for checking the signature.

5.3.1 Hash Function

In DAG DLTs mining is not present. However, in order to avoid spam in the network, each device is required to perform a little PoW in order to send its data to the network. This PoW process is much less expensive than Bitcoin's PoW. However, for lightweight devices, it can still be too costly. We discuss this aspect more in detail in Sect. 5.4.

Therefore, we propose to further improve this aspect by using a lightweight hash algorithm such as Quark [14], which offers about five times more throughput that Spongent, while consuming just 1.77 more μW.

5.3.2 Digital Signature Scheme

Digital signatures are used to verify the integrity of the data. The first blockchain, Bitcoin, uses Elliptic Curve Digital Signature Algorithm (ECDSA) to generate public and private keys for its crypto-currency wallets. For the sake of lightweight cryptography adoption in our architecture, we propose to employ the Edwards-curve Digital Signature Algorithm (EdDSA) digital signature scheme, which was designed to be faster than the current digital signature schemes while maintaining the same level of security. Its key advantages for lightweight devices are higher performance and straightforward, secure implementations [15].

5.4 Consensus

In a DAG DLT, when a node issues a new transaction, it has to solve a cryptographic hash puzzle similar to that of Bitcoin blockchain (i.e., PoW), in order to avoid malicious nodes spamming the network. Subsequently, the issuer has to approve two other unconfirmed transactions, and perform the rest of the consensus process. In this work we focus on improving the most heavy and inefficient aspect of the aforementioned consensus process for lightweight IIoT devices. In order to achieve this, we propose a custom reputation based mechanism that would drastically reduce the amount of PoW that IIoT devices must perform before issuing transactions.

We define a reputation score between zero and ten (0–10). Each IIoT device starts with a medium reputation of five. The reputation score increases or decreases based on the behaviour of the device. The aforementioned score is inversely proportional to the amount of PoW that a device has to perform (i.e. the more reputation score, the less PoW is required). The behaviour of a device is defined by **the number of valid transactions** that it emits, as well as **the amount of transactions** of that device. These two variables are independent from each other. Thus, even if the transactions that a device issued recently were valid, its amount of PoW would grow in order to issue more transactions. Using this mechanism we can effectively avoid network spamming, while reducing the computational burden for lightweight devices.

6 Conclusion and Future Work

We proposed a holistic multi-layer and multi-DLT architecture for an Industry 4.0 case scenario. In this paper we presented the complete design of the first layer, which is a DAG DLT for the machine and production line levels. We proposed several improvements to the DAG DLT regarding the participation of lightweight devices in the network, the cryptography, the storage and the consensus.

As future work we need to (1) perform the implementation and evaluation of the DAG data source layer and also (2) design and evaluate the remaining two layers of the architecture.

References

1. Xu, L.D., Xu, E.L., Li, L.: Industry 4.0: state of the art and future trends. Int. J. Prod. Res. **56**(8), 2941–2962 (2018)
2. Sisinni, E., Saifullah, A., Han, S., Jennehag, U., Gidlund, M.: Industrial internet of things: challenges, opportunities, and directions. IEEE Trans. Industr. Inf. **14**(11), 4724–4734 (2018)
3. Seok, B., Park, J., Park, J.H.: A lightweight hash-based blockchain architecture for industrial IoT. Appl. Sci. (Switz.) **9**(18), 3740 (2019)
4. Liu, Y., Wang, K., Lin, Y., Xu, W.: Lightchain: a lightweight blockchain system for industrial Internet of Things. IEEE Trans. Industr. Inf. **15**(6), 3571–3581 (2019)
5. Yu, Y., Liu, S., Yeoh, P., Vucetic, B., Li, Y.: LayerChain: a hierarchical edge-cloud blockchain for large-scale low-delay IIoT applications. IEEE Trans. Ind. Inform. **3203**(c), 1 (2020)
6. Cui, L., Yang, S., Chen, Z., Pan, Y., Xu, M., Xu, K.: An efficient and compacted DAG-based blockchain protocol for industrial Internet of Things. IEEE Trans. Industr. Inf. **16**(6), 4134–4145 (2020)
7. Tuptuk, N., Hailes, S.: Security of smart manufacturing systems. J. Manuf. Syst. **47**(February), 93–106 (2018). https://doi.org/10.1016/j.jmsy.2018.04.007
8. Mohamed, N., Al-Jaroodi, J.: Applying blockchain in industry 4.0 applications. In: 2019 IEEE 9th Annual Computing and Communication Workshop and Conference, pp. 852–858 (2019)
9. Schlechtendahl, J., Keinert, M., Kretschmer, F., Lechler, A., Verl, A.: Making existing production systems industry 4.0-ready: holistic approach to the integration of existing production systems in industry 4.0 environments. Prod. Eng. **9**(1), 143–148 (2015)
10. Pincheira, M., Vecchio, M., Giaffreda, R., Kanhere, S.S.: Exploiting constrained IoT devices in a trustless blockchain-based water management system. In: IEEE International Conference on Blockchain and Cryptocurrency, ICBC 2020 (2020)
11. Wang, Q., Wang, T., Shen, Z., Jia, Z., Zhao, M., Shao, Z.: Re-tangle: a ReRAM-based processing-in-memory architecture for transaction-based blockchain. In: IEEE/ACM International Conference on Computer-Aided Design (ICCAD), pp. 1–8 (2019)
12. Mourtzis, D., Vlachou, E., Milas, N.: Industrial big data as a result of IoT adoption in manufacturing. Procedia CIRP **55**, 290–295 (2016). https://doi.org/10.1016/j.procir.2016.07.038
13. Shen, J., Li, Y., Zhou, Y., Wang, X.: Understanding I/O performance of IPFS storage: a client's perspective. In: Proceedings of the International Symposium on Quality of Service (2019)
14. Aumasson, J.P., Henzen, L., Meier, W., Naya-Plasencia, M.: Quark: a lightweight hash. J. Cryptol. **26**(2), 313–339 (2013)
15. Bernstein, D.J., Duif, N., Lange, T., Schwabe, P., Yang, B.Y.: High-speed high-security signatures. J. Cryptogr. Eng. **2**(2), 77–89 (2012)

A Secure Blockchain-Based Solution for Management of Pandemic Data in Healthcare Systems

Arav Dalwani(✉)

Dhirubhai Ambani International School, Mumbai, India

Abstract. This paper tackles the issue of data manipulation in pandemics wherein often data is manipulated or underreported by certain government bodies for various reasons. This problem came to light during the first few months of the COVID-19 pandemic and had a major effect on the public health response in countries. To solve this issue, a blockchain based system and implementation is proposed to allow hospitals and testing centers to transmit their data while also minimizing the risk of manipulation from government bodies. The paper makes use of the United States CDC framework for data reporting by testing labs and hospitals in order to develop this proposal for a blockchain system. To accomplish this, the paper then delineates the implementation of private data collections and other features of the Hyperledger software. A potential web-app interface is also presented and the transaction logic has been described in detail. Along with this, an evaluation of the risks, limitations and potential extension of the system has been performed.

Keywords: Smart contract · Private data · Manipulation · Hyperledger Fabric · Consortium data exchange

1 Introduction

The new coronavirus disease 2019 (COVID-19) has spread and continues to spread rapidly throughout various regions around the world. At the time of the writing of this paper, on August 3rd, 2020, there have been more than 91 million confirmed case and 1.95 million deaths worldwide. Yet, 9 months after the World Health Organization declared a pandemic, one of the biggest obstructions to preventing the spread remains COVID-19 tests. While blockchain technology can be one of the greatest tools in combatting the virus, the adoption and proposals so far have been limited. The entire healthcare organization will eventually have to adjust to blockchain technology applications to support and fight these outbreaks.

While the transparency of testing data is often regarded to be of critical importance, such transparency is often not maintained. As explored in Sect. 2, manipulations and misreporting by governments have only widened the disconnect. Greater transparency with testing data and other metrics would greatly aid the healthcare community. Unfortunately, in the absence of a transparent, accountable, and permissioned system for reporting such vital information, data remains almost completely controlled by governments.

J. Prieto et al. (Eds.): BLOCKCHAIN 2021, LNNS 320, pp. 189–198, 2022.
https://doi.org/10.1007/978-3-030-86162-9_19

Additionally, information from different testing labs and hospitals remains scattered or withheld in many regions, with negative consequences. Consider the vital information such as testing delays, possible deaths and individuals who need to be retested, which may never come to light due to the lack of a system to identify these gaps. All information passes through governments in most cases, who take on the role of a third party that could very possibly abuse their role, as shall be discussed in this paper.

Digitizing such information on a blockchain would not only help the heath sector in the long run but also reduce the influence exerted by governments. This paper presents a novel approach to a blockchain based system for storing test data and immutable way that can be publicly accessible. The contributions of this paper are:

- It describes the construction of a network responsible for handling access-control and data storage of COVID-19 testing data and its architecture
- It details the flow of transactions over the blockchain network and the advantages of each for our healthcare agencies
- It delineates a proof-of-concept implementation of the same, existing limitations and risks, and potential further extension and integration.

2 Problem Statement

COVID-19 cases and testing data should be fully transparent to minimize errors discrepancies and should be freely reported from hospitals or testing centers themselves. Governments of course have to also be involved as they are the decision-making authority on the deployment of resources to various locations.

However, in many cases this sensitive data gets tampered with and underreported for various purposes and this can be highly damaging to a public health response. States within countries or countries as a whole often misreport and manipulate this data for their advantage as has been done in various places [1]. In Brazil during the pandemic, the national government completely stopped publishing the critical COVID-19 data and death information, an action that many experts warned would undermine the health response [2]. India has witnessed numerous cases of municipal corporations in Mumbai and Vadodara [3] or state governments such as West Bengal and Bihar purposefully undercounting infections and deaths [4]. A prominent Indian newspaper went as far to blame the government for the uncertainties in the COVID-19 numbers [5]. This became a significant issue of concern during India's second wave where multiple states were alleged to have manipulated the rising infection and death numbers according to many media reports. Even in the United States, there were worrying reports of states combining separate pieces of test data and in September, the CDC was even bypassed altogether in July, raising the issue of transparency [6]. Countries consequently face the issue of two possible extremes:

- Placing complete trust in governments at the local, state and national level to accurately report testing numbers. But as we have seen, even in democratic countries these numbers are frequently manipulated, and this is only discovered much later.
- Rely only on the testing centers and hospitals to handle the data and manage the response completely without government involvement. This may lead to haphazard and misdirected responses to the pandemic due to governmental absence.

This paper proposes an application of a blockchain based system that ensures that (a) COVID testing data is largely controlled by testing centers and hospitals thus making government involvement more responsible, (b) easy sharing of this critical data to various other parties in locations within the country.

3 Overview

Accomplishing the two criteria mentioned above in the problem statement requires the setup of a blockchain network that would allow testing centers to add test data, and hospitals to add hospital data to the blockchain, which can then be accessed by health-care departments. As a healthcare platform would deal with sensitive data and personal information, permissioned blockchains have desirable features for implementing this Electronic Health Records System.

CovDataApp is the prototype for handling of COVID data and for maintaining the Electronic Health Records. In this particular situation, we have chosen Hyperledger Fabric as the software to be used for this particular use case. This is because of the high transaction speed and the light open-source framework provided by Hyperledger [7]. The smart contract and transaction logic are all coded in Javascript, along with the access control rules. In this particular instance, we make use of the IBM Blockchain Platform for Hyperledger, however, the prototype can be run using Docker as well. The client application and user interface is created in React.js, which allows us to create a framework for the different organizations on the blockchain network.

The system proposed is based on the United States CDC framework for reporting data, wherein testing labs and hospitals send in their data to the state or local public health departments according to state/or local law or policy. This is mainly because the local health departments here carry out contact tracing rather than the national health unit. Similar systems of reporting test data and hospital data are present in India, Germany and the UK, where the government shifted contact tracing responsibilities to local councils in August. The CDC framework (Fig. 1) for reporting test data calls for specific and important details to be shared by the testing laboratories with the local governments, and then anonymized data to be sent to the CDC, in order to conduct analytics and evaluation of the data [8].

What to report

Laboratories should make every reasonable effort to provide the following data elements to state and jurisdictional health departments.

For more information on the data elements included in the June 4 HH: guidance, as well as technical specifications that support implementation, see HHS' COVID-19 Lab Data Reporting Implementatio Specifications.

1. Test ordered - use harmonized LOINC codes provided by CDC
2. Device Identifier
3. Test result–use appropriate LOINC and SNOMED codes, as defined by the Laboratory In Vitro Diagnostics (LIVD) Test Code Mapping for SARS-CoV-2 Tests provided by CDC
4. Test Result date (date format)
5. Accession # / Specimen ID
6. Patient age
7. Patient race
8. Patient ethnicity
9. Patient sex
10. Patient residence zip code
11. Patient residence county
12. Ordering provider name and nonpharmaceutical interventions (as applicable)
13. Ordering provider zip code
14. Performing facility name and CLIA number
15. Performing facility zip code
16. Specimen Source - use appropriate LOINC, SNOMED-CT, or SPM4 codes, or equivalently detailed alternative codes
17. Date test ordered (date format)
18. Date specimen collected (date format)

The following additional demographic data elements should also be collected and reported to state or local public health departments.

1. Patient name (Last name, First name, Middle Initial)
2. Patient street address
3. Patient phone number with area code
4. Patient date of birth
5. Ordering provider address
6. Ordering provider phone number

To protect patient privacy, any data that state and jurisdictional health departments send to CDC will be deidentified and not include some patient-level information. The deidentified data shared with CDC will contribute to understanding COVII 19's impact, case rate positivity trends, testing coverage, and will help identify supply chain issues for reagents and other

Fig. 1. Data variables collected by the CDC

From the CDC Website: "To protect patient privacy, any data that state and jurisdictional health departments send to CDC will be deidentified and will not include some patient-level information. The deidentified data shared with CDC will contribute to understanding COVID-19's impact, case rate positivity trends, testing coverage, and will help identify supply chain issues."

Therefore, the organizations who will be a part of this blockchain network are the Testing Labs/Centres, Hospitals, Local Health Departments and the National Health Department such as the CDC. In addition, to this news media outlets can be given access to join the main channel of the blockchain network in order to view some of the test data that is added to the main channel. Each organization carries their own certificate authority and peer, and the blockchain network runs an ordering service that validates all the transactions.

The participants of the blockchain network, all have access to the main channel. There is a common registration page where the user has to enter the Organization MSP ID, after which they would login and be directed to their separate, individual web app. This system was implemented as a prototype and not present a final implementation; therefore, the network was simplified so that it could be set up and deployed quickly. In a final implementation of this blockchain system, there would be many more participants – labs and hospitals – as that would be the requirement of the healthcare system.

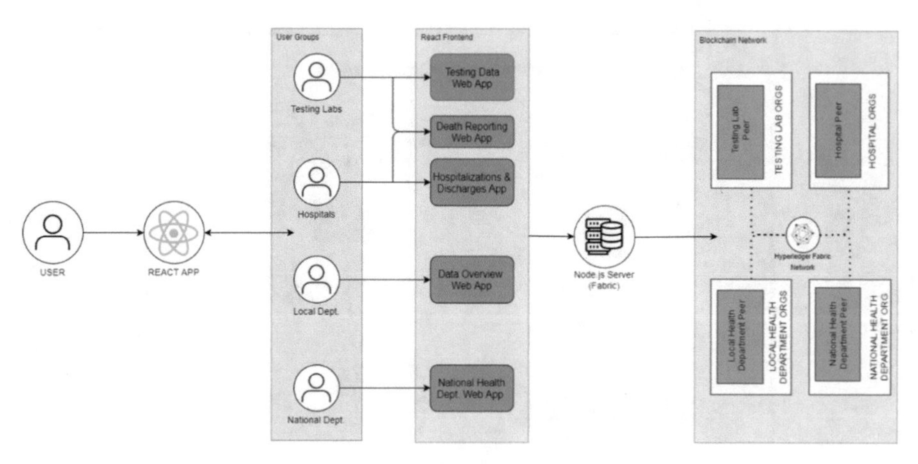

Fig. 2. Proposed system architecture

4 System Design and Architecture

4.1 System Components

There are many components of the proposed system, as it is complex and emulates the pandemic data reporting process across many stakeholders in the healthcare system.

As seen in Fig. 2, all members of the consortium exist as organizations on the main channel, running independent peers and CAs. This channel is the primary blockchain channel, and the only one that can authorize access requests. Each peer on the main channel has the necessary chaincode installed required to vote in favour or against an access request and submit data to the network. The specific contracts and transactions involved in the system are further explained later in the paper.

The second component is private data collections between two or more entities on the main channel, because of the need for preserving anonymity. The collections are instantiated such that the national health department is sent anonymized data while the local health department is given all required data for their operations. This is done by using private data collections, that are set up between the Testing lab orgs and the Hospital orgs on the network with the local department org. For example, all the Testing Lab orgs and Hospital orgs in New York would be connected through a private data collection with the org for the New York Health Department. At the same time, the orgs for all the testing labs and hospitals would form a private data collection (PDC) with the org for the national health department on the Blockchain network.

This is to ensure that patient and testing data is kept confidential, given that there are many stakeholders on the main channel, including News Outlets who have access. As a result, the data needs to be shared separately using a PDC with the local health department org. Then, under the CDC protocol that this system is built around, there needs to be a separate PDC relationship with the national health department org as only certain pieces of data are sent to ensure that it is anonymized. Further PDC relationships can be created individually between labs together and hospitals if deemed necessary to aid collaboration over the pandemic related data (Fig. 3).

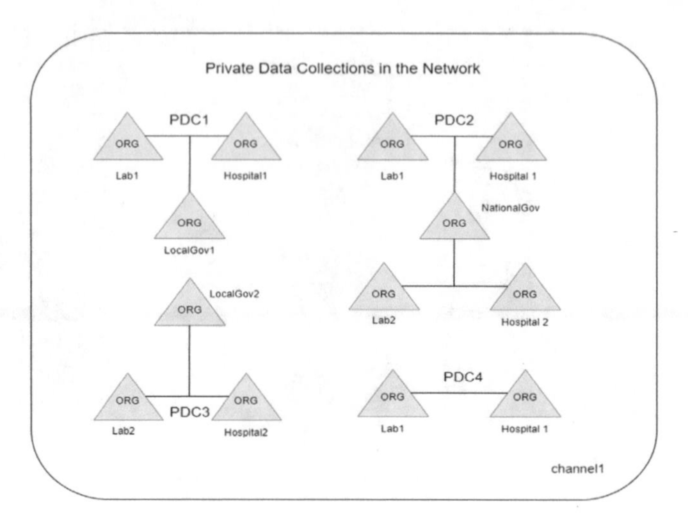

Fig. 3. Private data collections in the network

When a testing center submits data for a test only the data for the i) Test ID or Number (like E1 or T10) ii) Test Results Date iii) Test Result and the iv) Test Type are stored on the main channel, while all other patient/testing data is shared using private data collections. Similarly, when a hospital submits data about a patient, the data regarding the i) Patient ID or Hospital assigned ID, ii) the Changed Status of the Patient – Hospitalization/ Discharge/Death, iii) Date of Status Change and iv) Hospital Name are stored on the main channel. Since we need to ensure anonymity, the name of the patient, their address and all the other pieces of data that have been outlined in the CDC framework such as gender, ethnicity, age, etc. are shared using the private data collections with whichever health department concerned.

The third component is the off-chain storage implementation where the testing lab and hospital orgs store some of the assets off-chain. This mixed-assets storage ensures that the web-app can also retrieve data from the off-chain storage, as this data would be specific to that hospital org and not be mixed with data from other orgs. This novel addition ensures that in case of a breach of the system or of the web-app, the data is also securely stored for each specific org off-chain. Some data may also be non-transactional, which means it is not required to be submitted, in which case it could also be stored along with the off-chain data.

4.2 Transaction Flow

Let us assume that the members of the healthcare system of the country, i.e. the consortium, are to make use of private data sharing for reporting of vital pandemic related data. In such a situation, the consortium members (the different stakeholders) would have to join the network. The initial setup can be performed under the supervision of representatives from all consortium members. In the setup, the entities and their respective nodes are all added to the network and the main channel.

For simplicity, we have to assume that out of a consortium of n members, there are 6 entities – Lab1, Lab2, Hospital1, LocalGov1, NationalGov, NewsOutlet1. When an organization is added to the network, the roles and access control are clearly defined and assigned based on the organization. Let us consider the situation where Lab1 needs to report testing data. In such a situation:

1. The entities on the blockchain, namely Lab1, Hospital1 and LocalGov1 are added together to a private data collection – *Region1PrivateCollection.* This definition for this collection is in the form a.json file where these organisations, with distinct mspIDs are added to the collection.
2. At the same time, the Lab1, Lab2, Hospital1 and NationalGov, entities are added together to a separate private data collection – *NationalDataPrivateCollection.* The Lab2 Org is added to this collection and not to the previous collection, because it is assumed to be under a different region than Lab1 and Hospital1.
3. When data needs to be submitted, all the data elements as shown in the CDC framework are filled out either by the Lab1 org or the Hospital1 org. if the data is submitted by Lab1, the *RecordTest* transaction of the *TestContract* is executed.
4. In this transaction, the data elements described in Sect. 4.1 are stored on the main channel, but these data elements are limited. The rest of the data elements such as ethnicity, age, gender (from the CDC framework) are added to the *Region1PrivateCollection* using the putPrivateData() method of the Node SDK.
5. A large portion of the data elements that were added to *Region1PrivateCollection* are also added to *NationalDataCollection,* except that it would be deidentified and not include some patient level information. In keeping with this, data such as the name and address of the patient would be removed while other data about the Zip code or the region along with the gender, age ethnicity would be kept.
6. Data is submitted in a similar fashion for any hospitalizations or deaths that have occurred. Either the testing labs or the hospitals can report deaths, as a testing lab or centre may have conducted a test posthumously on the individual.
7. When data is submitted to the private data collections here, the actual private data is only accessible to those orgs authorised to be on the collections. A hash of the data is instead written to the ledgers of all the peers of the main channel
8. A separate transaction – *RecordRetestRequirement* is specifically invoked and executed only when the test data submitted shows that the Test Type is an Antigen Test and the result is negative. The CDC and many healthcare departments recommend the retesting of individuals who have tested negative on an antigen test using a full RT-PCR test. A contract-event *RetestEvent* is also emitted at the same time to notify the orgs of this requirement for that specific patient.
9. Another transaction – *RecordTestingFault* is invoked and transacted if there is a difference of 6 days between the data element for when the Test Results were Reported and when the Test was carried out. When the transaction is executed, the *FaultEvent* is also emitted and triggers a contract event listener. The event details are then shown on the web-app interface for NationalGov and LocalGov MSPs.
10. In the case of a hospital discharge being reported, the procedure is slightly different: Firstly, the web-app for Hospital1 gathers the data for all currently Hospitalized patients in the hospital from the off-chain data source. The names of these patients

then show up as a list on the web-app UI for the Hospital, and a specific name is selected to indicate a discharge of the patient. Only the discharge date and the final condition of the patient are submitted as data elements.

4.3 Interface

Figure 4 shows how the proposed system could be set up as an application to collect data from the various stakeholders. In this case, the testing page has been shown containing a portion of the data elements that were listed in the CDC framework. This proposal was designed to present only a sample of the interface that could be used for the sharing and transmission of data and not a final implementation. React along with HTML and CSS were used to create the interface shown, which contained a few other pages. We used the IBM Blockchain Platform to run and test the interface.

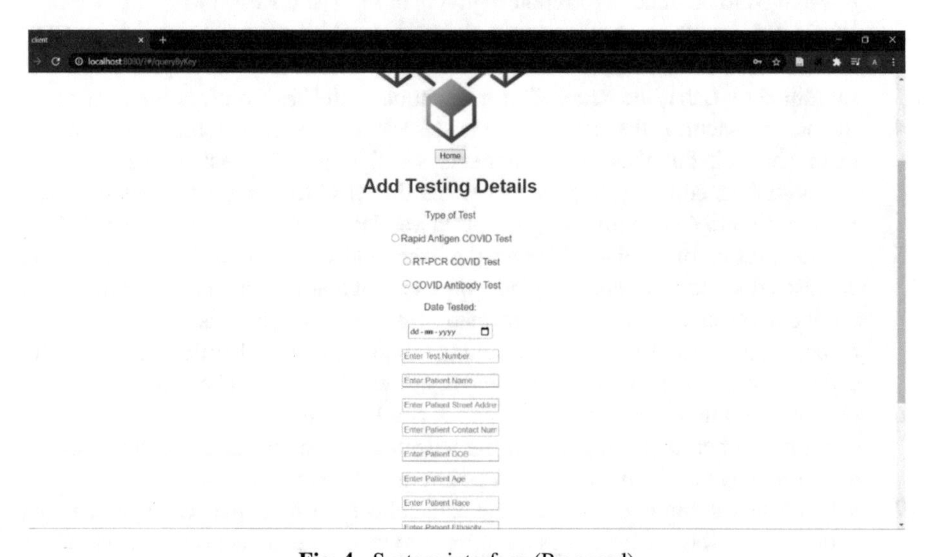

Fig. 4. System interface (Proposed)

We thus used the simplified set up that we described in Sect. 4.2. The NationalGov and the LocalGov MSPs that were added were given Reader permissions in the Access Control that was designed for this channel. The reason for this was that the health departments would still receive health data without manipulating the data that being sent or committing any unrequired transactions to the ledger. However, this does not have to be the case and the permissions given to these orgs can be changed if necessary. Additionally, while we only conducted the basic set-up on the IBM Blockchain Platform, the same can be replicated using standard Hyperledger Fabric.

4.4 Risk Assessment

The following are the potential risks associated with the implementation of the system described in this paper, along with a brief analysis of the likelihood of risk:

- Majority of Nodes Compromised: If more than 50% of the channel nodes are compromised simultaneously, data integrity is at risk. However, this is improbable in a large consortium with industry-grade security added to prevent breaches.
- Data collection operation compromised: Data is fetched directly from the data sharing entities' - hospitals and testing labs. These could be compromised if they are illegally infiltrated, however this is quite improbable, especially on the large scale that it would have to occur.
- Malicious code in the system: Ill-intentioned organizations that manage to join the network can embed malicious code in the network. Sandboxing the operation to an extent and running suitable malware checks can resolve this, so the risk is remote.
- Errors in Data Submission: Since this is a new system, the individuals handling the submission of data may be unfamiliar with it. However, they can invoke the UpdateTest transaction to rectify errors, and these are likely to be few in the long run.

The severity of a data breach would be significant and there could be many public health and epidemiological complications. However, it is not likely for a data breach to occur on a notable scale. The risk can be further mitigated with the implementation of audit measures that keep the consortium informed about unauthorized operations.

5 Potential Extension

This system would not just be handy in the case of COVID-19 but in any other pandemic as well. Most of the data reporting and sharing protocols for hospitals and testing labs would expectedly remain the same and the data elements that are to be collected are also expected to not differ much. A potential addition to this system would be features of contact tracing, boosting the public health response to the pandemic.

Additionally, the system could be coupled with epidemiological Machine Learning (ML) models that could be applied to data given on a region-by-region basis or a national basis. Taken together, the numbers provided by the testing sites and hospitals would be added up (based on test type and other metrics etc.) and can then be used to conduct data analysis. This would provide highly useful information to aid the public health response as this AI based modelling and analytics can prove to be a useful method of identifying the most vulnerable and hard-hit regions among other uses.

6 Limitations

The proposed system requires a combination of blockchain with manual data entry methods, from the respective orgs sending the data. This one limitation of the system, while it would help in digitization of testing and hospital records on the chain, would also require data to be entered manually, which can prove to be an arduous process.

Another potential issue is that the blockchain network could potentially become too large. While Hyperledger does not have a limit itself on network size or the number of orgs/MSPs that can join, a very large network may become difficult to manage.

A further limitation is that this system follows the protocols set up by the CDC for data reporting and collection. Some other countries may have very separate methods of doing the same, even though many nations follow practices similar to the United States. If the methods of data reporting differ vastly, then the proposed system may not be usable. However, we believe this design to be quite robust in nature and usable, barring the implementation limitations that could come up.

7 Conclusion

In this paper, a framework for secure, verifiable yet transparent collection of pandemic data has been described. The proposal of this system utilized a main or common channel with all the stakeholders added, along with separate private data collections to ensure protection of data. The specific operation and the transaction flow were delineated, such that the data could be effectively shared with other orgs. We showed how a web-app and an interface could work in synchrony with the chaincode and contracts to achieve the goal of such a system. The problem of data manipulation and intentional misreporting would thus be tackled through the proposed system in this paper.

References

1. Pawariya, A.: Why We Shouldn't Trust The Official Covid-19 Data Anymore; Most States Are Fudging The Numbers. Swarajyamag, 27-May-2020. https://bit.ly/36f7rMs. Accessed on 14 Jan 2021
2. Brazil stops publishing coronavirus data; Bolsonaro says website 'not representative' of reality as experts warn against undermining pandemic's gravity. Firstpost, 08-Jun-2020. https://bit.ly/3Au7OR6. Accessed on 14 Jan 2021
3. VMC faces Covid data manipulation flak, mayor says it 'could be to avoid panic'. The Indian Express, 19-Sep-2020. https://bit.ly/3hjuP1I
4. Singh, S.S.: Coronavirus: The mystery of the low COVID-19 numbers in West Bengal. The Hindu, 02-May-2020. https://bit.ly/3wfmyzT
5. Panneerselvan, A.S.: The uncertainties over COVID-19 numbers. The Hindu, 20-Aug-2020. https://bit.ly/3ypN8Yw. Accessed on 14 Jan 2021
6. Stolberg, S.G.: Trump Administration Strips C.D.C. of Control of Coronavirus Data. The New York Times, 14-Jul-2020. https://nyti.ms/36btXpm
7. Asenjo, R., Fuentes, L., Munoz, D., Constantinescu, D.: ClinicAppChain: a low-cost blockchain hyperledger solution for healthcare. Adv. Intell. Syst. Comput. **1010** (2019)
8. How to Report COVID-19 Laboratory Data. Centers for Disease Control and Prevention, 05-Jan-2021. https://bit.ly/36f7Mic. Accessed on 14 Jan 2021
9. Ladia, A.: Privacy centric collaborative machine learning model training via blockchain. Adv. Intell. Syst. Comput. **1010** (2019)
10. Private data. Hyperledger. https://bit.ly/3xkQeNk

Investigation on Vulnerabilities Location in Solidity Smart Contracts

Mirko Staderini(✉) and Andrea Bondavalli

Department of Mathematics and Informatics, University of Florence, 50134 Florence, Italy
{mirko.staderini,andrea.bondavalli}@unifi.it

Abstract. Smart contracts had a very fast increasing development in the last years. Once a smart contract is deployed on a blockchain due to code immutability, its residual vulnerabilities cannot be patched. Reducing the number of residual vulnerabilities becomes thus very important and normally is achieved through static analyzers. This paper investigates the physical position (location) of vulnerabilities in Solidity smart contracts. To this purpose, we use a language-independent systematization of vulnerabilities and we consider the outputs of a set of static analyzers processing a representative set of smart contracts. We analyze the distributions of the locations where tools find positive outcomes. We create the ground truth of vulnerabilities for a subset S of smart contracts through manual inspection and we first perform a comparison of the distributions within this set. Then we generalize our findings by comparing the distributions between the manually inspected subset and the full set. Such comparison allows us to identify where certain classes of vulnerabilities are located, suggesting specific areas in Solidity smart contracts where the search for vulnerabilities should focus.

Keywords: Smart contracts · Solidity · Vulnerabilities · Locations

1 Introduction

Smart contracts are one of the most important innovations of the second generation of the Blockchain. The basic idea is to execute automated computerized transactions. Their diffusion has allowed the development of applications in different areas (e.g., financial, medical, insurance, gaming, betting). However, as for any other kind of software, design and coding faults can cause weaknesses that may lead to exploitable vulnerabilities. In this context, this problem is even more critical, considering that developers cannot patch smart contracts once they have been deployed on the Blockchain.

Ethereum is one of the most used platforms, and it offers Solidity as the main (and Turing-complete) programming language. Thus, the analysis of vulnerabilities of Solidity smart contracts is extremely important for the platform security.

Static analysis is widely used to discover vulnerabilities in the early stages of the software life cycle. It can cover 100% of the code at a low cost, despite its incomplete fault coverage [1]. In this work, we use outcomes of some selected static analysis tools (SATs) to identify the *physical position* of vulnerabilities in smart contracts. In the rest of the

J. Prieto et al. (Eds.): BLOCKCHAIN 2021, LNNS 320, pp. 199–211, 2022.
https://doi.org/10.1007/978-3-030-86162-9_20

paper, we use the term *location* to indicate the physical position. Each line of a contract tested by a static analysis tool can result in two different outcomes depending on whether the tool identified a vulnerability or not. A positive result (P - *positive* – tool identified a vulnerability) can be a *true positive* (TP – correct detection of an existing vulnerability) or a *false positive* (FP – wrong detection of a non-existing vulnerability). Clearly, this is only a partial view as a negative result (N – negative - the tool did not identify a vulnerability) can be a true negative (TN – correct assessment of no vulnerability) or a false negative (FN – missed detection of an existing vulnerability). For our purpose, however we seek for the location of positives (as these are the possible vulnerabilities they can be automatically found).

Contribution. In this paper, we try to address the following main research question: *Where in smart contracts are vulnerabilities located, and where tools have to focus on?*

Motivation. Although vulnerabilities and analysis tools are a widely debated topic, the characterization of the position of vulnerabilities in Solidity smart contracts is surprisingly less investigated compared to other programming languages. Actually locating a specific class of vulnerabilities into smart contracts can help in different ways. On one side, tool developers can be guided for improving the vulnerability detection capabilities of the tool, while on the other side, software developers can produce more secure contracts focusing on the specific areas where such vulnerabilities are more likely located. The work develops as follows:

- As preparatory steps, a set of Solidity vulnerabilities are identified and mapped to a taxonomy based on the *Common Weakness Enumeration* (CWE) following the methodology developed by [2] and a set of SATs is selected according to specific guidelines – Sect. 2.
- Next, a representative set of smart contracts constructed through a random extraction from Etherscan. 300 smart contracts have been selected so to avoid over-representing similar repeated characteristics to form the *reference dataset*. Each tool processes the whole dataset. As smart contracts normally contain a relevant number of vulnerabilities getting smart contracts from Etherscan does not allow having a ground truth on their vulnerabilities. In order to understand better, we selected within the reference dataset a *pilot set* of 15 contracts, which have been manually inspected to determine their ground truth. We then run the SATs on the 15 contracts of the pilot set, and we could determine the location of TPs, FPs – Sect. 3.
- We first provide a deep analysis of the pilot set. Following, for each class of the taxonomy, we determine the location of the vulnerabilities by analyzing the true positives. Then, we compare the true positives against all the positives found (which also include false positives), we analyze their distribution and develop some considerations. Finally, we try to generalize our findings by comparing the distributions of positives between the pilot and reference set – Sect. 4.

The paper concludes by discussing the limitations of our analysis (Sect. 5), summarizing related works (Sect. 6), and providing conclusions (Sect. 7).

2 Taxonomy and Static Analyzers

2.1 Taxonomy

Several vulnerability databases exist, e.g., BugTraq, and *Common Vulnerabilities and Exposures* (CVE). The *National Vulnerability Database* (NVD), the U.S: Government repository for vulnerabilities, uses the *Common Weakness Enumeration* (CWE) to classify CVE entries. CWE [5] is a wide-spread used list of software weaknesses based on a hierarchical structure. The entire CWE list has three main hierarchical representations (*software development*, *hardware design*, and *research concepts*) that focus on different aspects. We use the research concepts, a language-independent representation that focuses on the behaviour of weaknesses. Each item of the list provides a weakness type at a defined level of abstraction. By proceeding top-down, the *root* is the most generic abstraction, and it represents the view. *Pillars* and *classes* are independent of any specific language. *Bases* and *variants* describe the weakness at a lower level of detail.

Given the central role of the CWE classification, in this work, we mapped the list of relevant vulnerabilities into the CWE language-independent representation.

To identify such a list of relevant vulnerabilities, we examined several research papers (e.g., [3, 4]), we investigated online documentation (e.g., the *Smart Contract Weakness Classification and Test Cases* – SWC), and some *GitHub* repositories. Finally, we considered only vulnerabilities that can be exploited in the Solidity releases >= 0.5, and we grouped vulnerabilities with a similar or overlapping definition. At the end of the selection process, we ended up with 32 vulnerabilities. To map them into our taxonomy, we use the methodology described by [2] and briefly summarized below. Mapping vulnerabilities into a CWE taxonomy is a manual process [6] that uses the vulnerabilities definition to identify, for each vulnerability, a list of CWE-id candidates. Subsequently, from the previous list, we select the best match (placed into a Pillar sub-hierarchy). Next, we proceed bottom-up, selecting the Class or Pillar (in the same Pillar sub-hierarchy) that groups vulnerabilities with common behaviours. Table 1 shows the mapping among the vulnerabilities and the related CWE.

2.2 Static Analyzers

We identified candidate Static Analyzers by starting with the survey of [7], then looking at GitHub repositories and searching for papers on static analyzers. The initial list of tools (38) to be selected has been refined using the following criteria: (i) processing of contracts in Solidity release 0.5 or up; (ii) analysis of contracts without assertions or user-defined properties; (iii) targeting to vulnerabilities detection and explicit identification of the row of detection (referring to original smart contracts). The first criterion selects the tools that support a Solidity release >= 0.5; the second one allows finding results independent of the user settings; the third one permits obtaining comparable locations.

The refinement (that excluded among others EthBMC [8] and SmartBugs [9]) ended up with eight tools. Table 2 lists the selected eight tools highlighting the release, the *input mode* (BC - *bytecode* or SC - *source code*), and the *internal representation* (CFG – *control flow graph* or AST – *abstract syntax tree*). A * indicates that specific options extend the support of the tool to the 0.5 release of Solidity.

Table 1. Mapping among Vulnerabilities and the Taxonomy CWE based.

CWE	Vulnerability	CWE	Vulnerabilities
CWE-20 Improper Input Validation	Ether Lost in Transfer [3] Requirement Violation [21] Short Addresses [4]	CWE-668 Exposure of Resource to Wrong Sphere	Blockhash usage [3] Malicious library [13] Secrecy Failure [4] Timestamp dependency [3]
CWE-284 Improper access control	Authorization through tx.origin [4] Unprotected Ether Withdrawal [21] Unprotected self-destruct [21] Visibility of Exposed Functions [4]	CWE-669 Incorrect Resource Transfer Between Spheres	Call to the Unknown [3] Delegatecall to the Untrusted Callee [4] DoS by External contracts [13]
CWE-330 Use of Insufficiently Random Values	Generating Randomness [3]	CWE-682 Incorrect Calculation	Integer Overflow or Underflow [21] Arithmetic Precision Order [12]
CWE-345 Insufficient Verification of Data Authenticity	Missing Protection against Signature Replay Attack [21] Signature Malleability [21] Typecast [3]	CWE-691 Insufficient Control Flow Management	Arbitrary Jump [21] Freezing Ether [4] Insufficient Gas Griefing [21] Reentrancy [3] Right to left override [21] Unexpected Ether Balance [21] Transaction Ordering Dependence [4]

(*continued*)

Table 1. (*continued*)

CWE	Vulnerability	CWE	Vulnerabilities
CWE-400 Uncontrolled Resource Consumption	DoS costly Patterns and Loops [4] Gasless send [3]	CWE-703 Improper Check or Handling of Exceptional Conditions	Exception Disorder [3] Unchecked send [4] Unchecked Call Return Values [4]

Oyente, based on symbolic execution, is one of the first developed tools to find vulnerabilities in the Ethereum smart contracts. Osiris extends Oyente by focusing on the integer overflow and underflow. Mythril is based on the symbolic execution of the *Ethereum Virtual Machine* (EVM) bytecode and detects vulnerabilities in EVM-compatible smart contracts. Slither converts smart contracts into an intermediate language; it uses AST and CFG as internal representation.

Table 2. Selected tools.

Tool	Release	Input mode	Internal representation
Securify2 (Sfy2) [10]	30th June 2020	BC	CFG
Securify (Sfy) [11]	5th June 2020	BC	CFG
Slither (Sli) [12]	0.7.0	SC	AST, CFG
SmartCheck (SmC) [13]	2.1	SC	AST
Remix-IDE (Rmx) [14]	30th June 2020	SC	AST
Mythril (Myt) [15]	0.22.6	BC	CFG
Oyente* (Oye) [16]	30 September 2020	BC	CFG
Osiris* (Osi) [17]	0.0.1	BC	CFG

Securify defines specific patterns; then, it determines whether some information inferred from the software violates the patterns (generating an alarm) or not. Securify2 extends Securify increasing the number of discoverable vulnerabilities. SmartCheck looks for specific code patterns to identify vulnerabilities. Finally, Remix-IDE consists of several modules, including a static analyzer (used in this work) for vulnerability scanning.

3 Experimental Settings and Methodology

We built a reference dataset (more than 300 smart contracts[1]): (i) first, randomly extracting real smart contracts from the public repository Etherscan using a java crawler; (ii) next, selecting smart contracts with Solidity release 0.5 and excluding those containing syntactic errors; (iii) finally, excluding contracts that generated some processing failures when processed by tools.

Each tool processed the whole reference dataset. Testing each row of a contract can result in two different outcomes: negative or positive detection. By focusing on positives, each tool delivers results with its own codes and specific format. For each tool, we identified the codes related to the vulnerabilities belonging to our taxonomy, excluding the other codes from the analysis. Finally, we harmonized results following the methodology described below.

Let consider the number of *lines of code* (LOC) of each contract and define the *location of detection* (LoD) as the line of a smart contract where a tool detects a positive. We define the *relative location of detection* (RLoD) as the ratio between the LoD and the LOC of the smart contract under analysis. The RLoD identifies where positives are located in the contracts. Thus, a tuple (*tool, address, RLoD, category*) represents a positive. *Tool* is the tool that identifies the positive, *address* is the smart contract address, *RLoD* is the relative location of detection, and *category* is the class of the CWE taxonomy the intrusion belongs to.

Determining whether each positive is a true positive or a false positive needs a massive amount of (manual) work. Thus, we extracted a subset composed of 15 contracts (referring to it as a *pilot set*) from the reference set. The pilot set is built as follows: (i) contracts must contain main features of the languages (e.g., functions that exchange Ether, low-level calls); (ii) the set contains (at least) one contract for each category (gambling, game, exchange, finance, properties) most represented in the reference set, which is also the most represented in the Ethereum platform [18].

We first analyzed the pilot set. The manual inspection permitted to determine the ground truth, i.e., to determine for each positive finding whether it is a TRUE positive or a FALSE one. In addition, it permitted to determine for each negative whether it is a TRUE negative or a FALSE one. For our purpose, however we sought for the location of positives. We used the RLoD (defined above) to determine the location. For the pilot set, we used a new tuple (*tool, address, RLoD, category, diagnosis*) adding the field *diagnosis,* which can assume values TP or FP. For each class of the taxonomy, we determined the location of the vulnerabilities by analyzing the TPs. Next, we analyzed our results by comparing the TPs to all the positives (including FPs). Finally, by comparing the distributions of positives between the pilot and reference set and having a clue on the true positives, we tried to understand and discuss the generality of our findings.

---

[1] The list of addresses can be retrieved at https://doi.org/10.5281/zenodo.5046761.

4 Analysis

4.1 Pilot Set Analysis and Datasets Comparison

By considering the sum of the number of positives that the whole set of SATs finds in contracts, we found that 65% of contracts in the reference set (55% in the pilot set) contain from 0 to 50 positives; 20% of contracts in the reference set (33% in the pilot set) contains from 51 to 100 positives; the remaining 15% of contracts in the reference set have more than 100 positives (12% in the pilot set). Thus, the two sets have a similar distribution of positives among the contracts.

We focus on the pilot set. First, we determine the *ground truth* (GT) for each class of our taxonomy. Then, we calculate the *coverage* as the percentage of detected vulnerabilities (TP) over all the vulnerabilities (TP + FN). Table 3 highlights the whole number of vulnerabilities (GT = 486), TP, FP, and the coverage for each tool. Table 4 details the analysis for each tool and class, showing the tool coverage and the number of vulnerabilities (GT).

Table 3. Pilot set analysis.

GT = 486	Securify2 (Sfy2)	Securify (Sfy)	Slither (Sli)	SmartCheck (SmC)	Remix (Rmx)	Mythril (Myt)	Oyente (Oye)	Osiris (Osi)
TP	320	40	229	182	43	27	6	8
FP	241	88	16	50	7	18	4	8
Coverage	0.67	0.08	0.48	0.38	0.09	0.06	0.01	0.02

We can observe that: (i) Mythril covers completely CWE-330 and CWE-669; (ii) Securify2 covers CWE-703 completely, and with a high percentage CWE-284 and CWE-20; (iii) Slither increases the coverage of CWE-691.

Whether all the tools process a smart contract, the result is positive if at least one tool has a positive finding. This way, a TP is a row in which one of the tools finds a positive and contains a vulnerability. The last row of Table 4 highlights the TPs that the whole set of SATs finds for each class (TPset): comprehensively, the whole set identifies 446 TPs over 486 vulnerabilities (92%).

As observed, tool findings and coverage change based on the class and then they depend on the class distribution. Figure 1a highlights the distributions of CWE classes in the pilot (red) and reference set (blue). As it can be observed, the classes have comparable frequency distributions in the two sets. Moreover, from Fig. 1b results that CWE-345, in grey, has no positives in either set; thus, we decided to exclude this class from further analysis.

4.2 Location of Vulnerabilities

This Section analyzes the location of specific classes of vulnerabilities. To perform a comparison of distributions, we use boxplots with the median and the *interquartile*

Table 4. Detailed Pilot Set Analysis. The table shows the tool coverage, the number of vulnerabilities (GT), and the number of TPs that the whole set of SATs finds (TPset) in each class (column) of the taxonomy CWE based.

	CWE 20	CWE 284	CWE 330	CWE 400	CWE 668	CWE 669	CWE 682	CWE 691	CWE 703
Sfy2	0.7	0.8	–	–	–	0.5	0.1	0.2	1.0
Sfy	0.1	0.1	–	–	–	–	–	0.2	0.4
Sli	0.1	0.7	–	–	0.3	0.5	–	0.7	0.6
SmC	–	0.7	–	0.4	–	0.2	–	0.1	–
Rmx	–	–	–	0.7	0.6	0.5	–	0.5	–
Myt	–	0.1	1.0	–	0.2	1.0	–	0.1	0.4
Oye	–	–	–	–	–	0.3	0.3	0.1	–
Osi	–	–	–	–	–	0.3	0.3	0.1	–
GT	118	257	4	12	24	6	12	34	19
TPset	90	257	4	8	20	6	8	34	19

range (IQR), the interval between the *upper hinge* (UH - 75th *lower. hinge* (LH - 25th percentile). First, we focus on the pilot set, in which we analyze locations of the true positives (TPs) and compare them with all positives. Then, whether the distributions are different, we look at the false positives (FPs). Finally, by comparing the distributions of positives between the pilot and reference set and having a clue on the true positives, we try to understand and discuss the generality of our findings.

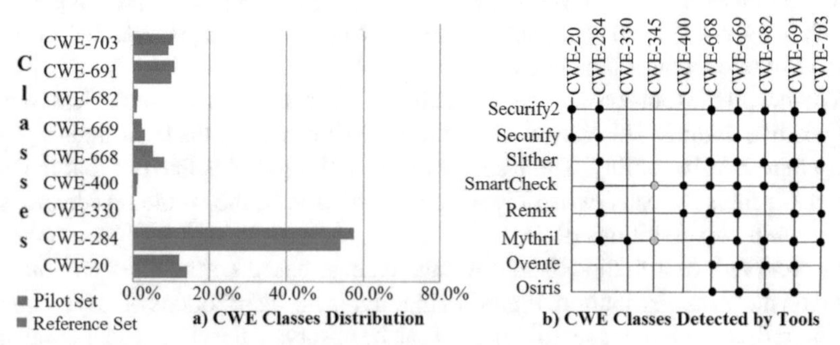

Fig. 1. Data overview. On the left, the distribution of CWE classes. On the right, which classes tools can detect in both sets.

Comparison in the Pilot Set. Figure 2 focuses on the distributions of the positives in the pilot set. As only four positives for CWE-330 have been detected, we decided to exclude this class from further analysis.

When we consider true positives, we can observe that:

- CWE-20, CWE-400, CWE-691, CWE-668, and CWE-703 have an LH greater than 50%; then, more than 75% of TPs locates in the second half of contracts. In particular, more than 75% of TPs locate in the last quarter of contracts for CWE-703;
- CWE-682 spreads the IQR (50% of TPs) between 65 and 40;
- CWE-284 and CWE-669 spread their IQR between 77 and 35.

Comparing the TPs to the positives, we determine that distributions of CWE-20, CWE-284, CWE-400, CWE-691, CWE-682, and CWE-703 differ in the upper and lower hinge, and median less than 5%. Consequently, TPs are representative of the locations of positives.

CWE-668 distributions differ significantly in the LH. FPs (at the right of Fig. 2) affect the distribution of positives. The manual inspection of the pilot set shows that Osiris and SmartCheck find 36 FPs and no TPs. Then, we analyze the distribution of positives without the findings of SmartCheck and Osiris (P_{SCO}) (the red boxplot in Fig. 2). TPs and P_{SCO} distributions do not differ significantly; therefore, locations of true positives are representative of locations of P_{SCO}.

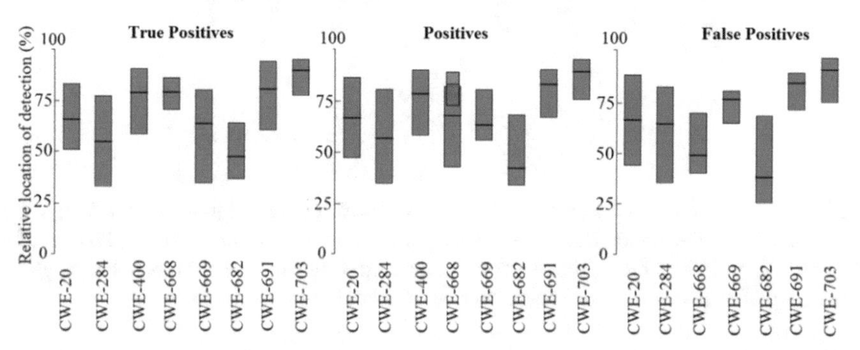

Fig. 2. Boxplots compare distributions of true positives (on the left), positives (in the middle), and false positives (on the right) in the pilot set. Each boxplot shows the interquartile range for a CWE-id of the taxonomy. A horizontal line within the box represents the median. The red boxplot highlights the distribution of positives without SmartCheck and Osiris for the CWE-668.

Finally, we focus on CWE-669. Again, the distributions differ in the LH. However, for values greater than the median, which has a value of 65, the distributions are comparable. Locations of the TPs represent locations of positives when values are over the median.

Generalization of Findings. Comparing the distributions of positives between the pilot and reference set after investigating the true positives permits discussing the generality of our findings.

As highlighted in Fig. 3, distributions of CWE-20, CWE-284, and CWE-682 differ in the upper and lower hinge, and median less than 5%. Consequently, locations of positives in the pilot set are representative of locations of positives in the reference set. From the previous Section, we know that locations of TPs in the pilot set are representative of the location of positives in the reference set (for these classes). We can then generalize findings: (i) concerning CWE-20, 75% of vulnerabilities locates in the second half of the contracts; (ii) CWE-682 locates mainly on the second third of contracts; (iii) CWE-284 spreads throughout all the length of contracts.

We analyze the distributions of positives of CWE-669 without the findings of SmartCheck and Osiris (red boxplots in Fig. 3). Distributions differ less than 10% in the UH median and LH, whereas the IQRs differ less than 5%. By generalizing findings from the previous Section, 75% of vulnerabilities are located over the 60 (LH) value.

The distribution of CWE-400 in the reference set has lower values than the pilot set for the LH, median, and UH; however, the difference between the values is less than 10. Therefore, the IQR among distributions does not differ significantly: we can affirm that distributions are similar.

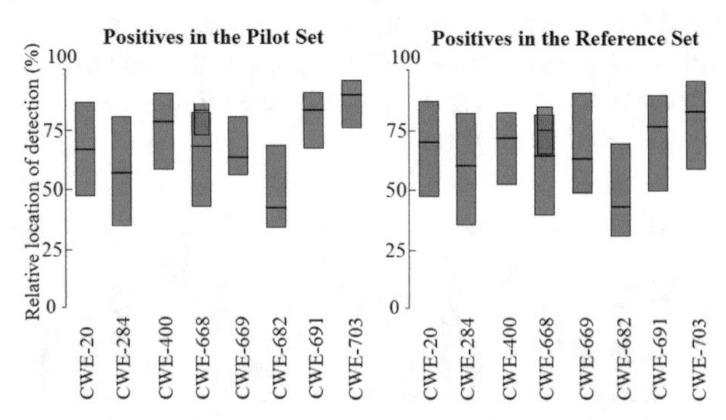

Fig. 3. Boxplots compare distributions between the positives of the pilot (on the left) and the reference set (on the right). Each boxplot shows the interquartile range for a CWE-id of the taxonomy. A horizontal line within the box represents the median. The red boxplot highlights the distribution of positives without SmartCheck and Osiris for the CWE-668.

Using the same previous arguments, we can generalize our findings: referring to the LH of the reference set, we can locate the 75% of vulnerabilities over the value of 50 (in the second-half of contracts).

CWE-691 and CWE-703 differ less than 5% in the UH and median, but differ significantly in the LH. So, the upper part of the distributions (over the 50th percentile) is representative of TPs. These observations allow us to locate the 50% of vulnerabilities in the last quarter of contracts for both distributions (median has a value greater than 75). No assumptions can be made for the lower part.

CWE-669 distributions have the same median. Analyzing the pilot set, we found that the distribution of TPs represents positives above the median. Thus, 50% of the vulnerabilities locates over the value of 65 (median of the distribution in the reference set).

5 Discussion and Limitations

The developers' skill has a significant influence on the quality of smart contracts. For the reference set, we have no information on the developers' skills. Different tools find positives in different classes. The difference has two main causes: (i) tools aim to detect different vulnerabilities; ii) tools can detect vulnerabilities with different capabilities. Each tool's outcome refers to a specific vulnerability (we remind that each vulnerability is mapped into the taxonomy CWE based). Tools detecting a vulnerability in a class can miss detecting other kinds of vulnerabilities belonging to the same class.

True positives are a subset of vulnerabilities. As a general statement, static analysis tools can miss detecting vulnerabilities: this leads to wrong locate the range of TPs in case of multiple missed detections. Using eight tools reduces this risk (processing the pilot set with all tools permits finding 92% of the vulnerabilities).

Our preliminary findings are based on a pilot set. However, we generalize results using the positives distributions in the reference set only in the case of comparable distributions. Consider two distributions. This work uses the lower hinge (LH), median, upper hinge (UH), and interquartile range (IQR) for comparison. We consider that two distributions have the same behaviour in two cases: i) the difference between LH, median, and UH values is less than 5%; ii) the difference is less than 10%, and the IQR ranges differ less than 5%.

This work locates vulnerabilities in a specific area of the contract, using the relative location of detection (RLoD). As an example, a range between 0 to 25 identifies the first quarter of a contract, a range from 26 to 50 identifies the second quarter, and so on.

6 Related Works

Vulnerabilities and tool analysis are active research areas in different domains, as static analysis is a traditional approach in various software dependability and security areas. Accordingly, research into applying static analysis to detect smart contracts' vulnerabilities started immediately after the appearance of Blockchain.

Atzei et al. [3] presented the first systematic investigation into Ethereum smart contract vulnerabilities and attacks. It provides a taxonomy of 12 kinds of vulnerabilities and analyzes attacks and possible countermeasures. More recently, Chen et al. [4] analyzed and compared vulnerabilities at the application level, data layer, and network layer. It also investigates the relationship between the vulnerabilities, their causes, their impacts, and possible exploitations.

Several works analyze and compare tools (mostly based on Solidity release 0.4). Parizi et al. [19] carried out an experimental assessment of static smart contracts security testing tools. It tested four tools on ten real-world smart contracts. Di Angelo et al. [7] independently analyzed 27 tools (static and dynamic) regarding the method used, maturity, availability, and detection aspects. Durieux et al. [9] retrieved quality metrics statistics for 47k smart contracts and compared tools analyzing a set of 10 vulnerabilities. Zhang et al. [20] focused on analyzing several bugs based on the IEEE framework. Using categories of bugs, it provided a benchmark between tools.

Our work uses some static analyzers but, differently from others, focuses explicitly on finding the location of vulnerabilities in Solidity (release up to 0.5) smart contracts. The investigation is based on a language-independent classification (CWE-based), permitting us to investigate the location of vulnerabilities having common behaviours. We evaluate the location of vulnerabilities for each class, starting from a manually inspected pilot set and generalizing findings using the reference set.

7 Conclusion

This work focused on determining the location of vulnerabilities in Solidity smart contracts. The analysis used the classification of vulnerabilities into a CWE taxonomy.

For all classes, we identified the tool coverage in the pilot set. Focusing on positives, we investigated where classes locate into contracts, first comparing true positives and positives distribution in the pilot set and then generalizing findings in the reference set. For some classes, we identified where a relevant percentage of vulnerabilities is located. Tool developers can use results to be guided for improving the vulnerability detection capabilities of the tool; software developers, on the other side, can produce more secure contracts focusing on the specific areas where such vulnerabilities are more likely located.

Starting from the analysis of this work, we plan to investigate the relation between the location of vulnerabilities and: (i) the use of Solidity constructs in some areas of contracts; (ii) the software complexity.

References

1. Okun, V., Guthrie, WF., Gaucher, R., Black, P.E.: Effect of static analysis tools on software security: preliminary investigation. In: Proceedings of the ACM Conference on Computer and Communications Security. ACM, Alexandria, USA (2007)
2. Staderini, M., Palli, C., Bondavalli, A.: Classification of ethereum vulnerabilities and their propagations. In: 2020 Second International Conference on Blockchain Computing and Applications (BCCA), pp. 44–51. IEEE, Antalya, Turkey (2020)
3. Atzei, N., Bartoletti, M., Cimoli, T.: A survey of attacks on ethereum smart contracts (SoK). In: Maffei, M., Ryan, M. (eds.) POST 2017. LNCS, vol. 10204, pp. 164–186. Springer, Heidelberg (2017). https://doi.org/10.1007/978-3-662-54455-6_8
4. Chen, H., Pendleton, M., Njilla, L., Xu, S.: A Survey on ethereum systems security: vulnerabilities, attacks, and defenses. ACM Comput. Surv. **53**(3), 1–43 (2020)
5. CWE Homepage. https://cwe.mitre.org. Accessed on 30 April 2021
6. CWE Mapping & Navigation Guidance. https://cwe.mitre.org/documents/cwe_usage/mapping_navigation.html. Accessed on 20 April 2021
7. Di Angelo, M., Salzer, G.: A survey of tools for analyzing ethereum smart contracts. In: 2019 IEEE International Conference on Decentralized Applications and Infrastructures, pp. 69–78. IEEE, USA (2019)
8. Frank, J., Aschermann, C., Holz, T.: ETHBMC: A bounded model checker for smart contracts. In: Proceedings of the 29th USENIX Security Symposium, pp. 2757–2774. USENIX Association (2020)
9. Durieux, T., Ferreira, JF., Abreu, R., Cruz, P.: Empirical review of automated analysis tools on 47,587 Ethereum smart contracts. In: Proceedings of the ACM/IEEE 42nd International Conference on Software Engineering, pp. 530–541. South Corea (2020)
10. Securify v2.0. https://github.com/eth-sri/securify2. Accessed on 31 Jan 2021
11. Tsankov, P., Dan, A., Drachsler-Cohen, D., Gervais, A., Bünzli, F., Vechev, M.: Securify: practical Security Analysis of Smart Contracts. In: Proceedings of the 2018 ACM SIGSAC Conference on Computer and Communications Security, pp. 67–82. Canada (2018)
12. Feist, J., Grieco, G., Groce, A.: Slither: a static analysis framework for smart contracts. In: 2019 IEEE/ACM 2nd International Workshop on Emerging Trends in Software Engineering for Blockchain (WETSEB), pp. 8–15. IEEE, Canada (2019)

13. Tikhomirov, S., Voskresenskaya, E., Ivanitskiy, I., Takhaviev, R., Marchenko, E., Alexandrov, Y.: SmartCheck: static analysis of ethereum smart contracts. In: Proceedings of the 1st International Workshop on Emerging Trends in Software Engineering for Blockchain, pp. 9–16. ACM, Sweden (2018)
14. Remix Project. https://github.com/ethereum/remix-project. Accessed on 30 March 2021
15. Mythril – repository. https://github.com/ConsenSys/mythril. Accessed on 30 March 2021
16. Luu, L., Chu, D.H., Olickel, H., Saxena, P., Hobor, A.: Making smart contracts smarter. In: Proceedings of the ACM Conference on Computer and Communications Security, vol. 24, pp. 254–269. ACM (2016)
17. Torres, C.F., Schütte, J., State, R.: Osiris: hunting for integer bugs in ethereum smart contracts. In: Proceedings of the 34th Annual Computer Security Applications Conference, pp. 664–676. ACM, USA (2018)
18. Oliva, G.A., Hassan, A.E., Jiang, Z.M.: An exploratory study of smart contracts in the Ethereum blockchain platform. Empir. Softw. Eng. **25**(3), 1864–1904 (2020). https://doi.org/10.1007/s10664-019-09796-5
19. Parizi, R.M., Dehghantanha, A., Choo, K.-K.R., Singh, A.: Empirical vulnerability analysis of automated smart contracts security testing on blockchains. Proceedings of the 28th Annual International Conference on Computer Science and Software Engineering, pp. 103–113. IBM Corp., USA (2018)
20. Zhang, P., Xiao, F., Luo, X.: A Framework and dataset for bugs in ethereum smart contracts. In: 2020 IEEE International Conference on Software Maintenance and Evolution (ICSME), pp. 139–150. IEEE, Australia (2020)
21. SWC Registry. https://swcregistry.io/. Accessed on 25 June 2021

Benchmarking Constrained IoT Devices in Blockchain-Based Agri-Food Traceability Applications

Miguel Pincheira(✉), Massimo Vecchio, and Raffaele Giaffreda

OpenIoT Research Unit, Fondazione Bruno Kessler, Trento, Italy
{mpincheira,mvecchio,rgiaffreda}@fbk.eu

Abstract. The adoption of blockchain technology combined with the Internet of Things in agri-food traceability scenarios is gaining momentum in terms of research and development. The intrinsic capability of blockchain to provide immutable and tamper-proof records is the perfect match for IoT systems comprising small sensing devices that can autonomously produce information across the entire process. However, there has been little discussion on the impact that blockchain has on the constrained sensing IoT devices in terms of their limited resources. These devices are the core of modern IoT systems providing the sensing layer for the entire application. In this paper, we focus on assessing the impact that a blockchain traceability system may have on such constrained sensing devices. To this end, we benchmark six IoT hardware platforms in terms of space and memory usage, processing time, and energy consumption as elements of a trustless event-based traceability application over two blockchain networks, namely Ethereum, as an example of permissionless network, and Hyperledger Sawtooth, as a permissioned counterpart. Our results and analysis provide an empirical reference for the study and development of blockchain-based traceability systems using constrained sensing IoT devices.

1 Introduction

The integration of the Internet of Things (IoT) is increasingly becoming a vital factor in blockchain systems for applications across several domains, such as finance, insurance, and logistics [2]. On the one hand, blockchain provides a trustless environment to digitize transactions smoothly and efficiently, making several processes leaner, faster, and more transparent [2]. On the other hand, IoT provides small and cost-effective sensing and actuation devices that can autonomously produce and consume information about the underlying process, with high granularity and reduced costs [7]. In Agriculture, blockchain-based IoT systems are generating considerable interest in terms of research and development for traceability systems, to cope with the increasing requirements of consumers that demand farm-to-fork visibility of agri-food products, i.e., a detailed history of where their food comes from, where it was grown or raised, how it has

J. Prieto et al. (Eds.): BLOCKCHAIN 2021, LNNS 320, pp. 212–221, 2022.
https://doi.org/10.1007/978-3-030-86162-9_21

been handled, and by whom [4]. Despite this growing interest, there has been little discussion on the impact that a blockchain software layer has on the more constrained sensing IoT devices. These devices are the core of the IoT system and the base for the entire blockchain-based applications.

In this paper, we focus on assessing the impact that a blockchain traceability system has on constrained sensing devices. To this end, we evaluate a set of IoT hardware platforms as part of an event-based traceability scenario using two different blockchain platforms. We benchmark six boards with different hardware architectures with limited program space and memory. We present the results in terms of space and memory usage, processing time, and energy consumption while discussing the implications of these results for the proposed scenario. Our experiments showed that even an 8-bit-MCU board is capable of directly interacting with both Ethereum and Sawtooth with minimal energy overhead. However, the differences between the protocol of both blockchain platforms have an impact on the constrained devices. We expect that our results and discussions provide an empirical reference for the study and development of blockchain-based traceability systems using constrained sensing IoT devices.

1.1 Related Work

A traceability system based on blockchain and IoT was proposed in [10] with a focus on food safety, based on Hazard Analysis and Critical Control Points (HACCP). The study describe the process but fail to provide a performance analysis. Authors of [6] focused on the trading portfolio of the blockchain-based traceability system. With a more practical approach, authors of [9] present a blockchain-based traceability system for the soybean. The work presents a detailed description of the implementation using the Ethereum network and focused on the smart contracts supporting the system. However, they fail to provide a detailed analysis of the IoT devices involved in the system. Overall, there has been little discussion on the impact that a blockchain may have on the more constrained sensing IoT devices [7]. If sensing devices become direct actors on the blockchain system, they can provide a *"root of trust"* for the sensed data. Moreover, sensing IoT devices as direct actors are one step closer to trustworthy oracles [5].

The rest of this paper is structure as follows: Sect. 2 presents a case study based on a simplified traceability scenario using constrained IoT sensing devices; Sect. 3 provides a detailed evaluation of the presented architecture with a focus on characterizing the constrained devices; Sect. 4 summarizes the insight from our evaluation and introduces possible future works.

2 Case Study: Traceability of Extra Virgin Olive Oil

To characterize the IoT devices, we consider a case study for traceability of extra virgin olive oil. Based on similar works for agri-food traceability [1,11] and particularly for olive oil [12], we simplified the traceability process into

4 stages: Farming, Manufacturing, Transportation, and Market. Each stage of the process is composed of various activities involving different actors. At each stage, IoT devices can monitor and autonomously register detailed information with high granularity. For our evaluation, we focused on event-based monitoring, i.e., a single record is stored on the blockchain instead of a detailed log. For instance, at the Farming stage, an event-based monitoring system registers the total production of the olives instead of each collected batch. At Manufacturing, the system records the transfer of ownership from the farmer to the processor. For Transporting, the system stores on the blockchain only temperatures anomalies (with respect to an agreed threshold) that might void a delivery agreement. Finally, during the Market stage, the system only records irregularities occurring during storage at the retailer.

2.1 High-Level Architecture

To implement the case study, we used an architecture that relies on constrained IoT sensing devices and blockchain to achieve transparency, auditability, and immutability of the stored records in a *trustless* environment for several unknown actors. In our architecture, as shown in Fig. 1, sensing devices have a unique blockchain identity and act as trustworthy data sources for smart contracts. Further, smart contracts provide a platform to implement business logic for the different stages of the traceability process using two types of smart contracts, i.e., TwinContract and AppContract. This architecture is described in [8] where it was used for a water-management system.

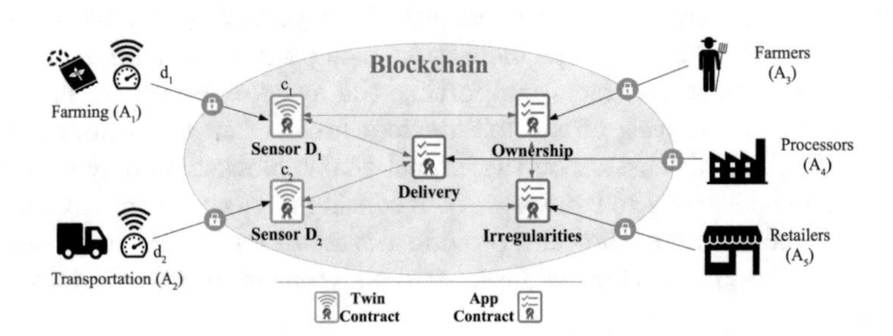

Fig. 1. High level architecture for blockchain-based traceability using constrained devices.

We conceptualized our architecture as a layered software framework, composed of three modules: Device, Gateway, and Blockchain. The Device module converts the sensed values into blockchain transactions that are later sent to the corresponding smart contract. The Gateway module is a simple relay component between the device and the blockchain layer. Finally, the Blockchain module gathers together all the smart contracts representing the sensors and the different process of the traceability system.

2.2 Implementation

The presented architecture is blockchain-agnostic, therefore, we tailored our implementation to two different blockchain platforms: Ethereum and Hyperledger Sawtooth. The reasons for choosing these implementations are two-folds. First, we have references for both public and private blockchain networks. Second, the selected blockchains provide different levels of customization for the records included on the ledger (transactions). Ethereum works with a single transaction structure, while Hyperledger Sawtooth allows the definition of a custom transaction structure. Nonetheless, migrating the architecture to other implementations should not present a major issue. In the following, we briefly described the implementation of the modules used for our evaluation.

2.2.1 Device Module

The objective of this module is to interact with the physical world and create blockchain transactions and has three components: Sensing, Blockchain Integration, and Communications. For our prototype, we implemented a Sensing component using a DHT11 temperature and humidity sensor and the open-source library to read the values from the sensor. For the communications component, we simply opted for serial communications at 115200 bps. The Blockchain Integration component converts the sensed values into a transaction, as the minimal blockchain data unit. We created a library for the Sawtooth based on our custom library for Ethereum, previously presented on [7]. We used C language within the Arduino development framework favoring cross-platform compatibility over code optimization.

2.2.2 Gateway Module

The main task of this module is to receive transactions from the lower layer, and to forward them to the upper layer, making the appropriate protocol conversions. For our prototype, we used serial communications at 115200bps with a simple relay agent written in Python 3.6 to connect to the Internet.

2.2.3 Blockchain Module

The purpose of this module is to provide a platform to implement the business logic of the application, using smart-contracts and their interactions. For our prototype, we implemented a SmartTwin for the temperature sensor and TwinApp for the transportation process (a detailed description of these smart contracts can be found on [8]). For Ethereum the smart contracts were implemented using solidity language and compiled for version 0.6.2. For Sawtooth, the contracts were implemented using Python 3.6.

3 Evaluation

3.1 Experimental Setup

To evaluate our solution, we tested our cross-platform prototype on six different microcontrollers boards (MCU) from the AVR, ARM, and ESP32 architectures. A comprehensive evaluation of all possible IoT boards is beyond the scope of this paper; however, the selected pool should provide a reference for other scenarios. Table 1 presents all the boards, detailing clock speed in megahertz (Mhz), program space in kilobytes (kb), memory size in kb, model, and a reference price (updated to May 2021).

Table 1. The hardware platforms used during our performance evaluation.

Device	Model	MCU	Architecture	Clock (Mhz)	Prog. space (kb)	Mem. size (kb)	Price (Eur)
UNO	Arduino Uno	ATMega328P	8-Bit AVR	16	32	2	18
EVERY	Arduino nano every	ATMega4809	8-Bit AVR	20	48	6	12
L031	STM32L031	Cortex M0+	32-Bit ARM	32	32	8	10
F303	STM32F303	Cortex M4F	32-Bit ARM	72	64	12	10
L452	STM32L452	Cortex M4	32-Bit ARM	80	512	96	12
ESP32	ESP DevKit	WRover-E	32-Bit ESP32	80	1024	320	10

For the Ethereum blockchain, we used the official Geth client (version 1.10.1-stable), while for Sawtooth we used the official client (version 1.2.6). The nodes run on separated virtual machines with 4 GB of Ram, 20 GB of SSD, and 4 vCPU on an OpenStack server using a clean Linux Ubuntu installation (version 18.04). The scripts that deploy and interact with the smart contracts were implemented using Python (version 3.6) and ran on a Lenovo T490s notebook, with 16 GB of Ram 256 SSD disk, and an Intel i-7 processor at 1.90 GHz over a clean Linux Ubuntu (version 18.04). The notebook and the nodes shared the same LAN connection.

3.2 Device Module Footprint

Using the statistics provided by the compilers, we estimated the footprint of each component of the device module. The results, in terms of disk usage (program space), are shown using Table 2a (absolute values) and Fig. 2a (normalized to the total available). The same approach was taken for memory usage, using Table 2b and Fig. 2b. Sensing (Sens) and Communications (Comms) components are the same for both implementations. The footprint of the Blockchain Integration component for Ethereum is represented by Eth, and the Blockchain Integration component for Sawtooth is represented by Saw. Our results show that an 8-bit board with 32 kb of disk space can fit the Ethereum implementation; however, the same amount of disk space is not enough on the 32-bit boards. Results also show that the Sawtooth implementation has a bigger footprint on the device, and only 4 of the 6 boards are capable of running the entire device module.

Table 2. Footprint for the device module (expressed in bytes)

Device	Available	Sens.	Eth	Saw	Comms.	Device	Available	Sens.	Eth	Saw	Comms.
UNO	32256	2014	26840	35214	2202	UNO	2048	207	1158	3214	188
EVERY	49152	1956	27428	35735	3105	EVERY	6144	196	611	2435	177
F303	65536	3040	29940	33404	14240	F303	12288	928	1540	3124	908
L031	32768	6168	30796	34432	14724	L031	8192	896	1508	3102	876
L452	524288	3032	33120	36568	17336	L452	163840	936	1560	3144	916
ESP32	1310720	1496	289318	293674	267270	ESP32	327680	13660	14996	16292	13612
(a) Disk usage						(b) Memory usage					

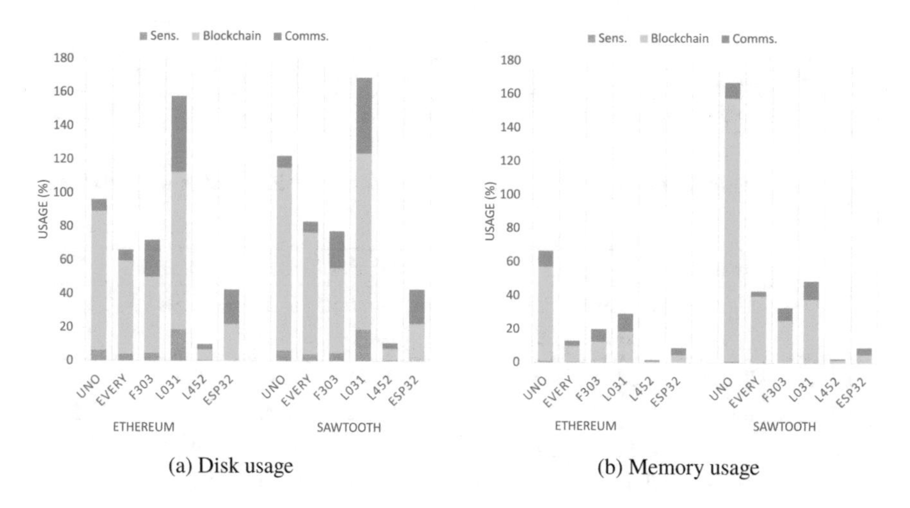

Fig. 2. Footprint of the device module (reference to the total available).

3.3 Processing Times and Transactions Size

For both blockchain platforms (i.e., Ethereum and Sawtooth), we measured the transaction size and processing time for 100 working cycles on the device module (i.e., sensing, blockchain integration, and communications). The average of all the experiments is shown in Table 3. As shown by the results, the processing times for creating a Sawtooth transaction are longer than Ethereum. Similarly, the size of the Sawtooth transaction is almost three times the average size of the Ethereum transaction.

3.4 Device Module Power Consumption

We measured the power consumption of each board when creating 100 transactions. We used a Otii device[1] that is able to provide current measurements with an accuracy of $\pm(1\% + 0.5\mu A)$ at 5V, with a rate of 1000 samples/s. As a reference, we used an idle state of 5s before each working cycle. It is important to notice that no low-power mode was used for the idle state. The average of

[1] https://www.qoitech.com/.

Table 3. Average processing times (in ms) and transaction size (in bytes) for the device module.

	Ethereum				Sawtooth			
Device	Sens.	Block.	Comms.	Total	Sens.	Block.	Comms.	Total
UNO	32	4245	23	4300	–	–	–	–
EVERY	32	4170	23	4225	32	8388	164	8584
F303	32	208	23	263	32	421	164	617
L031	–	–	–	–	–	–	–	–
L452	32	160	23	196	32	278	163	473
ESP32	32	83	23	139	32	146	163	341
	Avg. Transaction size: 134 bytes.				Avg. Transaction size: 925 bytes.			

Table 4. Power requirements (at 5V) during working and idle states on the different IoT devices.

	Idle state				Working State			
Device	Current (mA)	Current (var)	Energy (J)	Time (s)	Current (mA)	Current (var)	Energy (J)	Time (s)
UNO	34.38	2.85	0.86	5.0	34.36	2.37	0.73	4.30
EVERY	29.46	0.17	0.73	5.0	30.02	0.23	0.63	4.22
F303	65.0	0.12	1.63	5.0	67.93	1.0	0.08	0.26
L452	48.79	0.0	1.27	5.0	50.15	0.08	0.04	0.19
ESP32	42.26	0.03	1.06	5.0	61.98	55.23	0.03	0.13

(a) Ethereum implementation

	Idle state				Working State			
Device	Current (mA)	Current (var)	Energy (J)	Time (s)	Current (mA)	Current (var)	Energy (J)	Time (s)
EVERY	29.4	0.44	0.73	5.0	29.99	0.58	1.28	8.58
F303	64.67	0.27	1.62	5.0	67.86	1.76	0.2	0.61
L452	51.87	0.0	1.35	5.0	52.95	0.06	0.12	0.47
ESP32	42.23	0.02	1.06	5.0	59.93	43.35	0.09	0.34

(b) Sawtooth implementation

all the experiments is depicted in Table 4a for the Ethereum implementation, and in Table 4b for the Sawtooth implementation. As shown by the results, the processing times for creating a Sawtooth are longer and thus, have higher power requirements on the underlying IoT device.

3.5 Daily Energy Budget

We estimated the energy impact that integrating blockchain has for constrained sensing device using the following energy budget:

$$\mathcal{E}_{budget} = \mathcal{E}_{idle} + \mathcal{E}_{sens} + \mathcal{E}_{comm} + \mathcal{E}_{blockchain}$$

where $\mathcal{E}_{idle}$ is the energy of IoT device in idle state, $\mathcal{E}_{sens}$ is the energy for sensing, $\mathcal{E}_{comms}$ is the energy for transmitting the sensed value, and $\mathcal{E}_{blockchain}$ is the energy used for creating a blockchain transaction. Though simple, this analytical model is aligned with the ones found in the current literature [3, 8]. Based on the results of the power consumption and processing times, we estimated the energy budget for 1 h of work. In our scenario, we considered that the sensor measures the temperature every 5 min, and only one of this measurements (the event) is sent to the blockchain. Figure 3 shows the energy budget for the EVERY board, that averages the highest energy consumption. Similarly, Fig. 4 shows the energy budget for both blockchain implementation in an ESP32 board, which has the lower energy consumption. Results shown that the energy required by the blockchain operations is minimal compared to the energy used on the idle state.

4 Discussion and Future Works

In this paper, we assessed the impact that a blockchain-based distributed application has on constrained sensing devices. We benchmarked six IoT boards, from different hardware architectures, with limited program space and memory in an event-based traceability scenario using two different blockchain platforms. Surprisingly, our results showed that even an 8-bit-MCU board is capable of directly interacting with both Ethereum and Sawtooth with additional power requirements.

Overall, our results showed that the Ethereum blockchain is less demanding than the Sawtooth counterpart. Its device module implementation can fit into 32 kb of space and 2 kb of memory. However, this is the very minimum space since using other communications technologies that require additional resources might overflow such constrained devices. Nonetheless, devices within the same price range can seamlessly accommodate the presented architecture, with enough space for additional functionalities. Conversely, the Sawtooth implementation is more resource-demanding as the protocol group the transactions into batches. This difference demands extra resources, increases the processing times, and uses an encoding scheme that adds additional overhead to the transaction size. This translates into a minimal blockchain data unit that is 6 times larger than the one used in Ethereum, which may contrast with low-power networks that have a limited payload for the communications component.

The additional overhead on the Sawtooth platform also has an impact on the energy budget for creating and transmitting a transaction. In a 1-hour work cycle, with measurements every 5 min, the Sawtooth implementations have more

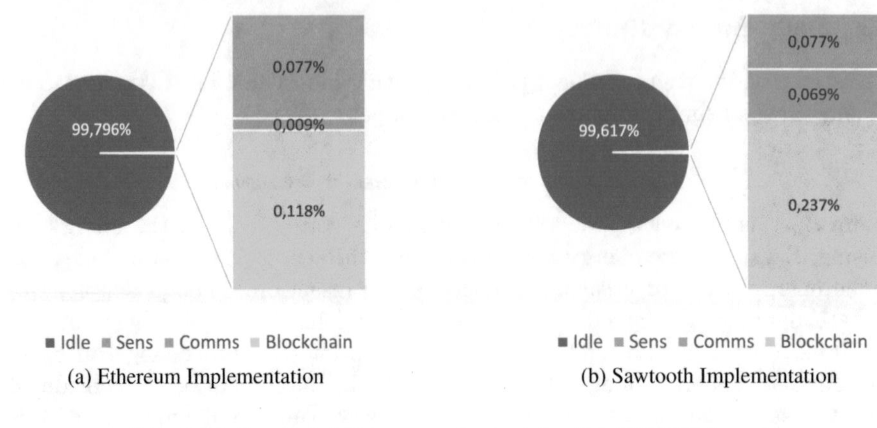

Fig. 3. Estimated daily energy budget for the EVERY board.

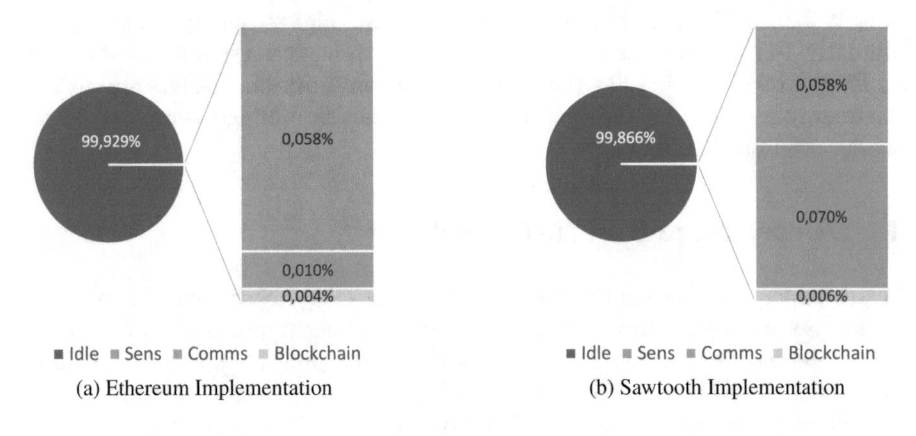

Fig. 4. Estimated daily energy budget for the ESP32 board.

energy requirements for both the blockchain operations and the communications. Still, these requirements are minimal, as the idle state uses more than 99% of the total energy budget of the scenario. Thus, constrained IoT devices can function as direct actors on a blockchain-based traceability system for Ethereum and Sawtooth. However, despite the similarities between blockchain platforms, small variations in the protocol will define different minimum requirements for the IoT devices to be used on the blockchain-based system.

We expected that this benchmark work provides an empirical reference for the study and development of blockchain-based traceability system using constrained sensing IoT devices. Future works includes evaluating power consumption on alternative scenarios, porting the software library to other blockchain implementations, and evaluating other communications technologies such as LoRaWAN, NB-IoT and 5G.

References

1. Accorsi, R., Bortolini, M., Baruffaldi, G., Pilati, F., Ferrari, E.: Internet-of-things paradigm in food supply chains control and management. Procedia Manuf. **11**, 889–895 (2017)
2. Ali, M.S., Vecchio, M., Pincheira, M., Dolui, K., Antonelli, F., Rehmani, M.H.: Applications of blockchains in the internet of things: a comprehensive survey. IEEE Commun. Surv. Tutor. **21**, 1676–1717 (2018)
3. Bouguera, T., Diouris, J.-F., Chaillout, J.-J., Jaouadi, R., Andrieux, G.: Energy consumption model for sensor nodes based on LoRa and LoRaWAN. Sensors **18**, 2104 (2018)
4. Caro, M.P., Ali, M.S., Vecchio, M., Giaffreda, R.: Blockchain-based traceability in agri-food supply chain management: a practical implementation. In: 2018 IoT Vertical and Topical Summit on Agriculture-Tuscany, pp. 1–4 (2018)
5. Heiss, J., Eberhardt, J., Tai, S.: From oracles to trustworthy data on-chaining systems. In: Proceedings of the IEEE International Conference on Blockchain (2019)
6. Mao, D., Hao, Z., Wang, F., Li, H.: Innovative blockchain-based approach for sustainable and credible environment in food trade: a case study in Shandong province, China. Sustainability **10**, 3149 (2018)
7. Pincheira, M., Vecchio, M.: Towards trusted data on decentralized IoT applications: integrating blockchain in constrained devices. In: 2020 IEEE International Conference on Communications Workshops, pp. 1–6 (2020)
8. Pincheira, M., Vecchio, M., Giaffreda, R., Kanhere, S.S.: Cost-effective IoT devices as trustworthy data sources for a blockchain-based water management system in precision agriculture. Comput. Electron. Agric. **180**, 105889 (2021)
9. Salah, K., Nizamuddin, N., Jayaraman, R., Omar, M.: Blockchain-based soybean traceability in agricultural supply chain. IEEE Access **7**, 73295–73305 (2019)
10. Tian, F.: A supply chain traceability system for food safety based on HACCP, blockchain and Internet of Things. In: Proceedings of the ICSSSM 2017, pp. 1–6 (2017)
11. Verdouw, C.N., Wolfert, J., Beulens, A., Rialland, A.: Virtualization of food supply chains with the internet of things. J. Food Eng. **176**, 128–136 (2016)
12. Zanoni, B.: Extra-virgin olive oil traceability. In: The Extra-Virgin Olive Oil Handbook, pp. 245–250 (2014)

Towards Micropayment for Intermediary Based Trading

Anupa De Silva(✉), Subhasis Thakur, and John Breslin

National University of Ireland, University Rd, Galway H91 TK33, Ireland
{anupa.shyamlal,subhasis.thakur}@insight-centre.org,
john.breslin@nuigalway.ie

Abstract. Involving an intermediary between producer-consumer environments is a common practice that reduces the managerial overhead on both parties. However, this mediation has both pros and cons. For example, it can overtake the power of producers in pricing, which could cause unfair revenue distribution. We present a novel micropayment protocol called MARI, which enables verifiable pricing and facilitates payment delegation for the prod-con environment. The proposed protocol is a blockchain-based unidirectional off-chain payment mechanism using a novel Merkel tree-based smart contract. MARI allows producers to communicate the fee structure directly to their consumers. It grants total control over producers' inbound payment flow, which an intermediary mediates and preserves the low overhead on the producers and consumers at the same time. Further, we explore how MARI meets the general requirements as a payment protocol and its scalability.

1 Introduction

Intermediary involvement between producer and consumer is a common practice in the real world. It decouples the consumers of a service from its provider and assists in managerial tasks, including access management, communication, and quality of service [4]. Hence, producers can minimize the cost and administrative overhead, allowing them to focus more on their core products. Furthermore, it is beneficial in providing recurring services where producers have to maintain persistent links with consumers. Also, resource-constrained producers cannot afford to manage a large number of consumers. For example, IoT-based systems are resource-constrained and produce data as a recurring service. Generally, they rely on a third party due to the lack of resources. Additionally, intermediaries can handle multiple such connections seamlessly and create highly scalable solutions.

However, having intermediaries is not always a desired feature in producer-consumer environments. The centralized nature of the intermediary role can cloud the views of both producer and consumer in terms of service and pricing [3]. As a result, consumers have to bear a higher cost for the services than the actual fee of the producer. Considering the pros and cons, we can realize that the presence of an intermediary is useful, but its position enables them to manipulate the link between producer and consumer.

J. Prieto et al. (Eds.): BLOCKCHAIN 2021, LNNS 320, pp. 222–232, 2022.
https://doi.org/10.1007/978-3-030-86162-9_22

This paper proposes the MARI payment protocol for a recurring producer-consumer service environment in which intermediaries perform managerial tasks. We deny the existence of an intermediary neither in service nor in payments. Hence, the intermediary also mediates in the compensation from consumer to producer in the same way. We facilitate payment delegation and define a verifiable mechanism to communicate the fee structure directly to the consumer. It makes sure that the producers get compensated through the delegation of intermediaries but without depending on it. First, we identify micropayment as a viable mechanism for recurring services. Then, we introduce a novel blockchain-based smart contract called Layered Multi-Lock Contract (LMLC), which can be released gradually by a third party who is not a contract participant. It allows making high-resolution micropayments with minimum on-chain cost. Then we present the flow of MARI followed by potential applications. Finally, we analyze and evaluate the performance, scalability and security aspects of the protocol. MARI demonstrates a strong form of authenticity where an entity can control an external payment flow towards itself with minimal additional overhead. Both consumers and producers can benefit from the transparency it creates. Further, it is scalable to facilitate multiple producers, consumers, and intermediaries with minimum additional cost.

2 Background and Related Works

2.1 Smart Contracts

Smart contracts are a digital form of real-life agreements created on top of blockchain technology. These pieces of software programs are subjected to be verified by the blockchain participants. The decentralized and distributed nature of blockchain makes them reliable with minimal trust on third parties [12].

MultiSig contract is one of the simplest yet powerful forms of contracts that can be implemented even using Turing incomplete scripts. They can enforce one or more participants to involve in spending. This nature enables MultiSig transactions to form offline payment channels [6] between two parties without making costly on-chain transactions for each payment. For example, when b needs to make periodic payment of v_δ to a, b initially funds an amount of v and publish a MultiSig transaction where both a and b has to sign to spend it. Afterwards, b keeps updating the balance and present the transaction to a (off the chain) as commitments. Once verified, a has the assurance of the payment although, it is not recorded on the chain as it is publishable at any time (i.e. if a dispute has occurred). There is no extra cost apart from the computation for signature verification and communication.

Apart from the signatures, we can design the MultiSig contract to integrate additional locks that need to be unlocked to move the funds. Time locks can mandate activation and expiration boundaries, whereas hash locks can be released by producing the pre-image of the hash value. The expiration time lock prevents funds to be locked forever because of an unresponsive counterparty in the contract. Activation time locks enable the transactions after a specific time, preventing old states from being published on the chain in offline payment channels. Hash locks are primarily valuable for the atomic swapping of assets between two parties [7]. It is the underlying technology used to create a network of payment channels [6] where payments are passed from one entity to another. The process is offline and atomic.

2.2 Micropayments

Trust is a fundamental requirement for trading unless transactions are atomic. In a transaction, whoever makes the first move (i.e. payment by the consumer or service by the producer) can lose if the counterparty avoids getting back. Either of the involved parties has to trust the other to get compensated or receive the service. If that is micropayment-enabled trading, the victim can minimize the loss by lowering the quantity at stake. Micropayment allows transactions in tiny amounts. Its resolution can vary depending on the application context and the nature of the payment system. Further, it implies the swiftness of making transactions and delivering service or product in exchange for the payment.

The concept of micropayment originated even before the World Wide Web (WWW). In 1960, Theodor Nelson [15] envisaged micropayments as a compensation mechanism for hypertext contents. Párhonyi [10] describes two generations of micropayment systems in the 1990s and 2000s. The first-generation micropayment emerged in the mid-'90s. However, it failed to provide a pragmatic solution for micropayment, although their innovative contribution is not trivial. The second generation started around 2000 with the advancement of internet technologies. Unlike the first generation, most of them survived for at least half a decade and even up until now. We argue that micropayment bounced back in its third generation with the emergence of Blockchain and, specifically, off-chain payment channels. Their decentralized nature eradicates the starting friction caused by the financial institutes and trust in third parties.

The contribution by Rivest et al. in cryptography-based micropayments protocols is notable. Payword [11] chain is one such innovation in commitment based micropayment protocols. A financial institute issues certificate for a PayWord chain, which is a chain of hashes. Then the recipient can sequentially release PayWords as payments. This method requires only one interaction with the broker during the payer-payee engagement, and it is an offline interaction. However, multiple Payword chains are needed for a client to engage with multiple merchants. Kim et al. [9] propose a method to mitigate this drawback where the client has to keep only one concatenated chain. In addition to the original system, the client has to embed a hash value given by the broker when making the payments. Therefore, the broker can keep track of multiple payments. However, the chain grows linearly along with new additions. Jutla et al. [8] propose an alternate solution using Merkle authentication trees, called PayTree.

In recent works, Zhi-Guo Wan et al. [14] developed MicroBTC based on Blockchain technology, inspired by PayWord. They modified the core Bitcoin implementation as the process demands programmable unspent outputs and loops. Galal et al. [5] proposed Merkel tree-based scheme, PayMerkel. Merkel tree root reduces the on-chain verification logarithmically. In essence, commitment-based schemes avoid costly signature verifications. It is particularly advantageous on Ethereum like blockchains that charge per operation. Otherwise, they make the payers restrain themselves to a particular frequency, whereas MultiSig based payment channel granularity is infinite. We exploit this limitation while designing the proposed protocol.

3 MARI Payment Protocol

MARI is a unidirectional payment protocol that supports payment from one entity to another through a third party. For example, a producer a is compensated by its service consumer c through an intermediary b. We call this setting a-b-c, and henceforth, we denote a producer, an intermediary, and a consumer by a, b, and c, respectively and superscript to denote a set of entities. For example, c-B-A denotes a consumer c connects with a set of intermediaries B to connect with a set of producers A. We first introduce a novel smart contract in Subsect. 3.1, followed by its usage, protocol flow and applications in Subsects. 3.2 and 3.3.

3.1 Layered Multi Lock Contract (LMLC)

Here, we present LMLC based on a modified Merkel Tree for the MARI payment protocol. The on-contract lock is LMLC's primary component, a Merkel Root encapsulating multiple locks from multiple parties. The complete Merkel Tree is a collection of subtrees and layered so that the below layer produces the leaves to the upper layer. For example, in the c-B-A setting, the on-contract lock is created at the top layer (θ), obtaining the locks from individual intermediaries as the leaves for the on-contract root. Similarly, the middle layer (β) consists of Merkel Trees based on the locks from the producers engaged with each intermediary. At the bottom layer (α), producers' hashed secrets act as the leaves to create producer-wise locks. We also embed different information, including monetary values and expiration times, alongside the locks using a prefix function at each layer. Further, we can independently release each lock at the α layer, making the on-contract lock openable multiple times. We define the LMLC and its components in detail as follows.

Definition 1. Let $\mathcal{L}$ be the recursive function which accepts n sized ordered lists of keys, K_n^λ to build the Merkel locks in each λ layer.

$$\mathcal{L}(K_n^\lambda) = \mathcal{F}^\lambda()\|\mathcal{H}(log(n)\|\mathcal{L}(K_n^\lambda[0:j])\|\mathcal{L}(K_n^\lambda[j:n])) \tag{1}$$

where $\mathcal{L}([k_i^\lambda]) = k_i^\lambda$ and j is the largest power of 2 less than n. $\mathcal{F}^\lambda$ is the prefix function and $\|$ denotes the concatenation. $K_n^\lambda[j_1 : j_2]$ indicates $[k_{j_1}^\lambda, k_{j_{1+1}}^\lambda, \ldots, k_{j_{2-1}}^\lambda]$ which is a sub-list of K_n^λ. $\mathcal{H}$ is a collision-resistant hash function.

Let $\mathcal{F}^\lambda$ be the prefix function that takes an ordered list of keys (i.e.leaves) K^λ at layer $\lambda \in \{\alpha, \beta, \theta\}$. When $\lambda = \alpha$, let $k_i^\alpha (\in K_n^\alpha) = v_i^\alpha \| e_i^\alpha \| \mathcal{H}(x_i^\alpha)$ where $v_i^\alpha \in V_n^\alpha$ is a monetary value, $x_i^\alpha \in X_n^\alpha$ is a random secret, and $e_i^\alpha \in E_n^\alpha$ indicates the expiration time. V_n^α and E_n^α are in ascending order. Then, $\mathcal{F}^\alpha(K_n^\alpha) \leftarrow v_{n-1}^\alpha \| e_{n-1}^\alpha$ (i.e. the last/maximum elements). The output is in the form of $v_n^\alpha \| e_n^\alpha \| l^\alpha$, where l^α is a hash value. We consider these outputs as the keys of the β layer (K^β). Then, we define $\mathcal{F}^\beta \leftarrow \sum v_i^\beta$ (i.e. the sum of all the last/maximum values at α). When $\lambda = \theta$, we consider the outputs of the β layer along with the address of the respective intermediary (add_b) as the input leaves and $\mathcal{F}^\theta \leftarrow \sum v_i^\theta$.

Definition 2. Let $\mathcal{W}$ be a recursion function that outputs the inclusion proof of ith item in K_n^λ when $\mathcal{W}(0, [k_0]) = []$.

$$\mathcal{W}(i, K_n^\lambda) = \begin{cases} if\ i < j, \\ \mathcal{W}(i, K_n^\lambda[0:j]), [(\mathcal{F}^\lambda(K_n^\lambda), \mathcal{L}(K_n^\lambda[j:n]))] \\ \\ if\ i >= j, \\ \mathcal{W}(i-j, K_n^\lambda[j:n]), [(\mathcal{F}^\lambda(K_n^\lambda), \mathcal{L}(K_n^\lambda[0:j]))] \end{cases} \tag{2}$$

Definition 3. Let i be the position of $k_i^\lambda \in K_n^\lambda$ and W^λ the output of $\mathcal{W}^\lambda$. Then $\mathcal{V}$ is the inclusion verification function, defined in Algorithm 1.

Definition 4. LMLC is a tuple, $(\tilde{l}, \tilde{v}, \tilde{e}, K^\lambda, \mathcal{F}^\lambda, \mathcal{L}, \mathcal{W}, \mathcal{V})$ where $\tilde{l}$ is the root lock which carries a total of $\tilde{v}$ and expires at $\tilde{e}$. It is created passing $K^\alpha, K^\beta, K^\theta$ to the function $\mathcal{L}$ in each layer separately.

Algorithm 1 Verification, $\mathcal{V}$

Input: $i, k_i^\lambda, W^\lambda$
Output: l'
Require: $i < n$
 Init : $l' \leftarrow k_i^\lambda$
 Init : $j \leftarrow 1$
1: **for all** $w_i \in W^\lambda$ **do**
2: **if** i is even **then**
3:   $l' \leftarrow w_i[0], \mathcal{H}(j \| l' \| w_i[1])$
4: **else**
5:   $l' \leftarrow l'[0], \mathcal{H}(j \| w_i \| l')$
6: **end if**
7: $i = \lfloor i/2 \lceil$
8: $j{+}{+}$
9: **end for**

Algorithm 2 Claim (On-chain)

Require: $\tilde{l}$, *claimed*
Input: $W^\alpha, W^\beta, W^\theta, x^\alpha, v^\alpha, e^\alpha, i^\alpha, i^\beta, i^\theta$
Input: sig, add_b
Output: payment settlement or error
1: Claimable check $\triangleleft$ i^β
2: Signature check $\triangleleft$ sig
3: $(v^\beta \| e^\beta, h^\beta) \leftarrow$
4: $\mathcal{V}(i^\alpha, v^\alpha \| e^\alpha \| \mathcal{H}(x^\alpha), W^\alpha)$
5: Expiration check $\triangleleft$ e^β
6: $(v^\theta, h^\theta) \leftarrow \mathcal{V}(i^\beta, v^\beta \| \mathcal{H}(h^\beta), W^\beta)$
7: $\tilde{l}' \leftarrow \mathcal{V}(i^\theta, v_\theta \| \mathcal{H}(add_b \| h_\theta), W^\theta)$
8: **if** $\tilde{l} == \tilde{l}'$ **then**
9: *claimed* $\Leftarrow (i^\alpha, v_i^\alpha)$
10: transfer v_i^α to add_b
11: **end if**

3.2 Design

Here we present the usage of LMLC in an A-B-C environment in different stages. They include initialization, verification, service, and closing.

Initially, producers connect with intermediaries, ideally with a MultiSig based payment channel. Producers generate an ordered list of random secrets (later take the hash value) for each segment of their services, along with the expiration time and monetary value. Altogether, they act as the producer's fee structure and the keys (or leaves) for the LMLC at layer α. The consumer picks them chronologically and builds a lock using $\mathcal{L}$, and it is verifiable (using $\mathcal{V}$) individually without revealing the producer secrets. There can be multiple such locks from different producers, managed by different intermediaries. Combining them layer by layer according to LMLC creates the root lock $\tilde{l}$, the overall expiry time $\tilde{e}$ and the total monetary value, $\tilde{v}$. Consumer place on the blockchain contract with funding $\tilde{v}$.

Both producers and intermediaries individually verify, 1) the existence of the contract, 2) the lock within (given the witnesses $\mathcal{W}^{\beta}$ and $\mathcal{W}^{\theta}$ by the consumer) using $\mathcal{V}$ which can verify inclusion of k_i in the lock l when $l^{\lambda} == \mathcal{V}(i, k_i^{\lambda}, W^{\lambda})$, 3) adequate $\tilde{v}$, and 4) $\tilde{e}$ is after the individual expiration. Successful verification indicates the consumer's willingness to pay intermediaries according to the respective producers' fee structure. However, they are not claimable by the intermediaries by themselves. The producers can also determine the number of consumers and their identities. It allows a to predetermine the charge which is to be claimed from its intermediary.

The service can begin after the verification. To compensate, c presents an off-chain commitment to the respective b in the form of a signed key $k_i \in K^{\alpha}$, that holds a value of v_i. b can claim it only if b is aware of x_i, the secret of k_i which belongs to a. Hence, b requests x_i from a in the form of an offline payment with the same lock where a has to reveal x_i to claim the payment, making the exchange atomic. Here, the intermediary can also include its brokerage fee. Similarly, b can handle multiple consumers and compensate a in bulk. Note that there is no direct interaction with the blockchain here. The signature verification (SV_o) assures the payment receipt.

Algorithm 2 gives the on-chain operation for an intermediary to claim a value. b should present the key with the secret along with its inclusion witness. The algorithm checks the inclusion level by level, and if successful, the respective amount is transferred to the given address (denoted by PT_c). The b's address (add_b) is also part of the verification. The contract allows several unlocks on different occasions by different parties. To avoid double-spending, we blacklist the already claimed positions at α layer along with their value. Contract owner can claim unspent amount after $\tilde{e}$ time.

Lemma 1. *Verification fails, provided* $\mathcal{V}(i', v_i^{\alpha} \| e_i^{\alpha} \| \mathcal{H}(x_i^{\alpha}), W^{\alpha})$ *where* $i \neq i'$.

Proof. First, we prove that the key position, i in the Algorithm 1 creates a unique sequence of odd and even that follows the binary sequence of value i. According to $\mathcal{V}$ and its notation, $i_j = 2 * i_{j+1} + r_j$ where r_j determines both the parity and j^{th} position of binary representation of the initial i (say i_1) such that $i_1 = 2^{n-1}r_{n-1} + 2^{n-2}r_{n-2}, + \cdots + 2r_1 + r_0$ (by basis representation theorem [1] and note that $|W| = log(n)$ and $i_1 < n$). The series, $\forall j\ r_j$, is the odd-even pattern the determines the control flow of the algorithm $\mathcal{V}$. From the same basis representation theorem, base two representation is unique. Therefore, we can derive that there exists a unique flow in $\mathcal{V}$, given i. i' gives a different flow and fails the verification because $\mathcal{H}$ receives a different set of inputs. □

3.3 Application

Our focus is on the compensation scenarios where an intermediary mediates between two parties in a continuous fashion. Such contexts are prevalent in the real world, although they were not distinguishably identified as such. For example, when a web platform mediates publishing videos for viewers, the producer becomes the publisher, consumers are viewers, and the platform becomes the intermediary. Here, the publisher's gain and platform revenue are generally decoupled and undisclosed. Another example is the intermediary involvement in data service domains due to the limited resources of data producers. However, it is not limited to digital services. For example, even a seamstress is a partial producer of a branded costume where the brand acts as an intermediary. MARI is applicable in these environments to improve the transparency of monetisation. The only limitation is that it is not pragmatic to apply it to services with no or low cost of reproduction without the producer's consent (e.g., videos in the first example).

MARI needs a blockchain with Turing complete scripting capability to implement the contract, defined in the Algorithm 2 as $\mathcal{V}$ consists of a loop. Hence, we used Ethereum to build the proof of concept of the protocol, which can be found here[1]. However, the MultiSig contract between producer and intermediary is not necessarily on the same blockchain. Therefore, it is an added advantage to integrate entities on different payment services. The rest of the functions are offline.

Table 1. Producer's perspective

	Setting I: a-C		Setting II: a-b-C		
	a	C	a	b	C
PT_c	$\|C\|$	$2\|C\|$	1	$\|C\|+1$	$2\|C\|$
DS_c	–	$3\|C\|$	–	–	$3\|C\|$
SV_c	–	–	1	$\|C\|$	–
HG_c	$log(\eta)\|C\|$	–	1	$log(\eta)\|C\|$	–
PV_o	$\|C\|$	–	$1+\|C\|$	$\|C\|$	–
SV_o	–	–	η	$\eta\|C\|$	–
SG_o	–	–	–	η	$\eta\|C\|$
DS_o	$\eta\|C\|$	–	η	–	–
HG_o	$2\eta\|C\|$	$\eta\|C\|$	2η	η	$\eta\|C\|$

[1] https://github.com/anupasm/MARI.

4 Analysis

Here, we analyse the amount of work required by each entity involved in the protocol from the positions of producers and consumers. The goal is to evaluate the overhead on a single entity. We examined it in different settings as well. In the first setting, there is a peer to peer engagement with a set of entities. It is analogous to the setting described in [5] when a payer pays multiple payees and vice versa. In the second setting, there is an intermediary in-between, as we described in the MARI protocol. Selected actions are high level and require notable effort both on-chain (c) and off-chain (o). They are, payment verification[2] (PV), signature verification[3] (SV), hash generation (HG), signature generation (SG), payment transfer[4] (PT), and data storage[5] (DS).

We assume payment resolution per cycle(denoted by η) is the same for all producers and consumers for simplicity. Further, we consider the worst-case where all sessions are consumed per cycle. Summary of the analysis is tabled in Table 1 and 2.

Table 2. Consumer's perspective

	Setting I: c-A		Setting II: c-B-A		
	c	A	c	B	A
PT_c	$2\|A\|$	$\|A\|$	2	$\|A\| + \|B\|$	$\|A\|$
DS_c	$3\|A\|$	–	3	–	–
SV_c	–	–	–	$\|A\|$	$\|A\|$
HG_c	–	$log(\eta)\|A\|$	–	$log(\eta)\|A\|$	$\|A\|$
PV_o	–	$\|A\|$	–	$\|B\|$	$\|A\| + \|B\|$
SV_o	–	–	–	$\eta\|A\|$	$\eta\|A\|$
SG_o	–	–	$\eta\|A\|$	$\eta\|A\|$	–
DS_o	–	$\eta\|A\|$	–	–	$\eta\|A\|$
HG_o	$\eta\|A\|$	$2\eta\|A\|$	$\eta\|A\|$	$\eta\|A\|$	$2\eta\|A\|$

5 Evaluation

MARI provides payment delegation, low overhead, verifiable pricing, and scalability. It inherits several security features by leveraging blockchain technology as an underline layer. However, they consist of both desirable and undesirable

[2] Contract verification process mentioned in Subsect. 3.2.

[3] Off-chain signature verification of commitment transactions and on contract signature verification in line 2 of Algorithm 2.

[4] On-chain monetary value transfer (i.e. initial funds transfer, claim funds by payee, or claim balance by payer).

[5] Storage for secret values.

points. Here, we evaluate how the proposed protocol achieves those features and satisfies the requirements of a state of the art payment protocol [2].

In general, blockchain-based payment systems provide pseudo-anonymity for the participants. However, transaction details, including monetary values and on-chain data, are publicly available for anyone to verify the details. From the perspective of providers and intermediaries, this can be beneficial to build up their reputation. However, consumers may concern about disclosing their consumed services even though pseudo-anonymity prevents revealing their true identity. In terms of the proposed LMLC, the map between consumer and service is not placed on-chain, and it is not exposed until the intermediary claims the respective values.

$\mathcal{H}$ is assumed to be a collision-resistant hash function. Hence, it can be proved that the Merkel tree of LMLC is also collision-resistant. Furthermore, we embed a level number inside hashing which prevents a second preimage attack. Therefore, given the key, an adversary cannot find its secret on its own. MARI payments are unidirectional, and both payment recipient has no incentive to publish old states whereas payment senders are bounded by expiration time. However, there is an opportunity for a double-spending in extended LMLC that allow multiple withdrawals. For example, suppose an adversary can provide a substitute number for the key position. In that case, the contract appends an invalid position to the claimed list, and the adversary can withdraw the amount multiple times. However, we prove that this is not possible in the LMLC in Lemma 1. Considering both facts, we can conclude that MARI preserves the integrity of the payments.

Distributed nature of blockchain-based systems provides reliability and availability for a system based on this protocol. One drawback of the system is that unresponsive clients of b can incur a loss. Although it is unavoidable without additional precautions, we argue that the loss is minimum in micropayment systems.

Fee structure is clearly stated for the agreed period and verifiable in MARI. It avoids the micropayment's mental cost as raised in [13]. In the event of a change of mind, incomplete or inaccurate decision to choose the service, a consumer can withdraw from the contract, which basically stops paying. It does not affect the compensation of the service provided by the producer thus far.

Producer's perspective in Table 1 demonstrates that the overhead on a has primarily moved to the intermediary with compared to the peer to peer engagement in on-chain operations. However, the signature verification incurs an additional cost. In the first setting, payments are solely based on hashing [5]. However, MARI occupies signatures and provides additional security for the payments compared to the hashes. Further, off-chain operations of a in the second setting are independent of the number of consumers, implying the ability to handle a higher number of consumers with static operations. Although it increases the overhead on the intermediary, off-chain operations do not cause a direct fee. Besides, intermediaries are supposed to be resource-rich entities. According to the consumer's perspective in Table 2, the on-chain operations of consumers is static. It means that consumers can engage any counterparty without additional on-chain costs. In general, out of all on-chain operations, the resolution value (η) is involved logarithmically, and it is only in hash verification. Therefore, Merkel tree-based micropayments can grow exponentially. It does not inflate the local storage as space can be recycled with new items.

Here, we can conclude that MARI largely preserves the low overhead on producers and the total on-chain cost compared to the first setting. As a result, off-chain operations are higher but primarily on intermediaries. Further, it can scale at a fixed cost for a consumer.

6 Conclusion

This paper presents MARI payment protocol, a novel approach to implement micropayment for intermediary-based trading. We create a novel blockchain-based smart contract where multiple parties can open the same lock multiple times and claim different monetary values, yet protected from double-spending. MARI facilitates producers to directly communicate the pricing structure to their consumers and assure the compensation even in the presence of an intermediary. The analysis of required operations shows that MARI is scalable and reduces on-chain cost and overhead on both producer and consumer. Additionally, we evaluate the protocol in terms of the security requirements of general payment protocols. We intend to integrate this protocol with existing communication protocols and develop an eco-system with the ability to update the connections dynamically.

Acknowledgements. This publication has emanated from research supported in part by a research grant from Science Foundation Ireland (SFI) and the Department of Agriculture, Food and the Marine on behalf of the Government of Ireland under Grant Number SFI 16/RC/3835 (VistaMilk), and also by a research grant from SFI under Grant Number SFI 12/RC/2289_P2 (Insight), with both grants co-funded by the European Regional Development Fund.

References

1. Andrews, G.E.: Number Theory. Courier Corporation (1994)
2. Asokan, N., Janson, P., Steiner, M., Waidner, M.: State of the art in electronic payment systems. Adv. Comput. **53**, 425–449 (2000)
3. Bessy, C., Chauvin, P.M.: The power of market intermediaries: from information to valuation processes. Valuat. Stud. **1**(1), 83–117 (2013)
4. Bhargava, H.K., Choudhary, V.: Economics of an information intermediary with aggregation benefits. Inf. Syst. Res. **15**(1), 22–36 (2004)
5. Galal, H.S., ElSheikh, M., Youssef, A.M.: An efficient micropayment channel on ethereum. In: Data Privacy Management, Cryptocurrencies and Blockchain Technology, pp. 211–218. Springer (2019)
6. Gudgeon, L., Moreno-Sanchez, P., Roos, S., McCorry, P., Gervais, A.: Sok: layer-two blockchain protocols. In: International Conference on Financial Cryptography and Data Security, pp. 201–226. Springer (2020)

7. Herlihy, M.: Atomic cross-chain swaps. In: Proceedings of the 2018 ACM Symposium on Principles of Distributed Computing, pp. 245–254 (2018)
8. Jutla, C.S., Yung, M.: Paytree: "amortized signature" for flexible micro-payments. IACR Cryptol. ePrint Arch. 2012, 10 (2012)
9. Kim, S., Lee, W.: A pay word-based micropayment protocol supporting multiple payments. In: Proceedings. 12th International Conference on Computer Communications and Networks (IEEE Cat. No. 03EX712), pp. 609–612. IEEE (2003)
10. Párhonyi, R.: Micropayment systems. In: Handbook of Financial Cryptography and Security. CRC Press (2010)
11. Rivest, R.L., Shamir, A.: Payword and micromint: two simple micropayment schemes. In: International Workshop on Security Protocols, pp. 69–87. Springer (1996)
12. Rouhani, S., Deters, R.: Security, performance, and applications of smart contracts: a systematic survey. IEEE Access **7**, 50759–50779 (2019)
13. Szabo, N.: Micropayments and mental transaction costs. In: 2nd Berlin Internet Economics Workshop, vol. 44, p. 44 (1999)
14. Wan, Z.G., Deng, R.H., Lee, D., Li, Y.: MicroBTC: efficient, flexible and fair micropayment for bitcoin using hash chains. J. Comput. Sci. Technol. **34**(2), 403–415 (2019)
15. Wikipedia contributors: Micropayment – Wikipedia, the free encyclopedia (2021). https://en.wikipedia.org/wiki/Micropayment. Accessed 29 June 2021

BlockFLow: Decentralized, Privacy-Preserving, and Accountable Federated Machine Learning

Vaikkunth Mugunthan, Ravi Rahman(✉), and Lalana Kagal

Computer Science and Artificial Intelligence Lab, Massachusetts Institute of Technology, Cambridge, MA 02139, USA
{vaik,r_rahman,lkagal}@mit.edu

Abstract. Federated machine learning enables multiple clients to collectively train a machine learning model without sharing sensitive data. However, without proper accountability mechanisms, adversarial clients can weaken the collective model. BlockFLow is a fully decentralized, privacy-preserving, and accountable federated learning system. It introduces an Ethereum blockchain smart contract to coordinate a federated learning experiment and to hold clients accountable. BlockFLow rewards clients proportional to the quality of their individual contributions, does not reveal the underlying datasets, and is resilient to a minority of adversarial clients. Unlike existing systems, BlockFLow does not require a centralized test dataset, sharing of datasets between the clients, or any trusted entities. We evaluated BlockFLow on logistic regression models. Our results illustrate that BlockFLow successfully rewards honest clients and identifies adversarial clients. These results, along with blockchain costs that do not scale with model complexity, demonstrate the effectiveness of BlockFLow as an accountable federated learning system.

Keywords: Blockchain accountability · Federated machine learning

1 Introduction

Machine learning models benefit from large, diverse training datasets. Organizations with sensitive datasets can only develop locally-optimal models as regulations can prohibit sharing datasets across organizational boundaries. Federated learning overcomes these limitations by enabling the development of globally-optimal models without sharing sensitive data [12,15]. In a traditional federated learning environment, individual clients train local models on their datasets and share their local models with a centralized server, which computes and shares the averaged model. However, this traditional setup is susceptible to accountability threats.

V. Mugunthan and R. Rahman—Denotes Equal Contributions.

J. Prieto et al. (Eds.): BLOCKCHAIN 2021, LNNS 320, pp. 233–242, 2022.
https://doi.org/10.1007/978-3-030-86162-9_23

For example, consider a logistic regression experiment where some clients invert their output labels and train and submit models based on this data. Such adversarial models can worsen the shared model [10]. Other clients would then use this weakened shared model in the next federated learning round, resulting in reduced overall accuracy [14,20]. To defend against such attacks, clients must be held *accountable* for their contributions. Evaluating all contributions can prevent adversarial contributions from being included in the shared model. Moreover, good contributors can be rewarded. Though it is trivial for a centralized server to evaluate models such as by measuring accuracy on a secret (test) dataset, it is not always feasible to agree upon a trusted, centralized server or test dataset (e.g. in international collaborations). Thus, a decentralized federated learning process, which is resilient to a minority of adversarial clients, can be preferable.

BlockFLow is a fully decentralized, privacy-preserving, and accountable federated learning system. It introduces an Ethereum blockchain smart contract to coordinate federated learning experiments and hold clients accountable. BlockFLow collects deposits (in Ether) from all clients, which the smart contract redistributes among the clients throughout the experiment to reward good clients and penalize adversarial clients. The system is resilient to a minority of adversarial clients. Unlike existing systems, BlockFLow does not require (secret) test datasets, sharing of datasets between the clients, or any trusted entities.

2 Related Work

Given the wide attack surface for federated learning, numerous techniques have been proposed to mitigate privacy and accountability threats. However, unlike our work, previous approaches use a semi-honest trust model, require trusted client(s), expose datasets, or do not defend against individual adversarial clients.

Blockchain platforms have been used previously for accountable federated learning. Some systems [9,13] evaluate models directly on the blockchain, via an initially-hidden but verifiable test dataset. While trust-less, these systems require datasets to be publicly revealed on the blockchain and incur significant blockchain computational expenses for on-chain model evaluation. These techniques are only applicable for small models with public datasets.

Instead of a public, on-chain evaluation, others [1,11] propose using trusted aggregators and evaluators. These systems assume that all clients agree to trust centralized server(s). While these proposals offer better privacy guarantees, where no data is publicly revealed, their weak trust model significantly limits their applicability. The systems proposed by [16,17] provide guarantees for only the semi-honest trust model and not for the adversarial setting. In addition, they don't provide any accountability guarantees.

The usage of a private blockchain (Hyperledger) by [4] introduces the drawback that it is enforced by fewer nodes, potentially making it a less accountable system. The frameworks proposed by [4,17,21] do not consider any privacy mechanisms in their protocol; hence, collusion among clients may result in model inversion attacks.

Finally, [3] proposes a robust, trust-less, gradient-based validation scheme, where only differentially private gradient updates are revealed on a public blockchain. It includes only the gradient updates most similar to the average update. While both accountable and privacy-preserving, error increases proportionally to the number of adversarial clients. It also does not scale well with respect to blockchain costs for large models, as gradients are averaged on the public blockchain.

BlockFLow offers numerous advantages compared to these existing works. First, it is both privacy-preserving and trust-less: it does not require any secret test dataset, trusted set of clients, or the revealing of data or weights on a public blockchain. It also supports a robust threat model. So long as there are a minority of adversarial clients, BlockFLow filters out all adversarial clients and thus preserves shared model quality.

3 System Design

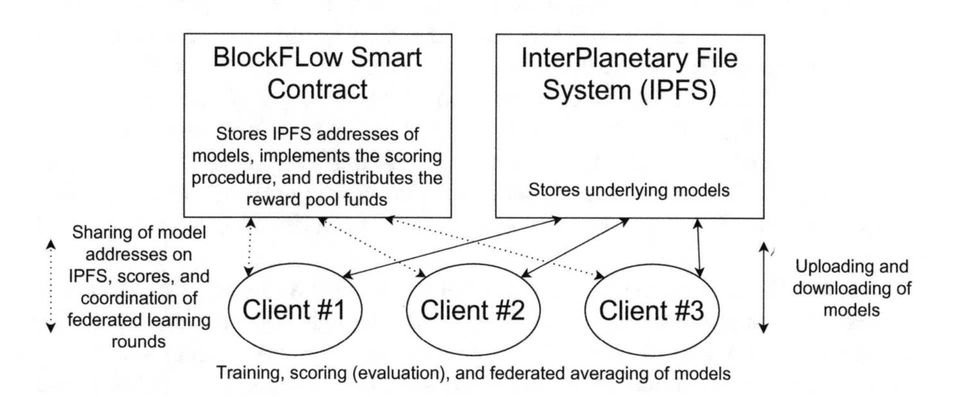

Fig. 1. BlockFLow System Architecture. Clients train and evaluate models locally. The BlockFLow smart contract, which runs on the Ethereum blockchain, stores models' IPFS addresses, calculates scores, and collects and redistributes reward pool funds. The InterPlanetary File System (IPFS) stores the model parameters. This architecture minimizes blockchain costs: no evaluation, dataset storage, or model storage is on the blockchain.

3.1 Client

The BlockFLow client, which each experimental participant runs locally, performs all machine learning training, evaluation and scoring of others' models, and federated learning averaging. It has a modular interface, which allows for federated learning experiments of any model type. Each experimental setup must define interfaces for training and serializing a model to a file, evaluating serialized models and producing numerical scores, and performing weighted federated averaging of models (given their scores). For privacy, the training interface may incorporate differential privacy [7]. The BlockFLow client handles the communication with the BlockFLow smart contract and the InterPlanetary File System (IPFS). Figure 1 illustrates the system architecture.

During the training stage, the BlockFLow client provides a deadline and starting model to the training script. It expects this script to return a serialized model before the deadline elapses. The client uses this serialized model as the client's submission for the federated learning round.

During the scoring stage, the BlockFLow client provides others' serialized models from the training stage to the scoring script. For each model, it expects the scoring script to return a score on $[0, 1]$, where 0 should be used to indicate errors or extremely poor models, and 1 represents perfect models. Higher scores must correspond to better models. The experimental setup must define the scoring metric that clients should use (e.g. for classification experiments, accuracy or F1 scores would suffice).

During the federated averaging stage, the BlockFLow client provides others' serialized models from the training stage and corresponding scores to the federated averaging script, and it expects this script to return the model to use as a starting point for the next invocation of the training script (in the next federated learning round). The scores provided to the federated averaging script incorporate all other clients' evaluations (not just one's own evaluation from the scoring stage). This script more heavily weighs models with higher scores, and scores below the 50th percentile are rejected due to the BlockFLow threat model (see Sect. 4).

Because BlockFLow requires all clients to retrieve and evaluate all others' models, total communication costs are $O(N^2R)$, where N is the number of clients, and R is the number of federated learning rounds. Storage costs are $O(N)$, since clients must evaluate all others models' in each round. For efficiency, our implementation performs batch downloads and evaluations. It is not necessary to store models from earlier federated learning rounds. To reduce storage costs to $O(1)$, one could download and evaluate models one at a time.

3.2 InterPlanetary File System (IPFS)

IPFS is a peer-to-peer file system, where anyone can host any content [2]. IPFS file locations, called addresses, are cryptographic hashes of the underlying content. Addresses are tamper-resistant, as clients can easily verify that they have the correct content by recomputing the cryptographic hashes of the data and checking that they match the associated addresses. Content disappears from the IPFS network when no IPFS node holds the content associated with the address. IPFS is free, as nodes are not compensated for holding content.

Unlike a blockchain, IPFS does not provide timestamping or prove ownership. However, recording an IPFS address in a blockchain timestamps the content with the blockchain's block number, and proves ownership with the cryptographic key pair of the blockchain transaction. BlockFLow stores underlying model data on IPFS to save on blockchain gas costs, and only records the IPFS address in the BlockFLow smart contract.

3.3 BlockFLow Smart Contract

The BlockFLow Ethereum smart contract coordinates a federated learning experiment. Before the experiment begins, clients must enroll by posting a deposit. The smart contract redistributes these deposits throughout the experiment in proportion to clients' contribution qualities. For each federated learning round, the contract is broken into six stages: train, retrieve, score commit, score reveal, tally, and refund. Each stage in each federated learning round has a deadline (defined in terms of block numbers). Clients who miss deadlines are eliminated from the experiment and cannot collect any future refunds. While more lenient policies are possible, this design simplifies the implementation and reduces gas costs. So long as there are always a majority of honest clients remaining, the smart contract and protocol continue to function correctly. If all clients are eliminated, then any remaining funds in the smart contract are burned.

Before a train deadline, clients submit their models' IPFS addresses to the smart contract. Before a retrieve deadline, clients report which other models they are able to retrieve. Clients who are unable to retrieve a majority of others' models, or those with models which are not retrievable by a majority of other clients, are eliminated. Before a score commit deadline, each client submits a cryptographic hash of a random salt and the scores they assigned to others' models. Only the remaining clients are evaluated. Before a score reveal deadline, clients reveal their assigned scores and the random salt. During a tallying stage, clients invoke the smart contract methods to provide gas to run the contribution scoring procedure (see Sect. 3.3). During a refund stage, clients collect refunds from the reward pool proportional to their score from the tallying stage. The smart contract reserves a fraction of the reward pool for future federated learning rounds. This process repeats for each federated learning round. After the final round, the remaining deposits are refunded among the remaining clients.

Contribution Scoring Procedure. The contribution scoring procedure, which is implemented in the BlockFLow smart contract, incentivizes clients to be honest without requiring any trusted third-parties or test datasets. Algorithm 1 describes how the smart contract computes clients' overall scores, which the smart contract uses to divide the reward pool among the clients and clients use for federated averaging. Clients' scores are the lesser of the quality of their models and evaluations. Section 4 discusses how this procedure incentivizes clients' honesty.

BlockFLow can handle non-extreme cases of non-independent and identically distributed (non-i.i.d.) datasets among the clients. For example, in a classification problem, the scores provided by clients with more data corresponding to a particular target variable than another will (negatively) reflect the bias.

Under extreme non-i.i.d. cases (e.g. each client has a different subset of labels completely), BlockFLow would perform poorly, as clients would be unable to accurately evaluate others' models.

Algorithm 1: Contribution Scoring Procedure.

for all client pairs $\{a, k\} \in \{N \times N\}$ **do**
  $s_{a,k} \leftarrow \texttt{eval}_a(k)$ {client a evaluates k's model off-chain using a's own dataset.}
end for
for all clients $k \in N$ **do**
  $m_k \leftarrow \text{MED}\{s_{a,k} : \forall a \in N\}$ {Compute the median model scores}
end for
for all client pairs $\{a, k\} \in \{N \times N\}$ **do**
  $t_{a,k} \leftarrow |s_{a,k} - m_k|$ {Compute the evaluation quality scores}
  $t'_{a,k} \leftarrow \max(0, \frac{0.5 - t_{a,k}}{0.5 + t_{a,k}})$ {Transform the evaluation quality scores}
end for
for all clients $a \in N$ **do**
  $d_a \leftarrow \min\{t'_{a,k} : \forall k \in N\}$ {Compute each client's least accurate evaluation}
end for
for all clients $k \in N$ **do**
  $m'_k \leftarrow \frac{m_k}{\max\{m_{k'} : \forall k' \in N\}}$ {Scale the median model scores}
  $d'_k \leftarrow \frac{d_k}{\max\{d_{k'} : \forall k' \in N\}}$ {Scale the least accurate evaluation scores}
end for
for all clients $k \in N$ **do**
  $p_k \leftarrow \min(m'_k, d'_k)$ {Compute the overall scores}
end for
return p

4 Threat Analysis

BlockFLow supports a threat model where it is resilient to attacks from up to 50% of the clients. Specifically, we assume that at least half of the clients are honest and follow the BlockFLow protocol. In addition, we assume that the Ethereum blockchain is secure and that no minority of BlockFLow clients control sufficient mining power to attack the underlying blockchain. Such attacks are beyond the scope of BlockFLow.

One such attack involves storage of models on IPFS. While IPFS is immutable [2], it does not guarantee that content is available on the network. An adversarial client could publish an IPFS address for unavailable content or neglect to download others' models that are actually available. To prevent these attacks, the BlockFLow smart contract eliminates clients who were unable to retrieve a majority of others' models or whose models were not retrieved by a majority of the other clients. Since the trust model guarantees a majority of honest clients, the honest clients would fulfill these conditions for each other.

Other attacks focus on the contribution scoring procedure. Consider the case where clients submit adversarial or low-quality models. When the other clients evaluate these models, the honest clients would score them poorly. As there are a majority of honest clients, the median score would come from an honest client. As such, adversarial clients are guaranteed to receive low model scores, and thereby, a limited share of the reward pool funds.

Clients may also collude during evaluation to attempt to steal the reward pool funds. Suppose a minority of dishonest clients report perfect 1.0 scores for their models and 0.0 scores for all others (e.g. models from honest clients). However, since strictly less than half of the clients are adversarial clients, the median score is guaranteed to come from an honest client. As any honest client's score is a fair evaluation, it is impossible for colluding clients to materially affect the evaluation scores.

Moreover, the fabrication of scores will only penalize those who attempt it. The contribution scoring procedure limits one's overall score with the evaluation on which one was furthest away from the median. Specifically, any client with an evaluation of more than 0.5 away from the median will receive an overall score of 0, and no share of the reward pool. As long as the median scores are bounded by the range of the honest clients' scores, which our threat model guarantees, fabrication is never optimal.

5 Evaluation

We evaluated BlockFLow using logistic regression models ($\alpha = 1.0$) on the Adult Census Income (Adult) [5] and The Third International Knowledge Discovery and Data Mining Tools Competition (KDD) [5,8] classification tasks. All discrete features are one-hot encoded, and continuous features were scaled globally and independently to be between 0 and 1. Two-thirds of the data was split in equal shares (unless otherwise noted) among the N clients, without overlap, and clients reserved 20% of their data for training validation. We applied differential privacy using the Laplace mechanism ($\epsilon = 0.01$) [6]. Clients used their entire datasets when evaluating others' models. The remaining one-third was reserved for testing. F1 scores were used for the scoring function.

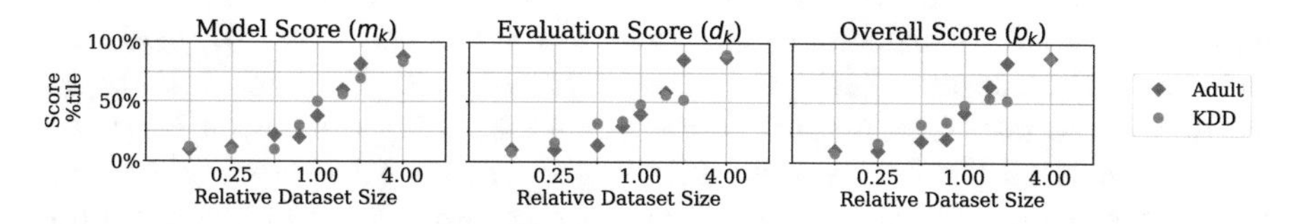

Fig. 2. Experiment 1: Rewarding Larger Datasets. Out of $N = 50$ clients, we varied the dataset size for 10 clients. These clients' model, evaluation, and overall scores (m_k, d_k, and respectively p_k from algorithm 1) are highly correlated with the relative dataset size. These results illustrate how BlockFLow can identify and reward those with large datasets while maintaining dataset privacy and without sharing underlying data.

In the first experiment, we evaluated how the contribution scoring procedure rewards those with larger (i.e. more robust) datasets. As shown in Fig. 2, clients' model, evaluation, and overall scores were strongly correlated with their dataset size ($\rho > 0.9$ for all evaluations on a logarithmic scale). As such, BlockFLow can identify and reward those with more meaningful contributions without requiring the underlying data to be shared.

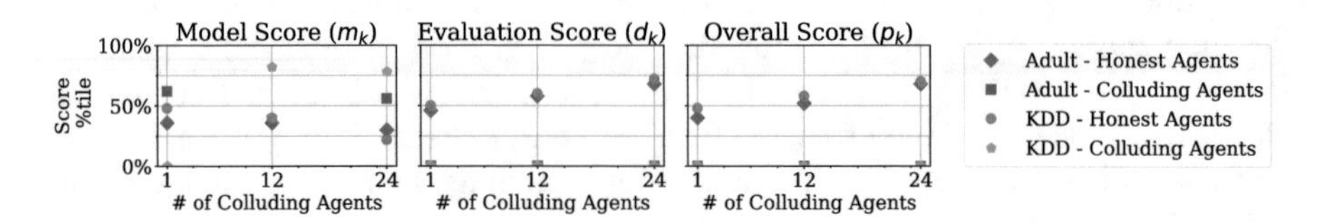

Fig. 3. Experiment 2: Discouraging Collusion. Out of $N = 50$ clients, a subset colluded by awarding other colluding clients perfect scores. While colluding clients obtained better model scores, note the statistically lower ($p < 10^{-31}$) overall scores for colluding clients compared to the honest clients. Hence, it is a sub-optimal strategy to fabricate evaluation scores.

In the second experiment, we examined collusion attacks. Specifically, consider the fabrication of evaluation scores where a minority subset of clients award perfect (1.0) scores to other colluding clients. Figure 3 illustrates how colluding clients' received lower scores.

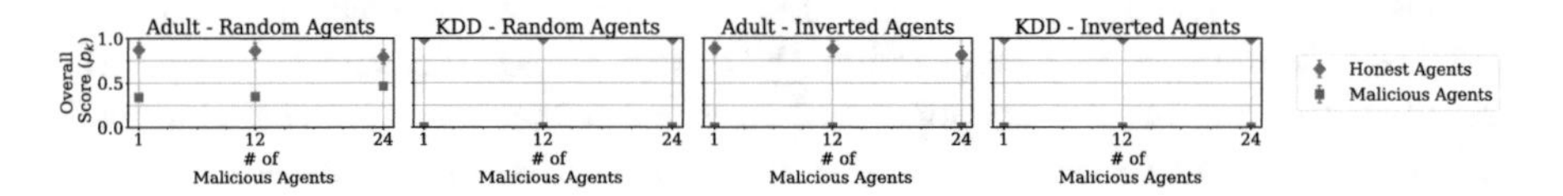

Fig. 4. Experiment 3: Penalizing Adversarial Clients. Out of $N = 50$ clients, a subset of adversarial clients trained on random data or inverted output features. For all $N > 1$ client experiments, adversarial clients' scores were statistically lower ($p < 10^{-22}$).

In the third experiment, we explored adversarial training setups, where a minority subset of clients submitted models trained on *random* or *inverted* data. Independently sampling each feature with probabilities from the real dataset distribution formed datasets for random clients. Flipping the output labels created the inverted clients' datasets. Like honest clients, adversarial clients split their datasets into 80% train and 20% validate portions. Figure 4 illustrates how the adversarial clients' scores were statistically lower than those of the honest clients.

Finally, we analyze the blockchain costs of our system. Each Ethereum opcode costs a specified amount of gas [19]. Gas is independent of market conditions, such as the price of gas or the exchange rate of Ethereum. A regression determined that the total gas consumption for N clients and R federated learning rounds is:

$$gas(N, R) = 6744N^2R + 52229NR + 89483N + 22336R + 2649601$$

Collectively, these experiments validate how the contribution scoring procedure can hold clients accountable in a federated learning setup without any trusted auditors or public datasets. Combined with the strong threat model for the Ethereum blockchain and support for differential privacy techniques, BlockFLow is resilient to attacks by an adversarial subset of clients while also being fully privacy-preserving. Moreover, blockchain costs that do not scale with model complexity illustrate how BlockFLow can be used with even the most complex of federated machine learning experiments.

6 Future Work and Conclusion

Future work for BlockFLow includes optimizing the BlockFLow smart contract to further reduce blockchain and communication costs to support experiments with many clients and federated learning rounds. Randomized evaluation can reduce costs. Instead of requiring each client to evaluate every other client's contributions, the BlockFLow smart contract could randomly select $Q << N$ clients to evaluate each client's work. A sufficiently large Q, with high probability, would lead to accurate results and still be resilient to a minority of adversarial clients.

Other improvements include modifying the BlockFLow smart contract to support ERC-20 tokens [18] for collateral. These tokens can eliminate Ethereum exchange rate volatility.

BlockFLow presents a fully decentralized, privacy-preserving, and accountable federated learning system. The evaluation demonstrated how our contribution scoring procedure runs efficiently via a blockchain smart contract, accurately rewards honest clients in proportion to their contributions, and penalizes adversarial clients who attempt to cheat. Unlike previous works, it does not require any trusted agents or public datasets. This combination of features broadens the reach of federated learning systems to environments where clients need not know nor trust each other but are rewarded for contributing high-quality models.

Acknowledgments. We thank Omar Dahleh for his contributions with comparing related works.

References

1. Awan, S., Li, F., Luo, B., Liu, M.: Poster: a reliable and accountable privacy-preserving federated learning framework using the blockchain. In: Proceedings of the 2019 ACM SIGSAC Conference on Computer and Communications Security, pp. 2561–2563 (2019)

2. Benet, J.: IPFS-content addressed, versioned, p2p file system. arXiv:1407.3561 (2014)
3. Chen, X., Ji, J., Luo, C., Liao, W., Li, P.: When machine learning meets blockchain: a decentralized, privacy-preserving and secure design. In: 2018 IEEE International Conference on Big Data (Big Data), pp. 1178–1187. IEEE (2018)
4. Desai, H.B., Ozdayi, M.S., Kantarcioglu, M.: BlockFLA: accountable federated learning via hybrid blockchain architecture. In: Proceedings of the Eleventh ACM Conference on Data and Application Security and Privacy, pp. 101–112 (2021)
5. Dua, D., Graff, C.: UCI machine learning repository (2017)
6. Dwork, C., McSherry, F., Nissim, K., Smith, A.: Calibrating noise to sensitivity in private data analysis. In: Theory of Cryptography Conference, pp. 265–284. Springer (2006)
7. Dwork, C., Smith, A.: Differential privacy for statistics: what we know and what we want to learn. J. Priv. Confidenti. **1**(2), 2 (2010)
8. Elkan, C.: Results of the KDD'99 classifier learning. ACM SIGKDD Explor. Newsl. **1**(2), 63–64 (2000)
9. Harris, J.D., Waggoner, B.: Decentralized and collaborative AI on blockchain. In: 2019 IEEE International Conference on Blockchain, pp. 368–375. IEEE (2019)
10. Kairouz, P., et al.: Advances and Open Problems in federated learning. arXiv:1912.04977 (2019)
11. Kim, H., Kim, S.-H., Hwang, J.Y., Seo, C.: Efficient privacy-preserving machine learning for blockchain network. IEEE Access **7**, 136481–136495 (2019)
12. Konečný, J., McMahan, H.B., Yu, F.X., Richtárik, P., Suresh, A.T., Bacon, D.: Federated learning: strategies for improving communication efficiency. arXiv:1610.05492 (2016)
13. Kurtulmus, A.B., Daniel, K.: Trustless machine learning contracts; evaluating and exchanging machine learning models on the ethereum blockchain. arXiv:1802.10185 (2018)
14. Li, L., Xu, W., Chen, T., Giannakis, G.B., Ling, Q.: RSA: byzantine-robust stochastic aggregation methods for distributed learning from heterogeneous datasets. In: Proceedings of the AAAI Conference on Artificial Intelligence, vol. 33, pp. 1544–1551 (2019)
15. McMahan, B., Ramage, D.: Federated learning: collaborative machine learning without centralized training data. Google Research Blog, p. 3 (2017)
16. Passerat-Palmbach, J., Farnan, T., Miller, R., Gross, M.S., Flannery, H.L., Gleim, B.: A blockchain-orchestrated federated learning architecture for healthcare consortia. arXiv:1910.12603 (2019)
17. Ramanan, P., Nakayama, K.: Baffle: Blockchain based aggregator free federated learning. In: 2020 IEEE International Conference on Blockchain (Blockchain), pp. 72–81. IEEE (2020)
18. Vogelsteller, V., Buterin, F.: Eip-20: Erc-20 token standard. Technical report 20, Ethereum Improvement Proposals, November 2015
19. Wood, G.: Ethereum: a secure decentralised generalised transaction ledger. Ethereum Proj. Yellow Paper **151**, 1–32 (2014)
20. Wu, Z., Ling, Q., Chen, T., Giannakis, G.B.: Federated variance-reduced stochastic gradient descent with robustness to byzantine attacks. arXiv:1912.12716 (2019)
21. Zhang, Z., Yang, T., Liu, Y.:. Sablockfl: a blockchain-based smart agent system architecture and its application in federated learning. Int. J. Crowd Sci. (2020)

COVID-19 Early Symptom Prediction Using Blockchain and Machine Learning

Sarada Kiranmayee Tadepalli and Ruppa K. Thulasiram(✉)

University of Manitoba, Winnipeg, MB, Canada
tadepask@myumanitoba.ca, tulsi.thulasiram@umanitoba.ca

Abstract. The COVID-19 outbreak has resulted in unprecedented and difficult times for world's population. Social distancing and self-isolation have become very important to reduce the spread. This called upon the creation of numerous applications that have used proprietary models for symptoms-tracking and contact-tracing around the world to mitigate the spread. In most of the applications data collected is stored in a centralized database without verification and hence, the data is not reliable. In this study, a decentralized application for COVID-19 symptoms tracking using Blockchain is proposed in order to enhance reliable data collection for training Machine Learning (ML) models. The Blockchain integration in this application will help in collecting COVID-19 symptoms data from the patients with trust. In addition to this, the data would be first verified by an entity of the decentralized network (e.g. a COVID-19 testing lab). Then, with the consent of the patient, this data is provided to the centralized system for retraining the ML. In short, the main advantage of this architecture is that the data from the users is collected and checked by a laboratory first and then provided to the ML model. The process helps in identifying the incorrect ML prediction and further train the ML model with reliable data for accurate prediction. Moreover, the trust of the users is earned as the data transfer happens with their consent and, besides, all transactions are recorded on the Blockchain, which is possible with the help of the Distributed Ledger Technology (DLT).

1 Introduction

Coronavirus (COVID-19) is a new virus outbreak that is broadly believed to have started in Wuhan, China, in December 2019 and has rapidly spread to different countries around the world [1]. COVID-19 was declared as a pandemic, by World Health Organization (WHO) on the 11th of March 2020, that left 166 countries affected by the end of March 2020 [2].

Currently, the data collected is centralized because of which the interoperability of electronic medical information between organizations is tedious. In the case of data collected for COVID-19, organizations like WHO require a platform where the data interoperability is smooth. Moreover, the COVID-19 data collected should be verified by a medically trusted authority so that incorrect conclusions by the authorities can be prevented. This calls for the need

J. Prieto et al. (Eds.): BLOCKCHAIN 2021, LNNS 320, pp. 243–251, 2022.
https://doi.org/10.1007/978-3-030-86162-9_24

of a unified platform, where governments, local authorities, and hospitals can share COVID-19 data with trust. This trust-worthy data collected would help in building COVID-19 related applications, which could provide accurate results for preventing further spread. It is to be kept in mind that applications developed on trusted and verifiable data should also guarantee the data privacy of the patients [3].

Early diagnosis of the symptoms will help in reducing the number of COVID-19 cases. Many applications [4,5] that are developed for self-reporting, collect data and store it in a centralized system to train the machine learning (ML) model. But in the current scenario, the symptoms reported by a patient are stored in the data storage first and are verified by a lab later. So, the ML models trained with this data may provide incorrect predictions.

Our Main Contributions Include

* In this study, we propose a Blockchain-based architecture for collecting the data pertaining to COVID-19 symptoms.
* We design a DApp with a ML model deployed on the client-side that instantly predicts COVID-19 case to be either positive or negative based on the symptoms entered by the patient. The model used for the DApp is SGDClassifier [6] which is shown to predict with 90% accuracy.
* A smart contract is triggered for sending the symptoms data with the ML prediction as a transaction over Blockchain to the lab.

The security and the privacy concerns of the patients' data are assumed to be taken care of by mobile service provider and is out of scope for this study. When this patient gets a swab test done, the lab will check the ML model prediction and correct it (if wrong). The lab will also trigger a smart contract to update the patient with the test result. All the data exchanged would be recorded on the Ethereum Blockchain as transactions. The reliable data on Blockchain is collected by the centralized storage and is used for re-training the ML model. This re-training, performed with verified data, updates the ML model and provides more accurate results.

The rest of the paper is structured as follows: In Sect. 2, related work provides the recent research work on Blockchain for data management and COVID-19 data analysis. In Sect. 3, the Machine learning algorithm and Blockchain used for the DApp are elaborated. Section 4, showcases the cost incurred if the DApp were deployed on the Ethereum Main network and Sect. 5 concludes this study followed by future work.

2 Related Work

To run all the Decentralized Applications an Ethereum framework is created [7]. A Turing-complete programming language is included in Ethereum so that users can create "smart contracts" - that is, execution of actions is specified by a set of rules. The inclusion of a complete programming language was a breakthrough that provided complete freedom to developers, which helped them to create applications. The Ethereum process is similar to Bitcoin. Algorithms help in regulating Proof-of-Work (PoW) puzzles difficulty, which helps in maintaining a constant frequency of block creation in every 12–15 s. The miner is rewarded 5 ethers that has these advantages - 1) circulation of ethers without a central authority and 2) transactions are validated in a decentralized manner. Due to the sheer computing power the miner hacking is significantly reduced. However, the Ethereum is now using Proof-of-Stake consensus algorithm.

ML models are highly centralized. The large datasets required for analysis are generally proprietary. If the data is not updated by acquiring more data, then the published models will be out dated. Harris and Waggoner [8] have proposed a framework with smart contracts to host a continuously updated model. This model is on a Blockchain so that it can be free to use on interface. Both financial and non-financial (gamified) incentive structures for providing reliable data is proposed, in order to maintain the model's accuracy with respect to some test set proposed. They also identified advantages (public and decentralized ML model, good data for model updation, and transparency and trust) as well as some disadvantages (overwhelming the Blockchain network, third-party intervention and non-familiarity for the collaborative and decentralized AI) of their model.

Zoabi et al. [5] suggested that the efficient and quick diagnosis of COVID-19 can reduce the burden on healthcare system. This can be achieved by efficient screening using prediction models. This paper proposed a ML approach that was trained on records from 51,831 tested individuals using only eight binary features: sex, age $\geq$60 years, known contact and five preliminary conditions (cough, fever, soar throat, headache and cold). Israeli Ministry of Health published a nation-wide dataset and this was used to develop a model that detects COVID-19 cases by asking simple questions from users.

Ahamad et al. [9] developed a model using supervised ML algorithms that helped in identifying the features and predicting COVID-19 disease with high accuracy. Features examined included details of the individuals concerned (age, gender, observation of fever, history of travel, and clinical details such as the severity of cough and incidence of lung infection). Several ML algorithms were implemented on the collected data and the XGBoost algorithm [10] performed with highest accuracy (>85%). Fever (41%), cough (30.3%), lung infection (13.1%) and runny nose (8.43%) were few of the highly predictive statistical features. However, 54.4% of people did not develop any symptoms that could be used for diagnosis.

3 Machine Learning and Blockchain

Our methodology starts with the proposal of an architecture presented in Fig. 1. The proposed architecture is divided into three phases:

ML Model Phase where the Israel early symptom dataset [12] is explored using python libraries for training ML model. The SGDClassifier [6] is used as the incremental ML algorithm for the DApp. Simple control statements cannot be used for large datasets.

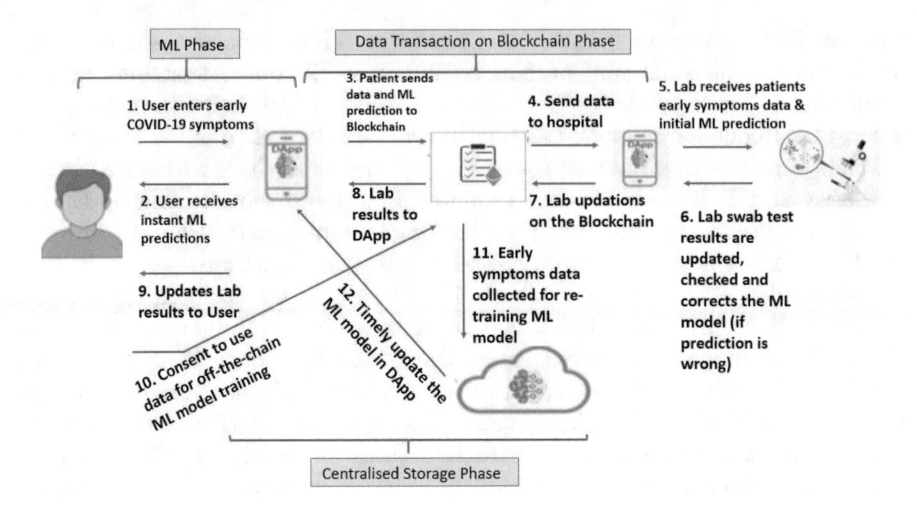

Fig. 1. Architecture of COVID-19 early symptoms prediction

Data Transaction on Blockchain, where the Decentralized Application (DApp) helps in the early symptom tracking. The DApp is developed using HTML, CSS and Web3Py. In this phase, the Ethereum Blockchain test network is used. The smart contracts are deployed to track the transfer between Ethereum accounts of the patients, labs and the centralized storage.

Centralized Storage Phase which, with the permission of the patient, the verified data is collected in the centralized storage for retraining ML model.

3.1 ML Model Phase

3.1.1 Data Exploration

The Israeli Ministry of Health released data of individuals who were tested for SARS-CoV-2 via Real-Time polymerase chain reaction (RT-PCR) using nasopharyngeal swab [12]. The dataset contains results of all the residents who were tested for COVID-19 nationwide. In addition to the test date and result there is a lot of information available, including clinical symptoms, gender and a binary indication as to whether the tested individual is aged 60 years or above. The dataset has over 1 million records from the period August 9th, 2020 through October 10th, 2020. The data is pre-processed by translating Hebrew to English for the exploratory data analysis and the description of the dataset is given in Table 1 and Table 2.

Table 1. Patient information

Sex (Male/Female)
Age ≥60 years (True/False)
Known contact with an individual confirmed To have COVID-19 (True/False)

Table 2. Symptoms

Fever (True/False)
Cough (True/False)
Sore throat (True/False)
Shortness of breath (True/False)
Headache (True/False)

The exploratory data analysis is performed first and is presented in the Fig. 2. Using various evaluation techniques the best ML model is then selected. The off-the-chain ML model training for the DApp and the data collected on the blockchain are used for retraining this model for accurate predictions as shown in steps 10–12 in Fig. 1.

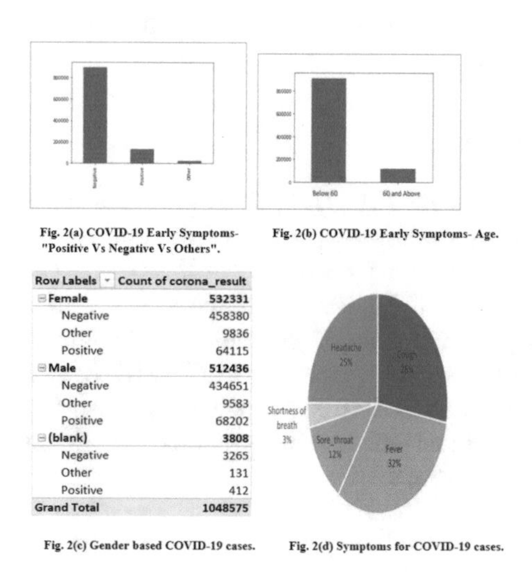

Fig. 2(a) COVID-19 Early Symptoms- "Positive Vs Negative Vs Others".

Fig. 2(b) COVID-19 Early Symptoms- Age.

Row Labels	Count of corona_result
Female	**532331**
Negative	458380
Other	9836
Positive	64115
Male	**512436**
Negative	434651
Other	9583
Positive	68202
(blank)	**3808**
Negative	3265
Other	131
Positive	412
Grand Total	**1048575**

Fig. 2(c) Gender based COVID-19 cases.

Fig. 2(d) Symptoms for COVID-19 cases.

Fig. 2. Exploratory data analysis

Figure 2(a) shows that in the dataset 80% of the cases reported were negative and fewer than 20% were positive and the rest were inconclusive. Figure 2(b) categorizes the COVID-19 patients' symptoms based on age. The analysis clearly shows that most of the people who reported COVID-19 symptoms were not older than 60 years. Figure 2(c) shows fever, cough and headache as the common symptom of all the COVID-19 symptoms reported in the majority of the cases. The data in Fig. 2(d) shows that male and female patients, who reported COVID-19 symptoms, were observed in equal proportions.

The "Known contact with an individual confirmed to have COVID-19 (True/False)" in Table 1 is not considered as data provided was insufficient. The understanding of the dataset is important for creating reliable ML model.

3.1.2 ML Model Selection

Predictions are made by the ML model deployed on the DApp, when a person enters the symptoms for preliminary testing of COVID-19. The ML model makes predictions on-the-fly and also gets retrained with data provided to it. Two incremental ML models considered are Perceptron model and SGDClassifier as the partial_fit() helps in retraining the model with new batch of data. The advantage of using such ML models is that it will keep a track on the evolving symptoms with different COVID-19 variants.

A. Perceptron Model

Perceptron is the most simple neural network model. This is a ML model, which is a binary or two-class classification. The hyperplane helps in learning for a decision boundary that separates two classes using a line in the feature space. This model will be appropriate for the problem where the classes (COVID-19 positive or negative) can be separated well by a line or linear model or linearly separable function [13].

B. SGDClassifier Model

Stochastic Gradient Descent (SGD) [6] is an optimization algorithm that minimize a cost function by finding the values of parameters or coefficients of functions. It is used for discriminative learning of linear classifiers under convex loss functions such as SVM and Logistic regression. The update to the coefficients is performed during each training instance, rather than at the end because of which it is used for incremental learning of COVID-19 symptoms. It is a classifier that implements a SGD learning routine supporting various loss functions and penalties for classification. Scikit-learn provides SGDClassifier module to implement SGD classification.

3.1.3 Evaluation

Accuracy is a ratio of correctly predicted observations to the total observations. A model is best when it has the high accuracy. Accuracy is a great measure but only when you have symmetric datasets where values of false positive and false negatives are almost same. Perceptron model is approximately 88% accurate where as SGDClassifier has 90% accuracy as evaluation results shown in Table 3 and 4. **Precision** is the ratio of predicted positive observations accurately to the total predicted positive observations. **Recall** is the ratio of accurately predicted positive observations to the all observations in actual class. **F1−Score** is the weighted average of Precision and Recall. In other words, $F1\text{-}Score = 2*(Recall*Precision)/(Recall+Precision)$. F1 is usually more useful than accuracy, especially if you have an uneven class distribution, i.e., if the cost of false positives and false negatives are very different. Further details of these metrics are available in [14].

Perceptron model has 94% F1-score where as SGDClassifier has 95% F1-score.

Table 3. Evaluation results for perceptron model

	Precision	Recall	F1-score	Support
Positive	0.88	1.00	0.94	268906
Negative	0.84	0.09	0.16	39802
Accuracy	–	–	0.88	308708
Macro average	0.86	0.54	0.55	308708
Weighted average	0.88	0.88	0.84	308708

Table 4. Evaluation results for SGDClassifier model

	Precision	Recall	F1-score	Support
Positive	0.91	0.99	0.95	268906
Negative	0.81	0.33	0.47	39802
Accuracy	–	–	0.90	308708
Macro average	0.86	0.66	0.71	308708
Weighted average	0.90	0.90	0.88	308708

Considering these metrics for ML models, Perceptron model showed 88% accuracy and F1-Score 94% where as SGDClassifier shows 90% accuracy and F1-score 95% as shown in Table 3 and 4. The SGDClassifer outperforms Perceptron model except for the Recall metric for positive COVID cases. Hence, SGDClassifier model is selected for the proposed architecture.

3.2 Data Transaction on Blockchain Phase

3.2.1 COVID-19 Symptoms Data Transaction and Predictions

The DApp shown in Fig. 1, consists of a questionnaire for the patient to enter the early symptoms data **D** in accordance with Tables 1 and 2. This data **D** is provided and the ML model deployed on the client machine for the initial predictions. The prediction **P** will help the patient with instant COVID-19 results. Next, the DApp provides the list of the nearest registered labs along with addresses in the network where the user can get a swab test. The early symptoms and ML prediction result is sent as a transaction **T(U, D, P)** on the Blockchain to the lab selected by the user address **U**. Steps 3 through 8 in Fig. 1 are done in Blockchain and important steps are described next.

3.2.2 Blockchain

The Blockchain used here is the Ethereum test network called Ganache. The smart contracts developed, using Solidity, save and send the patient's symptoms data as a transaction to the lab account **L**. After the lab performs the swab-tests and when the results for the patient are ready, the prediction **P** made by the ML algorithm are checked by the lab technician. If it is wrong, it is negated and the test results **Q** are updated otherwise the result is retained as **Q**. This correction is sent as a transaction **T(L, Q)** on the Blockchain and the result **Q** is also updated to the patient.

Additionally, the smart contracts are called by the centralized storage account **C**, then the aggregated data **F(U, L, D, P, Q)** consisting of the User and Lab addresses, patients symptoms data, ML model predictions and the verified ML results by the hospital respectively, is appended to the centralized dataset as a record. This record **F** helps in contributing reliable dataset for good ML model predictions in the future.

3.3 Centralized Storage Phase

The ML model in this architecture is trained off-chain in the centralized storage. More the amount of reliable data collected, the better will be the retraining of the model. The ML model is updated on the client side machines on day-to-day basis.

4 Ethereum Cost Details

The proposed system is developed on the Ethereum test network. Table 5 shows details related to the expected Ethereum gas spent when this proposed DApp is deployed on the Main network. Currently, 1 Gwei = 0.00000170 USD [15]. For the smart contract deployment on the Ethereum Main network, the gas spent is 2019063. Every time, when data is added either by a lab or by a patient, there is a gas cost for a transaction.

Table 5. Ethers spent description

Description	Gas spent	USD
Smart contract deployment	2019063	3.4255 USD
User symptoms data	217378	0.3688 USD
Hospital verification	70671	0.1199 USD

5 Conclusion and Future Work

The decentralized application (DApp) designed will help in the direct transfer of data to/from patient and lab. The predictions of machine learning used in DApp are constantly checked by the laboratory with the test results. This reliable data on Blockchain will in turn help retraining the ML model and provide more accurate predictions. Moreover, this reliable data can also be used instantly for the purpose of research and development by the government for prompt decision-making.

In future, the COVID-19 result report can be stored in the InterPlanetary File System (IPFS), which is a peer-to-peer network for storing and sharing data in a distributed file system. In this way, DApp can be extended to utilize the Deep Learning algorithms for numerous other applications, like COVID-19 predictions by Chest X-Ray Images [16]. Cough recordings can be transformed and inputted into a Convolutional Neural Network (CNN) [11] to detect COVID-19 symtoms through cough.

References

1. Huang, C., et al.: Clinical features of patients infected with 2019 novel coronavirus in Wuhan. China. The lancet. **395**(10223), 497–506 (2020). https://doi.org/10.1016/S0140-6736(20)30183-5
2. Khafaie, M., Rahim, F.: Article history: cross-country comparison of case fatality rates of COVID-19/SARS-COV-2. Osong Public Health Res. Perspect. **11**, 74–80 (2020). https://doi.org/10.24171/j.phrp.2020.11.2.03
3. COVID-19: How to fight disease outbreaks with data. https://www.weforum.org/agenda/2020/03/covid-19-how-to-fight-disease-outbreaks-with-data. (21-03-09)
4. Menni, C., et al.: Real-time tracking of self-reported symptoms to predict potential COVID-19. Nat Med. **26**(7), 1037–1040 (2020). https://doi.org/10.1038/s41591-020-0916-2. Epub 2020 May 11. PMID: 32393804; PMCID: PMC7751267
5. Zoabi, Y., Deri-Rozov, S., Shomron, N.: Machine learning-based prediction of COVID-19 diagnosis based on symptoms. NPJ Digit. Med. **4**(1), 3 (2021). https://doi.org/10.1038/s41746-020-00372-6. PMID: 33398013; PMCID: PMC7782717
6. Ruder, S.: An overview of gradient descent optimization algorithms. ArXiv:1609.04747 (2016)
7. Buterin, V.: A next-generation smart contract and decentralized application platform. White paper, 3(37) (2014)
8. Harris, J.D., Waggoner, B.: Decentralized and collaborative AI on blockchain. In: 2019 IEEE International Conference on Blockchain (Blockchain), pp. 368–375 (2019). https://doi.org/10.1109/Blockchain.2019.00057
9. Ahamad, M.M., et al.: A machine learning model to identify early stage symptoms of SARS-Cov-2 infected patients. Expert Syst. Appl. **160**, 113661 (2020). https://doi.org/10.1016/j.eswa.2020.113661
10. Chen, T., Guestrin, C.: XGBoost: a scalable tree boosting system, pp. 785–794 (2016). https://doi.org/10.1145/2939672.2939785
11. Laguarta, J., Hueto, F., Subirana, B.: COVID-19 artificial intelligence diagnosis using only cough recordings. IEEE Open J. Eng. Med. Biol. **1**, 275–281 (2020). https://doi.org/10.1109/OJEMB.2020.3026928
12. COVID-19-Isreal Govt. Data. https://data.gov.il/dataset/covid-19. (21-03-09)
13. Perceptron Model. https://machinelearningmastery.com/perceptron-algorithm-for-classification-in-python. (21-03-23)
14. Liu, Y., Zhou, Y., Wen, S., Tang, C.: A strategy on selecting performance metrics for classifier evaluation. Int. J. Mob. Comput. Multim. Commun. **6**, 20–35 (2014). https://doi.org/10.4018/IJMCMC.2014100102. Evaluation Metrics
15. Gwei to USD convertor. https://www.curvert.com/en/eth-usd/. (21-03-23)
16. Tabik, S., et al.: COVIDGR dataset and COVID-SDNet methodology for predicting COVID-19 based on chest X-ray images. IEEE J. Biomed. Health Inf. **24**, 3595–3605 (2020). https://doi.org/10.1109/JBHI.2020.3037127

A Smart Contract Architecture to Enhance the Industrial Symbiosis Process Between the Pulp and Paper Companies - A Case Study

Ricardo Gonçalves[1](✉), Inês Ferreira[2], Radu Godina[2], Pedro Pinto[1,3], and António Pinto[4,5]

[1] Instituto Politécnico de Viana do Castelo, Viana do Castelo, Portugal
rmiguelgoncalves@ipvc.pt, pedropinto@estg.ipvc.pt
[2] UNIDEMI – Department of Mechanical and Industrial Engineering, NOVA School of Science and Technology, NOVA University of Lisbon, Lisbon, Portugal
id.ferreira@campus.fct.unl.pt, r.godina@fct.unl.pt
[3] ISMAI and INESC TEC, Porto, Portugal
[4] CIICESI, ESTG, Politécnico do Porto, Porto, Portugal
apinto@inesctec.pt
[5] CRACS and INESC TEC, Porto, Portugal

Abstract. Pulp and Paper Companies collaborate to monitor and monetize waste and create value from their by-products. This process of Industrial Symbiosis requires the creation and maintenance of trusted and transparent relationships between all entities participating in these networks, which is a constant challenge. In this context, a blockchain-based system can help in establishing and maintaining these networks, serving as a ground truth between companies operating at a national or a global scale. This paper proposes a scalable and modular blockchain architecture design using smart contracts to enhance the industrial symbiosis process of the Pulp, Paper, and Cardboard Production Sector companies in Portugal. This design comprehends all entities participating in the network. The implementation of this design assumes the use of a permissioned ledger built using Hyperledger Fabric to provide the required trust and transparency between all entities.

Keywords: Blockchain · Smart contracts · Industrial symbiosis · Hyperledger fabric

1 Introduction

The term "symbiosis" refers to "a close connection between different types of organisms in which they live together and benefit from each other" [1]. In an industry context, an industrial symbiosis can be considered as a mutually beneficial relationship among companies to achieve a productive usage of by-products

J. Prieto et al. (Eds.): BLOCKCHAIN 2021, LNNS 320, pp. 252–260, 2022.
https://doi.org/10.1007/978-3-030-86162-9_25

and waste [2]. The industrial symbiosis has a collection of approaches to gain an advantage by involving physical exchanges of materials, through local and regional economies [3]. This happens due to the trade of by-products and utility sharing that occurs between industries, which include the reuse and commercialization of waste that can be used as secondary raw material [4].

An example of an industrial symbiosis process for Pulp, Paper and Cardboard industry companies in Portugal can be found in [5], where companies create the Fluidized Bed Sands (FBS) as a by-product and require normal sand for combustion. Figure 1 describes the network that is formed within this industrial symbiosis process, considering the different entities and their relationships. Five actors are depicted with direct or indirect flows: Pulp and Paper companies, Sand Producers, Mortar Producers, the organizations that use the resulting material mixture, and Environmental Portuguese Agency (EPA). Pulp and Paper companies use fluidized bed boilers for energy production. In fluidized bed boilers, a small amount of sand (regular sand) is used to maintain the energy production, this generates processed sand named FBS. The agency that represents every company in the pulp and paper industry is CELPA[1], the Portuguese Association of Pulp, Paper and Cardboard Producers. The mortar and sand industries have the potential to use this FBS in their industrial process. Finally, industrial manufacturers of fertilizer, building materials, and other industries, can use the sand mixture to manufacture their products. The dashed lines indicate that the EPA monitors the trades between each entity.

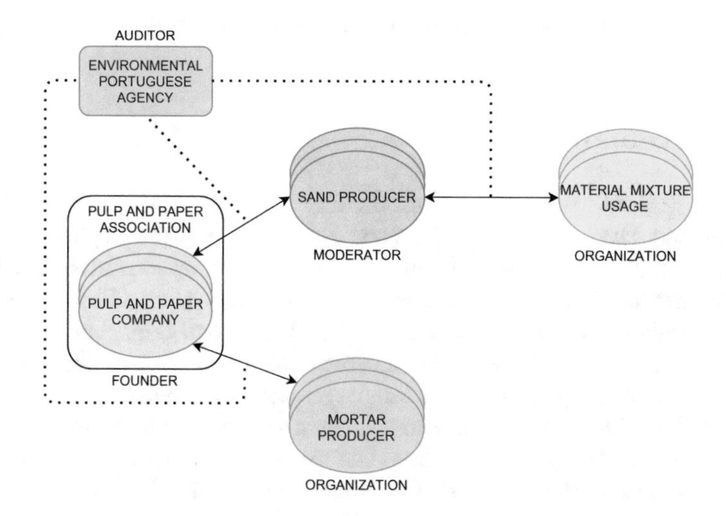

Fig. 1. Industrial symbiosis network for the FBS case study

For the industrial symbiosis network to succeed in this scenario it requires that all these companies create and maintain a relationship with each other; the

[1] http://www.celpa.pt/en/.

bigger the network, the hardest it is to maintain and create the relationships. Incoming companies must trust and should be trusted by the companies already in the network, and the relation and transactions between companies in the network should be clearly defined and auditable. The Blockchain technology can meet this requirement by design since it allows Smart Contracts to run automatically, executing all or parts of an agreement [6] and, at the same time, each transaction that is recorded on the blockchain cannot be changed once created, and can be verified or audited. The use of a cryptocurrency was not considered, mainly because most countries still do not legally recognize cryptocurrencies as a legal form of money.

This paper presents an architectural design of a blockchain network using Smart Contracts to enhance the industrial symbiosis network for the FBS of the Portuguese Pulp, Paper, and Cardboard industry. The proposal is based on Hyperledger Fabric and its chaincode as a basis for the Smart Contract development.

This proposal is intended to reduce the entry barrier of a new company that wants to join the system. In this case, there is no need to build a trust relationship with all the entities; the only requirement is to trust the blockchain network. The proposal also enables an auditor to manage and monitor the transactions, a role that the EPA can perform, monitoring all transactions.

The organization of this paper is as follows. In Sect. 2 the fundamentals of blockchain are described. In Sect. 3 the related work is presented. Section 4 describes the design, specification, and implementation of the proposed solution. Finally, in Sect. 5, all conclusions are drawn, including the identification of future work.

2 Background

Blockchain technologies comprising Smart Contract functionalities such as Ethereum and Hyperledger can be used to implement the depicted scenario. Ethereum was released in 2015 [7] and is a public and open-source, blockchain-based, decentralized software platform with the ability to run Smart Contracts. It also comprehends its cryptocurrency, Ether, the second-largest cryptocurrency currently in the market[8]. With the platform able to run Smart Contracts, written in Solidity or Go programming languages, developers can create their apps in a decentralized way. The ability to run Smart Contracts requires a fee, named "gas", which is what the network pays the miner for his work. The requirement of a fee for the execution of Smart Contracts avoids individuals from spamming the Ethereum network. Ethereum's consensus algorithm is the Proof-of-Work (PoW), this consensus requires the use of computer power by miners to validate the network's new block.

The Hyperledger project [9], founded in 2015 by the Linux Foundation is a suite of open-source frameworks, tools, and libraries to implement and create blockchain networks. One of the projects within Hyperledger is Fabric, which is software that allows a business to create their customized private decentralized ledgers. Unlike other blockchain-based technology, Fabric does not have a built-in cryptocurrency, although developers can implement it using Chaincode, the name given to the Smart Contracts of this framework.

Table 1 compares both solutions [10]. Ethereum is permissionless, offers a built-in currency, and its transactions are transparent and open to everyone. While on the other hand, Hyperledger Fabric is permissioned and does not have any cryptocurrency built-in, as stated in [11].

Table 1. Differences between hyperledger fabric and ethereum

	Hyperledger fabric	Ethereum
Purpose	Business to business	Business to contract and generalized applications
Consensus	Pluggable consensus algorithm	Proof of work
Access	Permissioned	Permissionless
Transaction visibility	Confidential	Transparent
Native currency	No	Yes, Ether
Programming languages	C++, C#, Go, Haskell, Java, JavaScript, Python, Ruby, Rust, Elixir, Erlang	Solidity, Go

Hyperledger Fabric is composed of three components channels, peers, and orderers. Channels are used to allow for privacy in the network and there are as many channels as needed. An entity can only view the transactions of the channels they take part in [12]. For example, assuming a system with three entities (A, B, and C), and that two channels were set up (C1 and C2) as follows: channel C1 connecting entities A and B; channel C2 connecting entities B and C. The transactions made by entity A are not visible to entity C, and consequently, transactions from C are not visible to A. These transactions are stored on blocks inside a peer ledger. As presented in Fig. 2, there are two types of peers, commitment peers, that only store the ledger, and endorsing peers, that also run chaincode. The Orderer triggers transactions and creates blocks containing such transactions. These blocks are then sent to the peer's ledgers where they are stored. The Orderer generates a new block when the number of transactions, or the amount of storage space it uses, reaches the limit set by configuration.

The Hyperledger Fabric is permissioned which means that a joining new user must have a valid certificate to access the system. The Certificate Authority (CA) have the responsibility of generating these certificates with the correct access rights. The Admin of each entity is accountable for ensuring that the users have network access. Chaincode is a program that runs on endorsing peers inside

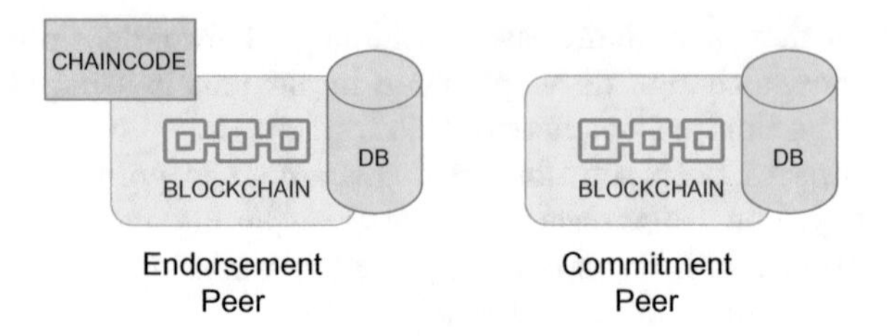

Fig. 2. Endorsement and commitment peer

docker containers. They are used to manage and manipulate the ledger. Chaincode permits two types of requests: queries and invocations. Queries are used to retrieve data, and invocations are used to modify the state of the ledger. Go, NodeJS and Java are examples of the currently available languages to program chaincode.

Comparing both Hyperledger Fabric and Ethereum one can conclude that, for the envisioned scenario, Hyperledger Fabric is the adequate solution. The required use of a cryptocurrency by Ethereum, and its associated costs of Smart Contract execution are clear drawbacks. These add complexity and introduce operational cost volatility due to the fluctuation of Ether's price. Another issue for the adoption of Ethereum relies on its permissionless nature; in the envisioned scenario only selected entities should be able to create transactions or alter the state of the ledger.

3 Related Work

Alexandris et al., in [13], proposed a blockchain-based mechanism as the basis for a collaborative circular economy business model that consists of assets transitioning between operators. The proposed mechanism enables assets monitoring by the involved entities, but also it's auditing by third parties such as regulators of the state. They concluded that there are benefits from the adoption of blockchain in regulated environmental jurisdictions.

Kouhizadeh et al., in [14], identified both blockchain and circular economy as the two new concepts that can positively impact the way we live in the future. In their work, they survey companies' cases and assess how blockchain will promote advances in the circular economy by linking blockchain to the multiple dimensions of the ReSOLVE model (regenerate, share, optimize, loop, virtualize, and exchange). In a previous work [15], the same authors focused on the particular problem of product deletion in the same circular economy context while using blockchain as a supporting technology.

The PlasticCoin cryptocurrency [16], proposed within the PlasticTwist H2020 research project, was developed to foster plastic reuse by citizens, promoting the circular economy while maintaining trust among its users. They make use of the Hyperlegder Fabric and developed a token following the concept described in the ERC-20 [17] standard of the Ethereum platform. These tokens are then printed and handed out, as scratchpads, to deserving people so that they can later claim their coins for reusing plastic materials.

From the research work surveyed, one can conclude that the circular economy in general, and industrial symbiosisin particular, can benefit from the adoption of blockchain and related concepts, such as cryptocurrencies and Smart Contracts. Moreover, the use of Smart Contracts or cryptocurrencies will be dependent on each specific case; while smart Contracts can be used for overall system automation, the cryptocurrencies can be used to attract citizens and foster their involvement.

4 The Proposed Smart Contracts Architecture

The proposed architecture assumes the following five types of roles: Founder, Moderator, Monitor, Organization Admin, and Organization User.

The Founder is the one responsible for the overall system management and comprises tasks such as the creation and deletion of organizations or moderators. The Moderator is responsible for creating new products and units as required. The Monitor can only read data from the network, for instance, accessing all transactions performed on the blockchain. Inside each organization, we can have regular users and administrators. The Organization Admin is the one that has the authority to manage new access certificates, thus creating or deleting users for its organization. The Organization User is the one that can manage both transactions and orders for its organization.

Based on the defined roles and their interactions, the proposed architecture design is presented in Fig. 3.

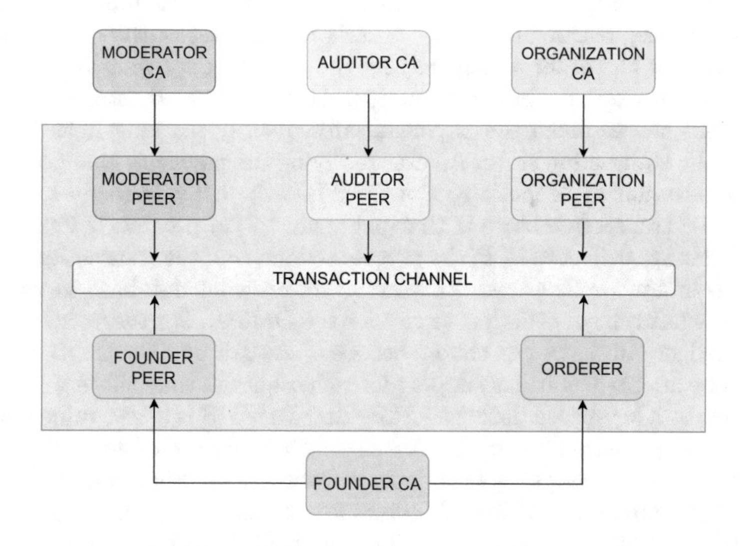

Fig. 3. Proposed architecture

The architecture assumes an implementation on a blockchain network built with Hyperledger Fabric and comprehends the following 4 peers: the founder peer (which is the network founder and maintainer), the moderator peer (that has the role to moderate the network), the auditor peer (who audits all the transactions and relations within the network), and the organizations' peer (a peer per any other company type in the network). Furthermore, the founder entity will also have an extra component, the orderer node.

The architecture also comprehends one CA for each entity, which it will provide their users with credentials. The last component in the design is the transaction channel, it comprises only one channel to convey transactions between all entities because there was no need for companies to keep their transactions confidential, or hidden from the other entities. Using only one channel means that all transactions are available to all entities, but also facilitates the existence of a monitoring agency, one of the identified requirements. Another identified requirement is that every company, except the monitoring agency, needs to create trade offers for their by-products. These offers then need to be disseminated within the network so that other entities will be aware of them, thus being able to buy the related by-product and registering a new transaction in the blockchain.

To access the system, users need a certificate proving their identity. Organization Admins generate these certificates for users inside each organization. The Founder entity generates a new certificate for each Organization Admin when it joins the network. Scaling the platform can be done easily by adding a new organization CA and peer and connect it to the transaction channel.

The steps to be performed by two entities A (the seller) and B (the buyer) are as follows. Firstly, both entities (A and B) must be registered and receive their corresponding certificates, which need to include the correct permissions. A can then create an order by identifying the product, its unit, quantity, and price. After this, B can browse the order's list and select the order issued by A. By selecting the order and specifying the amount they want to receive, will trigger the creation of a new transaction. At this point, B receives an invoice from A and has to make the payment, subsequently marking it as "paid" by changing the transaction state. A, after receiving the payment, changes the state of the transaction to "deposit received". When the cargo is ready to start the delivery, the transaction state is changed again to "in progress". Finally, when the products are delivered to B, the process is done, and the transaction is closed.

In this setup, we have two decisions to make, what database to use on the peers and what type of ordering service for the Orderer. For the peer's database, in Hyperledger Fabric we can choose between LevelDB and CouchDB. CouchDB was the one adopted as it has support for richer queries than those gets and sets of LevelDB's, a key-based database. Note that LevelDB is a key, value database. Regarding the ordering service, Raft was selected because of being the one recommended in Hyperledger Fabric documentation. The other two options available are Solo and Kafka. Solo only allows for the use of a single Orderer. Kafka is a distributed platform with multiples functionalities other than data storage, which would increase the overall system complexity, hindering its manageability.

For access control and authorization, the proposed solution will use the Hyperledger Fabric Attribute-Based Access Control. This method depends on attributes that are added to the certificates of the users when these certificates are created. These attributes are created by concatenating data type and the type of access to be authorized, separated by a dot. Taking the example of product creation, only users with the attribute "products.create" will be allowed to create new products in the proposed solution.

5 Conclusions and Future Work

This paper proposes a blockchain architecture design to enhance the industrial symbiosis process of the Pulp, Paper, and Cardboard Production Sector Companies in Portugal, providing the required trust and transparency to the network in order to enhance their industrial symbiosis process. The proposed architecture uses a permissioned ledger built using Hyperledger Fabric. Chaincode and Attribute-Based Access Control of Hyperledger Fabric are the building blocks used to implement the required logic that allows users to create orders and transactions between each other and exchange their by-products.

Using a blockchain-based network was found to have the following advantages facing the traditional systems. First, it enables fast transactions and quick integration of a new organization into the network. Then, the fact that every transaction is recorded in the blockchain makes it easier to audit. Finally, the blockchain-based system makes it easier to scale as it just requires starting up one peer and one CA node.

Despite already having advantages, the design can have some improvements to make it even better, beginning with a simple task like a front-end application to prevent users from using the Hyperledger Fabric's command-line interface. The Industrial symbiosis field is trying to help the world by preventing the waste of by-products and make industries greener. This type of blockchain-based architecture in Industrial Symbiosis systems can help the field reach a global scale, making the world more sustainable.

Acknowledgements. This work is partially financed by European Regional Development Fund (ERDF) through the COMPETE2020 Programme, within the STVgoDigital project (POCI-01-0247-FEDER-046086), and partially financed by National Funds through the Portuguese funding agency, FCT - Fundação para a Ciência e a Tecnologia, within project UIDB/50014/2020.

References

1. Smith, D.C., Douglas, A.E., et al.: The Biology of Symbiosis. Edward Arnold (Publishers) Ltd. (1987)
2. Chopra, S.S., Khanna, V.: Understanding resilience in industrial symbiosis networks: insights from network analysis. J. Environ. Manage. **141**, 86–94 (2014)
3. Chertow, M.R.: Industrial symbiosis: literature and taxonomy. Ann. Rev. Energy Environ. **25**(1), 313–337 (2000)

4. Jiao, W., Boons, F.: Toward a research agenda for policy intervention and facilitation to enhance industrial symbiosis based on a comprehensive literature review. J. Clean. Prod. **67**, 14–25 (2014)
5. Ferreira, I.A., Barreiros, M.S., Carvalho, H.: The industrial symbiosis network of the biomass fluidized bed boiler sand-mapping its value network. Resour. Conserv. Recycl. **149**, 595–604 (2019)
6. Mohanta, B.K., Panda, S.S., Jena, D.: An overview of smart contract and use cases in blockchain technology. In: 2018 9th International Conference on Computing, Communication and Networking Technologies (ICCCNT), pp. 1–4. IEEE (2018)
7. Torres, C.F., Steichen, M., et al.: The art of the scam: demystifying honeypots in ethereum smart contracts. In: 28th {USENIX} Security Symposium ({USENIX} Security 19), pp. 1591–1607 (2019)
8. Katsiampa, P.: Volatility co-movement between bitcoin and ether. Financ. Res. Lett. **30**, 221–227 (2019)
9. Dhillon, V., Metcalf, D., Hooper, M.: The hyperledger project. In: Blockchain Enabled Applications, pp. 139–149. Springer (2017)
10. Valenta, M., Sandner, P.: Comparison of ethereum, hyperledger fabric and corda. Frankfurt School Blockchain Center, vol. 8 (2017)
11. Androulaki, E., et al.: Hyperledger fabric: a distributed operating system for permissioned blockchains. In: Proceedings of the Thirteenth EuroSys Conference, pp. 1–15 (2018)
12. Benhamouda, F., Halevi, S., Halevi, T.: Supporting private data on hyperledger fabric with secure multiparty computation. IBM J. Res. Dev. **63**(2/3), 3–1 (2019)
13. Alexandris, G., Katos, V., Alexaki, S., Hatzivasilis, G.: Blockchains as enablers for auditing cooperative circular economy networks. In: 2018 IEEE 23rd International Workshop on Computer Aided Modeling and Design of Communication Links and Networks (CAMAD), pp. 1–7 (2018)
14. Kouhizadeh, M., Zhu, Q., Sarkis, J.: Blockchain and the circular economy: potential tensions and critical reflections from practice. Prod. Plann. Control **31**(11–12), 950–966 (2020)
15. Kouhizadeh, M., Sarkis, J., Zhu, Q.: At the nexus of blockchain technology, the circular economy, and product deletion. Appl. Sci. **9**(8), 1712 (2019)
16. Koscina, M., Lombard-Platet, M., Cluchet, P.: Plasticcoin: an ERC20 implementation on hyperledger fabric for circular economy and plastic reuse. In: IEEE/WIC/ACM International Conference on Web Intelligence-Companion Volume, pp. 223–230 (2019)
17. Fabian Vogelsteller, V.B.: EIP-20: ERC-20 token standard. Ethereum Improvement Proposals, vol. 20 (2015)

Establish Trust for Sharing Data for Smart Territories Thanks to Consents Notarized by Blockchain

Mongetro Goint(✉), Cyrille Bertelle, and Claude Duvallet

Normandie Univ, UNILEHAVRE, UNIROUEN, INSA Rouen, LITIS, 76600 Le Havre, France
{mongetro.goint,cyrille.bertelle,claude.duvallet}@univ-lehavre.fr

Abstract. Many territories are implementing development strategies strongly rooted in ambitions in terms of digital innovation. These strategies aim not only regional planning, but also economic development and people's quality of life. This refers to the concept of smart territories. To achieve their objectives, smart territories generally use citizens' data to offer them services and support them in better individualized decision-making. However, a major problem arises in data management, as it is necessary to have the consent of users to access and use data.

In this article, we propose a generic blockchain-based model for consent-based data sharing in the context of smart territories, while putting users at the center of control of their data. We take advantage of ADA-M (Automatable Discovery and Access Matrix), a matrix that provides a standardized way to unambiguously represent conditions related to discovery and access to data. We couple the use of ADA-M with a blockchain anchor system, which helps secure consent management for data access. This helps to develop an environment of digital trust for sharing data between users. Then, through a digital token, we offer a system of rewards for users in cases where their data is used for commercial purposes.

1 Introduction

The concept of smart territories has been developing for several decades. More and more territories are setting up strategies, in particular to manage in an integrated way the services offered to citizens, territory planning, economic development and quality of life by promoting the sensitivity of citizens and businesses to the environmental aspects. All these are essential for any development thought in terms of sustainability. One of the most popular variations of smart territories is that of the smart city. A smart city is characterized by its ability to connect physical infrastructure, Information Technologies (IT) infrastructure, social infrastructure and business infrastructure in order to take advantage of the collective intelligence of the city (Harrison and al. 2010). However, development of smart cities can be effective only if critical privacy and security issues are

J. Prieto et al. (Eds.): BLOCKCHAIN 2021, LNNS 320, pp. 261–271, 2022.
https://doi.org/10.1007/978-3-030-86162-9_26

resolved (O'Grady and O'Hare 2012). (Elmaghraby 2013) considers security and confidentiality as two main challenges that require special attention in smart territories. Security and confidentiality aspects are related according to the author, because lack of security can lead to invasion of privacy. These two strategic aspects are to be considered in particular in the implementation of data platforms for smart territories, as these platforms are essential for data integration. Otherwise, traditional data platforms are known to suffer numerous inefficiency problems (security, confidentiality, transparency, etc.) (Ball 2013, Goel 2014). They are therefore not typical models capable of responding to the security and confidentiality issues raised by Elmaghraby. They are generally based on trusted third parties. So, "digital trust" becomes a major stake.

Digital trust can be seen as inssurance placed by data owners in an actor empowered to manage their digital data. This means that data owners feel secure with their data, by securely controlling their distribution. Their consent is required to access this data. Especially since the problem of consent for access to data is legally recognized (GDPR 2016).

Blockchain has been proposed as an effective solution for consents management in several research studies, and it has been also suggested to build trust in smart cities (Kundo and Kundo 2019). In this paper, we present a generic blockchain-based data platform model, to make users adopt digital trust for data sharing in smart territories, while the blockchain secures the consent between different users of the platform for data access.

The rest of this article is organized as follows: in Sect. 2, we present blockchain technology and how it works. Also, the concept of smart contract is presented. The works related to our work are presented in Sect. 3. Then, in Sect. 4 we describe the architecture of the proposed model. Finally, Sect. 5 concludes the article and presents the perspectives and future works.

2 Blockchain Technology

Blockchain was first proposed in 2008 by Satoshi Nakamoto with Bitcoin (Nakamoto 2008). A blockchain can be thought of as a distributed and secure ledger to which all participants have access. The public register shared between its participants contains all the transactions that have been carried out there since its creation, while participants keep a copy of the ledger. Secure, verified by the entire network using cryptographic protocols and consensus algorithms, transactions in a blockchain are put into blocks and then arranged chronologically to form a chain.

Currently, blockchain technology is going well beyond its original use, cryptocurrencies. It allows data sharing in a decentralized and secure manner. This new approach was initiated with the Ethereum Blockchain (Buterin 2015), thanks to smart contract (Ethereum 2020a). A smart contract refers to a program that runs on blockchain when the clauses that have been defined there are met. Ethereum promotes development of decentralized applications (DApps). With this new approach, blockchain has been used for years in many fields such as finance (J.P.Morgan 2019), insurance (AXA 2017), health (Mamo et al. 2019), Internet of Things (Rantos et al. 2019), ect.

3 Related Work

Several works have already been carried out in several areas using blockchain for consent management for data access. In current section, we present some works deemed relevant in the context of our work. Also, we present ADA-M, an automatic discovery and access matrix that allows the production of structured metadata profiles according to regulatory conditions, in order to facilitate their sharing.

3.1 Existing Works on Consent Management with Blockchain

In several research works, blockchain has been proposed for the management of consents for data access.

Blockchain has been proposed by (Aldred et al. 2019) for consent management for data access. Further, Nicholas (Mano et al. 2019) have proposed an approach for automatic management of consents using the blockchain for biomedical data. In their work (Jaiman and Urovi 2020), authors have proposed a blockchain-based dynamic consent model. In addition, (Agarwa et al. 2020) proposed an approach with Blockchain in Consentio. In another work (Rantos et al. 2019), the ADvoCATE platform, based on the Ethereum blockchain has been proposed. Otherwise, few works focus on management of consents for data access in smart territories. Very recently, in their works (Makhdoom et al. 2020), authors proposed PrivySharing. In this work (Michelin et al. 2018), the authors proposed a blockchain-based model (SpeedyChain) for smart cities. In another work (Biswas and Muthukkumarasamy 2016), authors have proposed a security framework that integrates blockchain technology for smart devices.

3.2 Automatable Discovery and Access Matrix (ADA-M)

ADA-M has been proposed by (Woolle et al. 2018) as a matrix providing a standardized way, to unambiguously represent conditions associated with discovery and access to data in healthcare.

The matrix is made up of 4 sections: **Header section** (to describe what data is); **Profile section** (to define who can request access to the data); **Conditions section** (for the description of the terms or conditions that a data requester must comply with in relation to the use of the data); **Meta-Conditions section** (to describe aspects of the data that have to do with how the matrix is to be used and other special conditions).

With ADA-M, data provider can participate in data sharing and collaboration, making the meta-information about their data machine-readable and therefore directly available for digital communication and research activities.

4 Contribution

In current section, we will present our solution: a generic blockchain-based platform model for sharing data based on user consent, in order to establish digital trust in a context of smart territories. In smart territories, there can be a diversity of services, which digitize more and more data, therefore which require trust between actors for their sharing in order to create synergy within different services of the territory. So, we can imagine several possible scenarios with several data providers such as: transport services; actors in the health field and so on. An explicit scenario can be that of researchers in the field of health, who will provide data to the platform to be shared between them subsequently. So, our contribution aims, on the one hand, to propose a consent management model that takes into account different use cases of smart territories; On the other hand, establish digital trust between users of the platform for data sharing. A remarkable aspect of our contribution is the fact that we use the ADA-M matrix, originally intended for health data, to adapt it in our context for data use of policy statement and data use of purpose statement. Coupling ADA-M with blockchain technology allows good consent management. In addition, through a digital token system, the model also allow user data owner to be rewarded from data accessors, in case their data is used for commercial purposes.

First of all, we would like to present, on the one hand, the various assumptions we have made regarding data management in a context of smart territories; so we present what is meant by "consent" in our case.

4.1 Hypotheses

Hypothesis 1: We can assume that our platform will receive several types of data: personal data collected via mobile applications or from IoT devices (data from electrical systems for example). In addition, we can receive data via a web portal (data provided by health institutions, research entities or any other public/private administrations), etc.

Hypothesis 2: The data supplied by the data providers could be used for non-profit purposes (such as for research), or even for profit purposes (by commercial companies for example). In the second case, data owners may demand payment from the access requester for their data to be used.

Hypothesis 3: The platform will be able to accommodate different categories of users: data providers, data access requesters and role assigners.

Hypothesis 4: Data access seekers may search a dataset in order to obtain the consent of its owner to access it, also a data owner may search for a potential data accessor (such as a research institution) for give him his consent for access to his data.

4.2 Consent

Indeed, consent refers to the agreement between two users to share data with each other. Consent implies which user can access which set of data of another user, and for how long. In our context, a consent consists of: Id_UDO + Id_UDOA + Id_Policy of use + Id_Declaration of objective + Data pointer + Expiry date + Consent State. Details of each attribute of consent are shown in Table 1.

Table 1. Representation of the attributes of a consent

Id_DOU	Data owner identifier
Id_DOAU	Identifier of the data owner and accessor user
Id_Policy of use	Identifier of the use policy linked to data
Id_Declaration of objective	Identifier of the purpose of use linked to the data
Data pointer	Unique pointer to identify data
Expiry date	Expiration date of consent
Consent State	Boolean which indicates if the consent has been revoked or not

4.3 Our Model

As shown in Fig. 1, our model uses several storage spaces: firstly, the blockchain (**BC**) to manage user identities, data use policy, data use purpose statement, and consent for data access. While usage policy is for data providers to set conditions that access requesters must meet, usage purpose itself is for requesters to state why and how they will use the data.

A question to ask is: what is the use of blockchain in all of this? Indeed, given its characteristics, blockchain comes in this case as a certified and tamper-proof device to secure data transactions. For example, a consent anchored in the blockchain will be immutable. It should be noted that as part of our work, we have chosen the Ethereum blockchain. Our choice is based on the fact that Ethereum is a completely public blockchain, and also one of the most tested and used blockchains today; especially since it allows the development of smart contracts. In addition, it has a set of easy-to-use tools that will be useful to us as we implement our solution. It also allows the development of digital tokens, which will be used to reward users as part of our solution. So, the Ethereum blockchain as a whole appears complete for the theoretical proposition of our model. However, there is a concern: currently, Ethereum allows to manage a limited number of transactions per second; on the other hand, the price of transactions on the public network remains very fluctuating. This can therefore have an impact in a context of smart territories with a diversity of services where we should manage thousands of transactions. So, an alternative for should be the use of polygon[1].

[1] https://polygon.technology/.

It is a scaling solution for Ethereum dApps, developed to address major issues, such as slowness and high transaction fees, without compromising on security.

Otherwise, concerning storage spaces, on the other hand, we have a Storage Service Outside the Blockchain (SSOB)(**SSOB**), made up of multiple databases, where all other data will be stored. Then, we have the users: Data Owner User (**DOU**), Data Owner and Accessor Users (**DOAU**) and Role Assignment Users (**RAU**).

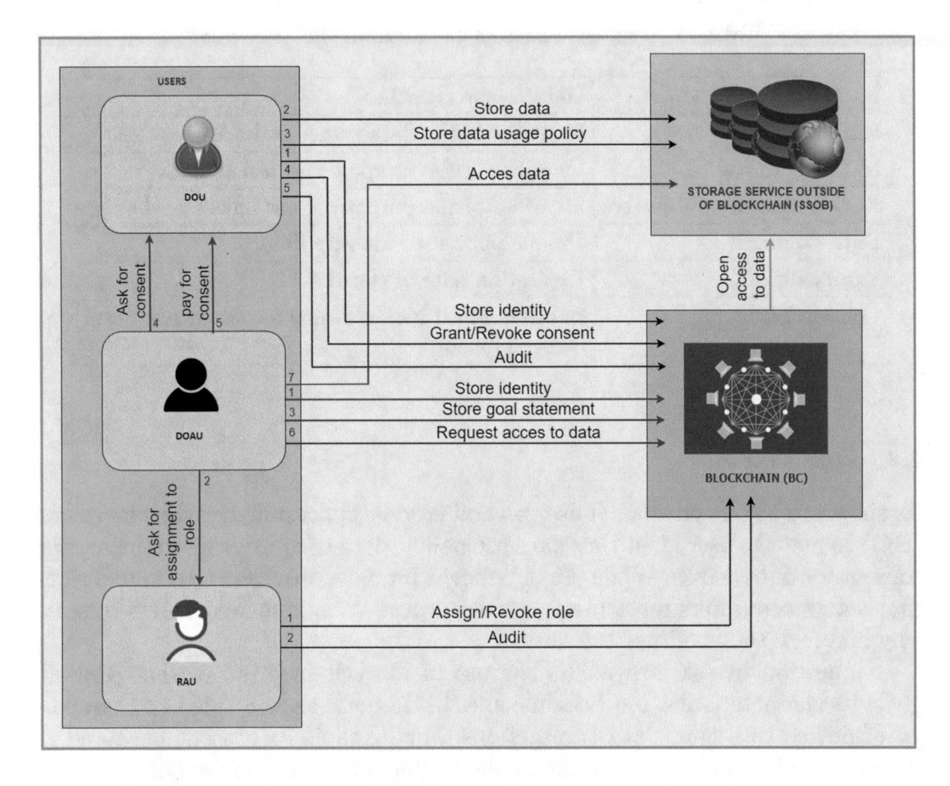

Fig. 1. Generic architecture of the consent management platform for smart territories

To begin with, we asked ourselves several important questions: "How will users get to the platform?" How will the data be recorded? and How will consent be established?". Indeed, we consider two different scenarios: a first scenario for sharing data between two users for biomedical research, and a second scenario for sharing transport data (commercial data) between two users.

4.3.1 Scenario 1

For the first scenario, each user registers on the platform and an identity (including their public key) is anchored in the blockchain. Blockchain technology uses the public key/private key system, using a digital wallet, to guarantee the identity of the sender of information and to ensure that the information is legitimate.

Once registered, the data provider will be able to choose what type of data he want to register (health data in this case), then send them to the platform which routes them to the respective database in SSOB as shown in Fig. 1. An access right (data fingerprint + user ID) is also registered in BC.

We start from the logic that each set of data will be linked to a usage policy declared by the user providing the data. To do this, we take advantage of the ADA-M matrix (J. Woolley et al. 2018). We adapt the ADA-M matrix representation to our context to allow DOU to define their data use policy for different use cases. Thus, during data recording, DOU is required to complete a usage policy declaration form which will be linked to his data and which will be recorded in SSOB (see Fig. 1). DOU will be able to choose from a set of proposals in order to complete the different sections of the form. In particular, a data pointer is included in the usage policy, which will allow you to know to which data set each usage policy is linked. Also, the data provider will indicate what type of data is it, who will be able to access it and under what conditions. It should be noted that, for the collection of data from IoT devices by the system, we can use the proposed approach by (Rantos et al. 2019) to require that every device be registered on the system before they can send data. The control of this data can then be ensured by the owner of the device via a mobile application.

On the other hand, once registered on the platform, the user requesting access to the data must be assigned a role. A role will determine which category of data can access a DOAU. Role grouping prevents any user from requesting access to any type of data. The assignment of users to roles is done by watchdogs, much as proposed in Consentio (Agarwal et al. 2020). In this case, we need another type of user: Role Assignment Users (RAU). An RAU can be, for example, the ministry of health (for the assignment of roles to researchers in the field of health). RAU will be a special user category, with some administrative right to be able to assign roles. However, it remains to be determined by what procedure a RAU will assign a role to a DOAU. This can be done by verifying an identity document, for example. Thus, once DOAU is assigned to a role, he will be able to register its purpose for using data in the blockchain, using a form based on the ADA-M matrix structure. In his statement of purpose, he may specify his status (medical researcher for example), the type of data to be sought, the purpose and the conditions of use of the data.

- *Assigning Consent for Scenario 1*

For granting/revocation of consent, two cases may arise:

1- A DOU decides to give its consent to a potential DOAU (a research entity for example), without request from the DOAU. This approach aims to allow users who provide data to advance research, for example.
2- A DOAU wants to search for data to use. Therefore, he will have to seek the consent of the DUO.

Considering the first case: DOU chooses a usage policy, then the system will search for all usage objectives correlated to it. DOU may subsequently choose to grant consent for access to a specific set of data. The consent (see the consent section described above) is therefore recorded in the blockchain using a smart contract. DOU may subsequently modify or revoke this consent.

Considering the second case: DOAU search in SSOB for data use policies corresponding to its declared purpose of use. They can then select a usage policy, then send a consent request to the owner of the data to which the usage policy is linked. The reference of the purpose of use in the blockchain is also sent to DOU when requesting consent so that he can verify it. Then, DOU can accept or reject the consent request. In the case of acceptance, the consent will be recorded in the blockchain immutable (see Fig. 1). Otherwise, DOAU will receive a notification that its request has been rejected.

4.3.2 Scenario 2

This scenario concerns the sharing of transportation data (business data) between two users. As for the first scenario, the user registration procedures, data registration, declaration of use policy remain the same. However, it can be assumed that this data is collected via a smart phone. These are data concerning the travel route of users, the means of transport used, etc.

In this case, as they are commercial data, it must be mentioned in the "Header" of the use policy by DOU. Also, in the "Profile" section of the usage policy, designed according to ADA-M structure, the data provider must specify what type of company will be able to access the data. And in the condition section, he has to define the conditions, such as the reward he wants to get for using his data. This reward will be made using a system of digital tokens that will be issued on the blockchain. The Ethereum blockchain makes it possible to issue tokens for such use cases (Ethereum 2020b). We assume that these tokens can be used as an asset on the platform and will rise in value to be able to be traded on exchange platforms. On the other hand, the RAU may in this case be the ministry of local authorities which will assign DOAU to a role. Once assigned to the role, DOAU can make its statement of purpose which will therefore be inked into BC, and then sent a consent request to the data owner.

- *Assigning Consent for Scenario 2*

When DOU will establish a consent to open access to its data to DOAU, the system will check the availability of funds (tokens) in the account (digital wallet) of DOAU in the blockchain, according to the amount specified by DOU. If the funds are available, consent is established, recorded in the blockchain and DOU is rewarded. Otherwise, the consent is rejected and each of the two parties will receive a notification indicating the failure to set up the consent.

Moreover, if DOU decides to update A consent, a completely new version of it will be created, because the blockchain being an immutable ledger does not allow the modification of data. This is, among other things, what makes the use of blockchain for consent management so strong: its immutability. We always have an authentic record of the consents that have been defined between users. As a result, we will search each time for the latest version for an established consent. On the other hand, if a DOU revokes access right to a DOAU on a dataset, the consent state will change to false, which will make the consent invalid (even if the expiry of the consent had not yet arrived). If there has been compensation for the consent, part of the funds received must be returned to the DOAU account. To calculate the value to return to DOAU, we can apply the formula: $\forall A_{rec}, A_{ret} = A_{rec} * D_{el}/D_{exp}$, where:

A_{rec}: Amount received by DOU
A_{ret}: Amount to be returned to DOAU
D_{el}: Elapsed Duration of the consent
D_{exp}: Expected Duration of the consent

It is good to note that the DOU can audit the blockchain to have control over all DOAU that have had access or/and still have access to its data.

- *Data access by DOAU*

Once connected to the platform, a DOAU can request access to a set of data belonging to a DOU. A smart contract will therefore verify in the blockchain whether there is consent allowing this user to access this data. The smart contract always checks the latest version of the consent for the user on a set of data, the expiration date as well as the status of the consent. If there is valid consent, the blockchain opens access to the data to the accessor in SSOB. Otherwise, access is denied and DOAU can send a consent request to DOU.

- *Data modification by DOU*

The possibility for a user of an information system to modify/rectify or even delete his data, is part of the strategies aimed to ensuring user confidentiality (Jaap-Henk 2014). It is also part of the principles of the RGPD (RGPD 2016). To comply with these principles, our platform will allow a user to modify or even request the deletion of his data. Indeed, for any modification or deletion of data from a DOU, a notification is sent to the DOAUs concerned by this set of data. In case of deletion, all concerning DOAU are therefore invited to delete the data on their side.

5 Conclusion and Future Work

Managing consent for data access remains a crucial aspect of ensuring the security and privacy of user data in data platforms. This helps to build digital trust between users for sharing their data. In this paper, we have presented a generic platform model based on blockchain technology to manage consent in a smart territories. We presented the architecture of our platform model and how it works. The blockchain allows us to make secure management of consent without having to resort to a trusted third party.

In future works, we will have to present the implementation part of our platform model. We will implement a proof of concept of the model to show its validity, and also to analyze its performance. Moreover, one of the perspectives for future work is to think about how we can secure the data stored in the database outside the blockchain, even if this is not part of the main objective of our work. Besides, an interesting idea is to create a better synergy between the different actors of the platform, would be to implement a smart contract for the notification of addition of new data. Being connected to the data use policy storage database, the smart contract could therefore notify potential DOAU whose data use purpose statement is correlated to a new use policy added by a DOU.

Acknowledgements. We would like to thank Le Havre Seine Métropole (LHSM), which supported this work as part of Mongetro Goint's doctoral thesis.

References

Agarwal, R.R., Kumar, D., Golab, L., Keshav, S.: Consentio: managing consent to data access using permissioned blockchains. In: 2020 IEEE International Conference on Blockchain and Cryptocurrency (ICBC), pp. 1–9 (2020)

Aldred, N., Baal, L., Broda, G., Trumble, S., Mahmoud, Q.H.: Design and implementation of a blockchain-based consent management system (2019). journals/corr/abs-1912-09882

AXA: AXA se lance sur la Blockchain avec fizzy (2017). https://www.axa.com/fr/magazine/axa-se-lance-sur-la-blockchain-avec-fizzy. Accessed 25 Jan 2021

Ball, J.: NSA's Prism surveillance program: how it works and what it can do. The Guardian (2013). https://www.theguardian.com/world/2013/jun/08/nsa-prism-server-collection-facebook-google. Accessed 31 Oct 2020

Biswas, K., Muthukkumarasamy, V.: Securing smart cities using blockchain technology. In: 2016 IEEE 18th International Conference on High Performance Computing and Communications, pp. 1392–1393 (2016)

Buterin, V.: Ethereum white paper: a next generation smart contract and decentralized application platform, pp 13–19 (2015)

Elmaghraby, A.S.: Security and privacy in the smart city. In: 6th Ajman International Urban Planning Conference AIUPC 6: "City and Security" At: Ajman, UAE, pp. 1–9 (2013)

Ethereum: Introduction to smart contracts (2020a). https://ethereum.org/en/developers/docs/smart-contracts. Accessed on 22 Jan 2020

Ethereum: ERC-20 token standard (2020b). https://ethereum.org/en/developers/docs/standards/tokens/erc-20. Accessed 10 Feb 2021

Gervais, A., Karame, G.O., Wust, K., Glykantzis, V., Ritzdorf, H. Capkun, S.: On the Security and Performance of Proof of Work Blockchains. In: Proceedings of the 2016 ACM SIGSAC Conference on Computer and Communications Security, pp. 3–16 (2016)

Goel, V.: Facebook tinkers with users' emotions in news feed experiment, stirring outcry. The New York Times (2014). https://www.nytimes.com/2014/06/30/technology/facebook-tinkers-with-users-emotions-in-news-feed-experiment-stirring-outcry.html. Accessed 31 Oct 2020

Harrison, C., et al.: Foundations for Smarter Cities. IBM J. Res. Dev. 1–16 (2010)

Jaiman, V., Urovi, V.: A consent model for blockchain-based distributed data sharing platforms. Inst. Electr. Electron. Eng. (IEEE) **8**, 143734–143745 (2020)

Morgan, J.P.: J.P. Morgan interbank information network grows to 300+ banks (2019). https://www.jpmorgan.com/insights/technology/news/iin-grows-to-300. Accessed 21 Jan 2021

Kundo, D., Kundo, D.: Blockchain and trust in a smart city. Environ. Urban. Asia 1–13 (2019)

Makhdoom, I., Zhou, I., Abolhasan, M., Lipman, J., Ni, W.: PrivySharing: a blockchain-based framework for privacy-preserving and secure data sharing in smart cities. Comput. Secur. 1–24 (2020)

Mamo, N., Martin, G.M., Desira, M., Ellul, B., Ebejer, J.: Dwarna: a blockchain solution for dynamic consent in biobanking. Eur. J. Hum. Genet.: EJHG **28**, 609–626 (2019)

Michelin, R.A., et al.: SpeedyChain: a framework for decoupling data from blockchain for smart cities. In: Proceedings of the 15th EAI International Conference on Mobile and Ubiquitous Systems: Computing, Networking and Services, pp. 145–154. ACM (2018)

Nakamoto, S.: Bitcoin: a peer-to-peer electronic cash system, pp. 1–4 (2008)

O'Grady, M.J., O'Hare, G.: How smart is your city? Am. Assoc. Adv. Sci. 1–3 (2012)

PricewaterhouseCoopers: Confiance numérique, pp. 1–4 (2016). https://www.pwc.fr/fr/assets/files/pdf/2016/10/confiance_num

Rantos, K., Drosatos, G., Kritsas, A., Ilioudis, C., Papanikolaou, A., Filippidis, A.P.: A blockchain-based platform for consent management of personal data processing in the IoT ecosystem. Secur. Commun. Netw. **2019**, 1–15 (2019)

RGPD: Règlement (ue) 2016/679 du parlement européen et du conseil du 27 avril 2016 relatif à la protection des personnes physiques à l'égard du traitement des données à caractère personnel et à la libre circulation de ces données, et abrogeant la directive 95/46/CE. J. Off. l'Union Eur. 1–88 (2016)

Woolley, J.P., Kirby, E., Leslie, J. et al.: Responsible sharing of biomedical data and biospecimens via the "automatable discovery and access matrix" (ADA-M). npj Genom. Med. **3**, 1–6 (2018)

Blockchain and AI in Art: A Quick Look into Contemporary Art Industries

Marko Suvajdzic[1(✉)], Dragana Stojanović[2], and Iryna Kanishcheva[3]

[1] Digital Worlds Institute, University of Florida, Gainesville, FL, USA
marko@digitalworlds.ufl.edu

[2] Faculty of Media and Communications, Singidunum University, Belgrade, Serbia
dragana.stojanovic@fmk.edu.rs

[3] CEO Public Art Platform, Monochronicle, Gainesville, FL, USA
contact@monochronicle.com

Abstract. In this exploratory text the authors review different ways in which Blockchain technology intersects with Artificial Intelligence (AI), and with art, and how it connects to a more and more frequently mentioned area such as contemporary art industries. These intersections are pointing at the two aspects worth exploring – the first one being a way in which technology (here Blockchain and AI) can be used in various fields and industries, and the other one following art as it opens its world to the new technological possibilities, enriching its forms, topics and manifestations, and questioning the status of the author as well. The art examples and case studies exhibited here will illustrate a couple of problems that can be solved and/or improved with Blockchain and AI technology. These include transparency, art data authenticity, art data monetization, smart contracts with artists, investment opportunities of NFT (non-fungible tokens), roles and activities of curators, psychology of aesthetics, and exploration of creativity.

Keywords: Blockchain · AI · Art · Intersection · Exploration · Technology · Art industries · Author

1 Blockchain, Artificial Intelligence and Art: Intersections and Reasons

Where are we, where do we want to be, how do we get there? To answer these questions we need data. Relevant, authentic, diversified and secure. It has been the case in different times, and different societies applied or developed different technologies to ensure the easiest, quickest and most efficient way to transfer data securely and safely. As for the technology in art, the times of doubt and the times of enthusiasm were interchangeably replaced with each other, producing rather negative or positive attitude towards technological world (compare Adorno and Horkheimer 2002 with McLuhan 1964 or Haraway 1991). However, the fascination with technology, as well as merging contemporary technology with art stayed the same. What artists, and the field of art can do is to offer a space and places for pure exploration of human creative potential, and of the ways in which

J. Prieto et al. (Eds.): BLOCKCHAIN 2021, LNNS 320, pp. 272–280, 2022.
https://doi.org/10.1007/978-3-030-86162-9_27

technology is used in creative acts and, subsequently, in creative industries. Technology, thus, has never been separate from art, and it has never been a "foreign body" to it. Even when technology is primarily emerging for other purposes than art, and even when it seems questionably useful or problematic to the world of art, the artists sought to In the words of Ruth Catlow, "they [the artists] know that a way to get to know something that doesn't yet exist is to collaborate with its possibilities and to do something/anything with it or about it. And by doing so they materialize and shape what it will be" (Catlow 2017, p. 22).

Back to contemporary times, it seems that technologies emerge faster and faster, and the two hottest technologies today, Blockchain and Artificial Intelligence (AI) can be effectively combined to use data in ways never before thought possible, not only in the arts, but many other industries. Data is the key ingredient for the development and enhancement of AI algorithms, and blockchain secures this data, allows to audit and make conclusions from the data by AI and monetize the produced data.

By definition, a blockchain is a distributed, decentralized, immutable ledger used to store encrypted data. On the other hand, AI is the engine or the "brain" that will enable analytics and decision making from the data collected, trace and determine why decisions were made in a particular way. Blockchain and its ledger can record all data and variables that go through a decision made under machine learning, which makes it particularly useful to the field of arts. In this way we can stress the importance of Blockchain technology in the matters of transparency, data authenticity, data monetization, in creating smart contracts with artists, in investment opportunities of NFTs (non-fungible tokens), and also in changing roles and activities of curators. Blockchain influences psychology of aesthetics, and create the open field for exploration of creativity. Adding AI into these intersections of Blockchain and art has already produced some impressive results (Suvajdžić et al. 2019), in the next chapters we will explore already mentioned areas, as well as some of the art projects where Blockchain and AI collide.

2 Blockchain Technology, AI and Its Usage in Art: Benefits and Innovations

2.1 Transparency

The Freedom of Information Act (FOIA) has provided the public the right to request access to records from any federal agency. It is often described as the law that keeps citizens in the know about their government (FOIA Government website). But how did this affect the art? Local Art Agencies, supported by federal funds, are required to disclose any information requested under the FOIA unless it falls under personal privacy, national security, and other exceptions. This could produce quite lengthy and complicated procedures, if it wasn't for the Blockchain technology. To simplify the access to information and secure each transaction, Monochronicle, public art platform suggests applying Blockchain technology, so all the transactions within the field of public art production are registered and visible automatically. This ensures the requested transparency in a rather simple, technologically-based way. Each block of a local art agency will contain a number of transactions (e.g. art purchases) and every time a new transaction occurs on the Blockchain, a record of that transaction will be added to every participant's ledger.

In this way all transactions are public and readable by everyone, but in the same time privacy of the individuals participating in selling/buying is kept safe to some extent, due to pseudo anonymity of the accounts. This is possible because the blockchain system itself defines private and public keys for every user. The private key is known only to the person involved in transaction and cannot be recovered if lost, and the public key is visible to the others, marking transparency of transaction. In this way Blockchain technology gives all the participants their privacy, and in the same time it enables transparency and visibility of every change in the (block)chain. This will keep all agencies updated on spending, decision making, and compliance with the law. The government agencies would have to use such a platform to record, store information, and also buy services to analyze and translate this data into presentable reports. Decentralized, transparent networks then can be accessed by anyone, around the world, including public interested in spending of taxpayers money and equality in decision making. (Authors, on Monochronicle website) Non-governmental entities, that are not required to reveal their records to the public, might be questioned on their transparency over time as well. Similarly, low-budget art commissions might need to be investigated and additionally funded.

2.2 Data Authenticity and the Authenticity/Value of the Artwork

The progress of AI is completely dependent on the input of data and is able to continuously improve itself. The better quality of the data, the better prediction. By collaborating with partners in art industries and providing highly secure, encrypted data ensures the future development of artificial intelligence and the industry. One of the problems pervading contemporary art industry is forging, or, better to say, copying the artwork. Especially digital art is prone to the issue of copying, and the matter is even more complicated if we have in mind that the copy of the artwork in this kind of sense proves itself to be absolutely the same as the original. Moreover, it is, and it might not be considered stealing, for even in the case of copying, the ownership of the original still stays with the previous owner. Also, it is hard to keep track with everything that happens in digital world, since the flow of information on the web is usually one-directional. Blockchain technology again comes useful in solving these kind of issues, and there are numerous platforms, such as Monegraph, Ascribe and so on which are dedicated to solving of these problems. These platforms seek to ensure successful tracking of an artwork, alongside with its potential illegal copies (Zeilinger 2016). Besides the problems with authenticity and ownership rights, once a digital work is copied, its value drops, for the things of value are marked with a specific quality – scarcity. Blockchain technology and AI would help in solving this by introducing the idea of "digital scarcity". Digital scarcity is obtained through issuing a limited number of copies (one or more), which are tied back to unique blocks proving ownership. This would solve the "elephant of the room in digital art", as McConaghy et al. are calling it (McConaghy et al. 2017, p. 463), referring to the fact that, up to Blockchain technology, we were not able to solve the matters of ownership and copying within digital art world. However, not all the artists have positive attitude towards this kind of an artwork protection, since they see it as further commodification of art (Zeilinger 2016). Some of the artists even created artworks that openly criticize this economic maneuvre, and one of the examples is Addie Wagenknecht, with her work Limited Editions of Unlimited (2012). Although Wagenknecht makes the internet series of signed copies of her works and invites everybody to share them and scatter them for free, we may say that even in this kind of an artivist move, the artist is still referring to Blockchain and AI technology, meaning that these two have already made their entrance within the artworld, and they cannot be ignored nor left aside.

2.3 Art Data Monetization: Blockchain and AI Technology

Besides mentioned problem of the authenticity of digital artwork, Blockchain technology can also help with the costs of developing and feeding AI for the companies that do not generate their own data. Through monetizing collected data there could be a huge revenue source created, coming from different kind of buyers (e.g. policy makers, brands). This also applies to the field of art, where artworks can also be tokenized and monetized. An example would be Eve Susman's 89 Seconds Atomized (2020), where she presents the potential of tokenizing her previous work, a video 89 Seconds at Alcazar, which is now sold through Blockchain and AI system (see Fig. 1). Every token of this artwork can be separately sold, bought and/or exhibited, which enables numerous reappearances of the work in all the new forms, and it also monetizes the artwork in a very innovative way.

Fig. 1. Eve Susman, 89 Seconds Atomized (2020)

2.4 Smart Contracts with Artists

The concept of smart contract was introduced in 1993 by Nick Szabo, a computer scientist (Seidler et al. 2017), and it was defined as a computer protocol that verify and enforce the performance of a contract without needing to inform or use human intermediaries. Surely, people still do have to enable the system, to feed it with the program and instructions, so it could develop itself independently afterwards, but it may be a matter of time before AI could create these programs and instructions by itself. This ability is already being explored through DAO/DAC systems (Decentralized Autonomous Organization/Decentralized Autonomous Corporation), where the protocols are transparent, but are enacted and governed by a non-human entity with agency (Seidler et al. 2017).

In the field of art, or contemporary creative art industries, there are more and more examples that show how this DAO/DAC system can be used to enforce the artwork that organizes, produces, reproduces, sells and governs itself. One of the examples listed here will be Plantoid by Primavera de Filippi (2016), and the other one will be a project terra0 by terra0 research group (2016).

According to the words of the author itself, Primavera De Filippi, Plantoid is a plant equivalent to an android, which would be a robot or synthetic organism designed to look, act and grow like a plant (De Filippi 2017). Currently there are several species of Plantoids around the world, which is the consequence of Plantoid's ability to reproduce within the digital world, art world and through actual, public spaces. Plantoid is, basically, a Blockchain and AI based life form, a hybrid creature that lives both in physical and digital world, and which reacts to inputs given in any of them. It uses algorithmic patterns to sustain itself, which makes it a kind of an algorithmic entity. Plantoid is fully autonomous, self-sustainable creature capable of reproducing itself through a combination of Blockchain code and human interaction. It is created as a bridge between technology and art and their often separately seen worlds, in order to illustrate how Blockchain and AI technology can benefit and innovate art, and also to show how art can respond to technology with a great flexibility (De Filippi 2017). In this way technology and art are not shown as conflicted areas, but as always merging areas that can benefit from each other. Once started, Plantoid cannot be shut down by any party, which makes it also a creature unto itself, opening up new questions about the object (or an objectification) of the artwork, if it is still possible at all in this case. Plantoid is, in the same time, an artwork, an artist, an art dealer and an art agent (O'Dwyer 2018). Plantoid seems to be both the subject and the object of the artworld, both an author of itself and its own owner. The material, physical human author, the artist, Primavera De Filippi, here appears more as co-author than the author of the work, which makes room for even more questions about the authorship in the field of art, when Blockchain and AI technologies are included. Plantoid still doesn't have a legal personality, but it stands as a materialized representation of DAO/DAC system in the art world (Fig. 2).

Fig. 2. Primavera De Filippi, Plantoid (2016)

As for terra0 project, the merging process of art and technology is similar as in Plantoid, although the idea is a bit different. Blockchain technology is here used as a technological and technical fuel, so to say, that keeps the project possible, sustainable and ongoing. Terra0 is neither a wood nor an artwork, and it is both. As in the case of De Filippi's Plantoid, it is both an object and a subject in the world of art – and economy. Although terra0 practically owns and regulates itself, hiring professionals to take care of it, it cannot be a legal subject yet, so it exists in a kind of a gray zone within technology, art and economy (Seidler et al. 2017). It stands for a concept of self-owned forest, and for a concept of an ongoing art-technology project. This forest itself pushes smart contracts for leisure activities, and so it accumulates capital for itself, which makes it self-sustainable (Fig. 3).

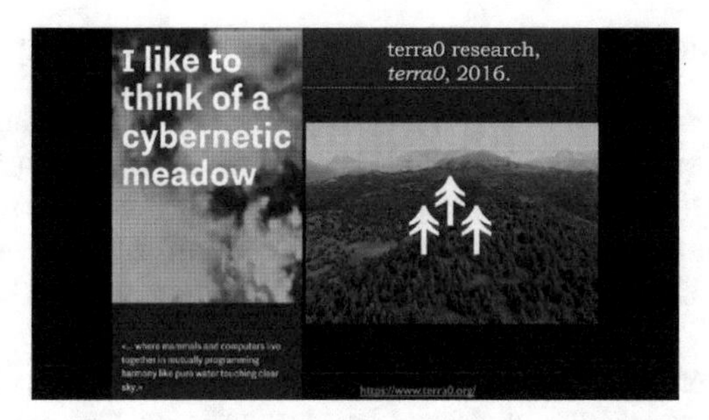

Fig. 3. terra0, terra0 research group (2016).

3 AI Art: Explorations of Creativity in Presenting and Making Art

3.1 Art and AI: Role of the Curator

In recent years, there are more and more initiatives that use or explore Artificial Intelligence in the field of art. The emergence of a new technology in art always has an impact on different aspects of artistic practice, such as art topics, media manifestations of art, the issue of the author, or value of an artwork, and it also has a deep impact on the institutions that encircle and produce art, and that determine its value (Dickie 1971). That means that a new technology always impacts the ways in which curators work – merging all these aspects into their daily practice.

One of the specific ways in which AI is used in art is a case of applying AI algorithms to create AI art. In order to create AI art, artists write algorithms for and AI which, in this way, "learns" a specific aesthetic by analyzing images fed to it and generates new images in adherence to the aesthetics it has learned.

This process is deeply connected to the process of curation. The first step in this process is pre-curation – selection of images. Then the output images are manually selected according to preferred qualities of an artwork, which is called post-curation. For example, when evaluating artists, local art agencies and curators develop a set of criteria, put weight, and apply to each participating artist. Evaluation is often abstract, based on personal perception of aesthetics and cultural background, while constantly under the risk of cognitive biases. The outputs of these manipulations reveal "top artists" that, in fact, are not always the best in the industry. This problem might be solved with the help of machine learning and AI. With more data and applied problems in the artist selection process, systems will learn and make recommendations. By using algorithms, the role of the curator is not undermined but rather enhanced. Throughout the process, the curator maintains an active hand: the feed or data selection must be pre- and post-curated, algorithms and tasks adjusted as needed, and outputs explained. Certainly, it is not yet clear if this is going to make the curators' work easier, or if it is going to be accepted as a good technological solution, but it is certainly a contemporary tool for art and worth exploration.

3.2 AI and Psychology of Aesthetics: Exploration of Creativity

Daniel Berlyne's work on thinking was succeeded by his development of a theory of aesthetic behavior, summarized in Aesthetics and Psychobiology (Berlyne 1971; Weiner and Craighead 2019). According to Berlyne, various aspects and/or properties are being compared and contrasted. During this process, organism collates (gathers) information from the stimuli available to it, and then selects a stimulus to which a behavioral response will be given.

In support of Berlyne's studies, an AI algorithm "GAN", that was being fed portraits, ended up producing a series of deformed faces (see Fig. 6). The generated portraits are certainly novel, surprising, and conflicting. The machine has failed to properly imitate a human face, because of artistic creative process and decision making of surprising deformities. By analyzing past artworks and modeling new artworks with AI, we can identify patterns and trends in aesthetic perception followed by consumer's demand.

All the new thoughts that originate in the mind are not completely new; they are based on our culture, all our knowledge and our experience. The greater the knowledge and the experience, the greater the possibility of finding an unthinkable relation that leads to a creative idea. Through AI, we can understand creativity like the result of establishing new relations between pieces of knowledge that we already have but not necessarily conscious about. Consequently, the AI has greater capacity to store and operate the information.

3.3 Investment Opportunities of NFT (Non-fungible Token)

The latest phenomenon in the investing world is the idea of non-fungible tokens, or NFTs, which represent a digital piece of a collectible online item. "Fungible" means something that can be easily traded for something else of the exact same value. Following this idea, it means that a fungible token (FT) has a kind of an interchangeable quality. Its value stays the same, but it is not necessarily important what the content is. In the case of non-fungible tokens (NFTs), no token is the same as the other, and in this aspect the content is what matters the most. Thus NFTs can be successfully used in the field of art, especially digital art. It may solve questions of digital art value and ownership, as well as digital art market problems such as scarcity, which is important in marketing and trading processes. If an artist, for example, posts a picture to Instagram, they will still own that photograph, but they grant a very broad license to Instagram to use that photo on their platform, and users to see the photo without any contribution to the artist (unless the user is a full-time influencer, paid every time followers shop through affiliate links).

Popular public artists (or so-called street artists) create digital collections to utilize new technology, and to explore the new medium. Their iconic street works (like murals) are trading into digital space.

Felipe Pantone says, "I created THE GRAFFITI COLLECTION as my very first NFT release as a way to best utilize this innovative medium and technology. Understanding graffiti's fleeting and ephemeral nature, only through blockchain technology do these pieces permanently establish ownership and remain in existence forever unlike many of the graffiti pieces I have created" (Pantone, official website, also see Fig. 7).

From April 12 to June 15, 2021, the number of sales involving non-fungible tokens (NFTs) in the art segment increased significantly. As of April 12, 2021, roughly 23.7 thousand NFTs were sold in the art segment during the previous 30 days. As of June 15, 2021, the aggregated number of sales over 30 days rose to approximately 93 thousand. Overall, most NFT sales in the art segment came from the primary market as of the period considered. Market capitalization of transactions globally involving a non-fungible token (NFT) from 2020 reached $338.04 million dollars (de Best 2021a, b, c, d; Statista Research Department 2021).

The future of NFTs for a wide range of stakeholders appears bright. Digital assets will have a marketplace in creative industries and enable artists to collect commission from every sale through Blockchin without the need for complex tracking systems.

4 Conclusion

All the possibilities of using Blockchain and Artificial Intelligence technology in the field of art that were mentioned in this paper are just emerging in the recent years. Exactly this makes them even more important, both for the artists and academics, as well as for art dealers, art curators and the audience too. It is important to explore options that Blockchain technology and AI are giving us, and to understand them from different points of view.

It is for sure that Blockchain and AI stand as relevant factors of enhancing the art word, especially in the field of digital media. We can conclude that technologies should be explored as both medium and subject of artistic practices, whether they might be conventional, enthusiastic, or critical towards the issues of using technology in art. This exploration can successfully happen only if all the actors within the artworld have a voice in how Blockchain and AI might be applied to the field they work in. In order for that to be possible, we, as academics and/or artists, or art workers, cannot just refuse contemporary technology, saying that it is not relevant to the process of creating and valuing art. We will have to say yes to technology, and to research the ways it could work for our future ideas.

References

Adorno, T., Horkheimer, M.: Dialectic of Enlightenment. Stanford University Press, Palo Alto (2002)

Berlyne, D.E.: Aesthetics and Psychobiology. Appleton-Century-Crofts, New York (1971)
Catlow, R.: Artists Re:thinking the blockchain: introduction. In: Catlow, R., Garret, M., Jones, N., Skinner, S. (eds.) Artists Re:Thinking the Blockchain, Torque Editions & Furtherfield, Signal, vol. 2395, pp. 21–37 (2017)
Danto, A.: Symposium: the work of art, the artworld. J. Philos. **61**(19), pp. 571–584 (1964). American Philosophical Association Eastern Division Sixty-First Annual Meeting
De Filippi, P.: Plantoid/the birth of a blockchain-based lifeform. In: Catlow, R., Garret, M., Jones, N., Skinner, S. (eds.) Artists Re:Thinking the Blockchain, Torque Editions & Furtherfield, Signal, vol. 2395, pp. 51–62 (2017)
Dickie, G.: Aesthetics, An Introduction. Pegasus, Cambridge (1971)
FOIA Government. https://www.foia.gov/faq.html
Godbole, O.: Ether Drops Below $2K, Bitcoin Wilts as China Tells Banks to Cut Off Crypto Transactions. https://www.coindesk.com/ether-drops-bitcoin-wilts. Accessed 30 Jun 2021
Haraway, D.J.: A manifesto for cyborgs: science, technology, and socialist feminism in the 1980s. In: Simians, Cyborgs, and Women: The Reinvention of Nature, pp. 149–182. Routledge, New York (1991)
de Best, R.: Market cap of NFT worldwide 2018–2020. Statista. Accessed 16 Mar 2021
McLuhan, M.: Understanding Media: The Extensions of Man. McGraw Hill, New York (1964)
McConaghy, M., McMullen, G., Parry, G., McConaghy, T., Holzman, D.: Visibility and Digital art: blockchain as an ownership layer on the internet. Strateg. Change **26**(5), 461–470 (2017)
Monochronicle. https://monochronicle.com/. Accessed 30 Jun 2021
NFT sales in the art segment worldwide in the last 30 days April–June 2021, by type. Statista Research Department, Statista (2021). Accessed 15 Jun 2021
O'Dwyer, R.: Limited edition: producing artificial scarcity for digital at on the blockchain and its implications for the cultural industries. Convergence: Int. J. Res. New Media Technol., 26, 1–21 (2018)
de Best, R.: Sales volume of NFT in different segments worldwide 2018–2020. Statista. Accessed 22 Apr 2021
Saxena, P., Srivastava, S.: An Entrepreneur's Handbook to Blockchain. https://appinventiv.com/guide/blockchain-guide-for-entrepreneurs/. Accessed 30 Jun 2021
Seidler, P., Kolling, P., Hampshire, M.: terra0 – can an augmented forest own and utilize itself?. In: Catlow, R., Garret, M., Jones, N., Skinner, S. (eds.) Artists Re:Thinking the Blockchain, Torque Editions & Furtherfield, Signal, vol. 2395, 63–72 (2017)
Storey, A.: How Much Does It Cost to Mint and NFT?. https://postergrind.com/how-much-does-it-cost-to-mint-an-nft/?utm_source=rss&utm_medium=rss&utm_campaign=how-much-does-it-cost-to-mint-an-nft. Accessed 30 Jun 2021
Suvajdžić, M., Stojanović, D., Appelbaum, J.: Blockchain art and blockchain facilitated art economy: two ways in which art and blockchain collide. In: 2019 4th Technology Innovation Management and Engineering Science International Conference (TIMES-iCON) (2019). https://doi.org/10.1109/TIMES-iCON47539.2019.9024403
Zeilinger, M.: Digital Art as 'Monetized Graphics': enforcing intellectual property on the blockchain. Philos. Technol. (2016). https://doi.org/10.1007/sl3347-016-0243-1. Published online under the terms of Creative Commons Attribution 4.0 International License
Felipe Pantone murals. https://pant1.tumblr.com/page/10, https://www.felipepantone.com/
Felipe Pantone NFT. https://niftygateway.com/collections/felipepantoneopen
de Best, R.: Market Cap of NFT Worldwide 2018–2020. Accessed 16 Mar 2021
de Best, R.: Sales Volume of NFT in Different Segments Worldwide 2018–2020. Accessed 22 Apr 2021
Statista Research Department: NFT Sales in the Art Segment Worldwide in the Last 30 Days April–June 2021, By Type. Accessed 15 Jun 2021

Machine Learning Powered Autoscaling for Blockchain-Based Fog Environments

John Paul Martin[1](✉), Christina Terese Joseph[2], K. Chandrasekaran[3], and A. Kandasamy[4]

[1] School of Computer Science and Engineering, Vellore Institute of Technology, Vellore, India
[2] Department of CSE, Indian Institute of Information Technology, Kottayam, Kottayam, Kerala, India
[3] Department of CSE, National Institute of Technology Karnataka, Surathkal, Mangalore, India
kchnitk@ieee.org
[4] Department of MACS, National Institute of Technology Karnataka, Surathkal, Mangalore, India

Abstract. Internet-of-Things devices generate huge amount of data which further need to be processed. Fog computing provides a decentralized infrastructure for processing these huge volumes of data. Fog computing environments provide low latency and location-aware alternative to conventional cloud computing by placing the processing nodes closer to the end devices. Co-ordination among end devices can become cumbersome and complex with the increasing amount of IoT devices. Some of the major challenges faced while executing services in the fog environment is the resource provisioning for the user services, service placement among the fog devices and scaling of fog devices based on the current load on the network. Being a decentralized infrastructure, fog computing is vulnerable to external threats such as data thefts. This work presents a blockchain based fog framework for making autoscaling decisions with the use of machine learning techniques. Evaluation is done by performing a series of experiments that show how the services are handled by the fog framework and how the autoscaling decisions are made.

Keywords: Fog computing · Resource provisioning · Service placement · Blockchain · Auto-scaling · Machine learning

1 Introduction

Cloud computing was introduced to deal with the demands of the ever increasing number of Internet Connected Devices. These devices collect data from the surroundings and send them to the cloud for processing. Handling such huge amount of data becomes difficult with the increasing number of Internet-of-Things (IoT) devices. The physical distance between the end users and the cloud

J. Prieto et al. (Eds.): BLOCKCHAIN 2021, LNNS 320, pp. 281–291, 2022.
https://doi.org/10.1007/978-3-030-86162-9_28

servers introduces a substantial amount of end-to-end delay, high latencyas well as high processing and communication cost. So, to deal with this problem, fog computing was introduced which reduces the distance between the computational devices and the actual end users. Fog computing proposes the use of idle computational resources available at the edge of the network, enabling services to be executed locally at the edge of the network. Fog devices constitute of routers, gateways, modems that provide some computation, networking and storage capabilities. Fog computing exploits the resource capacities of networking components to allocate services or to store data. Initially, services are closer to the clients and, data do not need to be transferred in its whole to the cloud. Both issues have a notable performance impact, reducing network latency and usage. One of the important tasks is to efficiently place the service requests onto the fog devices. The decision to place the services on a particular device depends on how the resources are distributed among the devices. The main goal of a resource provisioning algorithm is to minimize the total deployment time of the service request. Various algorithms are proposed for resource provisioning in the fog environments.

To realize a secure and ideal fog computing platform, there is a need for trust among the nodes. Providing trust in the distributed environment of fog can be done using the distributed ledger technology or the blockchain. Using blockchain, it is possible to provide trust and security in a decentralized system. This work employs using machine learning as a way to predict the resource usage in the fog environment and based on that taking the decisions to scale up or scale down the computational resources. The main contribution of the work is to provide a fog computing framework as previously proposed conceptually providing: (i) definition and creation of a blockchain based fog computing framework incorporating the security features of the blockchain technology, (ii) a fog computing framework implementing two resource provisioning techniques: first-fit algorithm and best-fit algorithm, (iii) an autoscaling mechanism utilizing machine learning methodologies that automatically scales the resources of the framework based on load on the devices. The rest of the paper is organized as follows: First we will look at state-of-the-art work done on resource provisioning & placement of services in fog environments and blockchain-based environments, in section (ii) and (iii) respectively, we will look at the infrastructure of the underlying fog framework in section (iv), how blockchain acts as a security measure in (v) and take a look at autoscaling with machine learning in section (vi). Afterward, we will look at the proposed work in sections (vii) and (viii) and evaluate the performance of the framework in section (ix). Finally paper concluded in section (x).

2 Related Work

The service placement problem in fog computing is addressed by multiple researchers. Finding the most appropriate location for module placement is an NP-Hard problem. Linear programming is a common approach to optimization. Major optimization techniques are categorized into subsections.

Heuristic Algorithms: Taneja et al. [5] proposed a module mapping algorithm which maps the network nodes and applications using their capacity and requirements in the form of key-value pairs. Hong et al. [1] formulated a module deployment algorithm that places services to the devices dynamically. Ashkan et al. [7] proposed two heuristic solutions to solve the dynamic Fog service provisioning problem. The problem is formulated as an Integer Non-Linear Programming task that is solved using the Min-Vol and Min-Cost heuristic to minimize the overall cost induced.

Genetic Algorithms: Another approach to solving the service placement problem is to model the problem and solving it using a genetic algorithm. Skarlat et al. [4] proposed a service placement plan that uses a Genetic Algorithm (GA) to solve the optimization problem. Wen et al. [6] devised a modified form of a genetic algorithm that maintains the composition of applications in parallel based on a number of constraints.

Other Approaches: Ni et al. [2] used priced timed Petri nets (PTPNs) for solving the resource allocation problem. Their proposal optimized price and time costs to complete a task. Singh et al. [3] has discussed various issues while balancing the load in fog environments.

3 Application Management in Fog Landscape

In this section, we will discuss how the application requests are processed by the fog devices, service images are shared and which devices are selected for deploying the services in the fog landscape. An IoT application is a set of services, once all the services are deployed, the application will begin executing. The services are either pre-defined or can be defined by the end user on runtime. The services are created in the form of service images and stored in the local database of fog control nodes and the cloud. Once the service images are defined, a user can send a task request in the form of a JSON application request. Then the fog control node will find a viable fog device to execute the service on.

Once the task request is received by the fog control node, it will provision the resources available to the application. The resource provisioning module is responsible for provisioning the resources to the incoming task requests. To make the fog landscape diverse, different roles are assigned for different fog devices. In total, four service types are defined, t1, t2, t3, and t4. In this work, we will compare two resource provisioning algorithms namely: first-fit algorithm and best-fit algorithm.

First Fit Algorithm: In this algorithm, the incoming task requests are provisioned in the first resource that satisfies the requirements. The devices are sorted based on the type of service they can host. The incoming task requests are also sorted and the algorithm iteratively finds the first fog device that can completely host the application based on the available resources present on the device.

Best Fit Algorithm: This algorithm works similar to the first fit algorithm with a minor difference that it will select the best possible resource to be provisioned for the incoming applications. From the available pool of resources, the one resource that maximizes the amount of resource left after resource provisioning will be selected.

Once the application is deployed in the fog devices, the fog control node will store the details of the running applications in the database.

4 Autoscaling with Machine Learning

We leverage machine learning as a tool for making autoscaling decisions. In our case, the dependent variable is selected as time. Now, in fog environments, prediction can be done using different resource patterns such as CPU, RAM or storage. Our experimental results showed RAM as a suitable candidate for the independent variable. Using CPU as an independent variable creates a discrete scatter plot making it difficult to predict the variable at a particular time. By looking at the device usage patterns, the RAM usage to time graph plot indicates a linear pattern. So, the linear regression model is used for modeling the variables.

When the RAM usage of a particular fog device in a fog colony crosses the 80% mark, the device is said to be overutilized and when it is below 40%, the device is said to be underutilized. Although these constants set the upper and lower limits for deciding whether a device is overutilized or underutilized respectively, there is no hard and fast rule to set these values as the limits, but in our experimental setup, we consider a device to be overutilized when its RAM usage crosses 80% and underutilized when it is below 40%. The autoscaling daemon first predicts the optimal RAM usage of a fog device using the machine learning model. It then periodically checks the aforementioned rules and accordingly takes the action. We use the linear regression model to train the dataset generated by the fog cells.

5 Proposed Work

In this work, we provide a novel way to run IoT applications on the fog framework providing data integrity with the use of blockchain technology and making autoscaling decisions with the use of machine learning to reduce the cost of maintaining the fog infrastructure. Figure 1 shows how the interaction between different components takes place in the fog landscape.

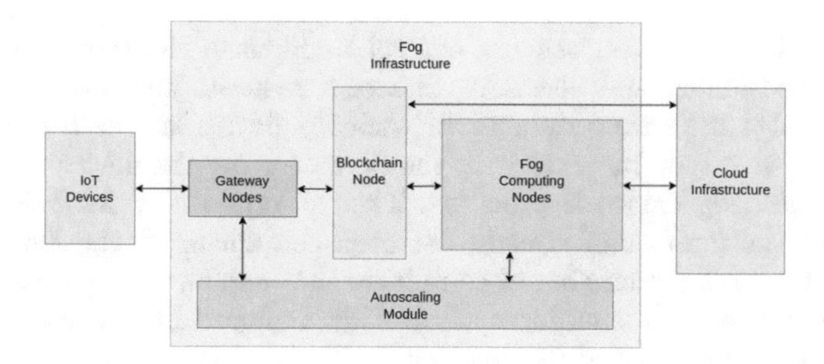

Fig. 1. High-level view of the relationship between different components.

6 Proposed Algorithm

Algorithm 1: First Fit Heuristic Algorithm

Input: Set of Fog Devices and Task Requests
Output: List of task assignments to fog devices and sorted requests

$assignments \leftarrow []$
$sortedRequests \leftarrow sortByServiceType(requests)$
$sortedChildren \leftarrow sortByServiceType(children)$
for $child \in sortedChildren$ **do**
  for $serviceType \in child.serviceTypes$ **do**
    for $request \in sortedRequests$ **do**
      if $serviceType == request.serviceType$ **then**
        $Utilization \leftarrow get.Utilization(child)$
        $containers \leftarrow getContainerCount(child)$;
        if $checkRules(Utilization)$ **and**
        $containers < MAX.CONTAINERS$ **then**
          $container \leftarrow sendDeployReq$
          $(child, request)$;
          $assignments.add(child,$
          $request.container)$;
          $sortedRequests.remove(request)$;
        end
      end
    end
  end
end
return $assignments, sortedRequests$

During the initial course, the fog control node sorts the devices eligible for resource provisioning and also sorts the task requests based on their service types. The algorithm then iterates for each fog device and each request type, and if the device type matches with the request type, the device utilization of that particular fog device is procured. Then it runs the *checkRules* function which finds the number of already deployed containers on the device. In our approach, we limit the number of containers that can be run on a fog device to be 15. If more than 15 containers are already deployed in any device, the task request will not be satisfied by that device. There are no resource limits added while creating an application container. From our experiments, we find that the resource usage in the form of RAM usage spikes and crosses the over-utilization mark when 15 applications are hosted on a single fog device. Similarly, if device utilization does not satisfy the predefined parameters, the fog device will not be able to execute the task request. Following parameters needs to be satisfied for the resource to be provisioned on a fog device:

$$\text{MAX CONTAINERS} < 15; \text{CPU USAGE} < 80\%; \text{RAM USAGE} < 99\%$$

The first algorithm used for resource provisioning in the fog framework is the first fit heuristic algorithm. It takes a set of fog devices (either fog nodes or fog cells) and a set of service requests as input and provides a list of task assignments as output. The task assignment provides the details of the fog devices and the task requests assigned to them. The resource provisioner iterates through all the child devices until it finds the first device that can handle the task request. The fog control node is responsible for resource provisioning in a particular fog colony in our fog landscape.

The second algorithm used in the fog landscape is the best fit heuristic. Unlike first-fit, it will select the best fitting fog device from the set of devices. In our approach, we use the RAM utilization as the best fitting measure. From all the available fog devices, the device having the least RAM utilization will be selected as the best fit device for resource provisioning. Using this algorithm for resource provisioning, the fog devices are better utilized.

Algorithm 2: Best Fit Heuristic Algorithm

```
Input: Set of Fog Devices and Task Requests
Output: List of task assignments to fog devices and sorted requests
assignments ← []
sortedRequests ← sortByServiceType(requests)
sortedChildren ← sortByServiceType(children)
for fogdevice ∈ sortedChildren do
    fd = NULL;
    leastRAMUtilized = MAX;
    for serviceType ∈ fogdevice.serviceTypes do
        for request ∈ sortedRequests do
            if serviceType == request.serviceType then
                Utilization ← get.Utilization(fogdevice)
                containers ← getContainerCount(fogdevice);
                if checkRules(Utilization) and
                containers < MAX.CONTAINERS and
                fogdevice.RAM < leastRAMUtilized then
                    leastRAMUtilized ← fogdevice.RAM;
                    fd ← fogdevice;
                    container ← sendDeployReq(fd,
                    request);
                    assignments.add(fd,
                    request.container);
                    sortedRequests.remove(request);
                end
            end
        end
    end
end
return assignments, sortedRequests
```

7 Evaluation

To evaluate the proposed fog framework, we will discuss how the service requests are handled by the fog landscape and how fog devices behave under different scenarios.

7.1 Experimental Setup

In this section, we will discuss the experimental setup used for evaluating the fog landscape. The fog controller runs on an Intel i7 7700 system. For the fog devices, Raspberry Pi 3b+ model is used with Hypriot as the operating system. For cloud resources, Amazon AWS is used. The fog landscape is implemented in Java 8 and Spring application framework. For communication between different fog devices, we use REST API and JSON messages.

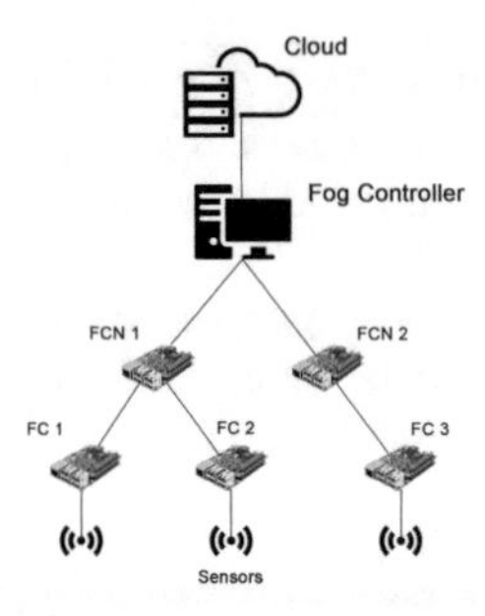

Fig. 2. Evaluation setup.

As seen in Fig. 2, the fog controller acts as a mediator between the cloud and the fog devices. FCN 1 has two children namely, FC 1 and FC 2 creating a fog colony. Similarly, FCN 2 has a single child called FC 3 creating the second fog colony. Sensors are connected to the fog cells to evaluate the fog application. The sensor used is a temperature/humidity sensor used for taking application readings.

7.2 Results and Discussions

We evaluate the resource provisioning algorithms with our evaluation setup. Different arrival patterns (random, constant, pyramid) for service requests are defined. We will measure the total service deployment time on the fog devices with those arrival patterns. The deployment results are shown in Table 1. We tested the deployment times by creating three test suites namely, random, constant and pyramid. The task requests are sent to the fog devices and only when all the task requests are satisfied, the application is deployed in the form of a container. The task requests are divided into four types, t1, t2, t3, t4. The t1 task request is made specifically for fog devices and can be sent to only fog cells. The task requests require the sensor equipment for functioning, t2 and t3 task requests are meant for both fog devices as well as the cloud. These task requests are used for accessing the local database and finally, t4 task requests are meant to be run only on cloud VM's. These task requests are responsible for accessing the cloud database. Random test suite sends random task requests of all the types at regular intervals of time, with random times of execution. Constant test suite sends task requests t1 and t3 only with different runtimes at regular intervals of time. Pyramid test suite gradually increases the number of task requests of type t3 and then again gradually decreases the number of the task requests sent at a regular interval of time. Initially, service requests are sent to the first fog colony containing fog cells FC1 and FC2. In the first-fit resource provisioning algorithm, the service requests and the fog cells are sorted. The incoming requests are sent to FC1 until it reaches the maximum capacity of hosted containers (i.e. 15). Only when the FC1 is overloaded, the incoming requests are placed onto FC2. So, the load is not equally balanced when using first-fit as the resource provisioning algorithm. On the other hand, the best-fit resource provisioning algorithm selects the best fog cell based on the ram usage of the devices. The device having the minimum ram usage will be selected as the potential fog device to place the service request. This way, the load is equally balanced amongst the fog devices in the colony as seen in Fig. 3. An evaluation test is also undertaken to view how the fog landscape behaves when the devices in a fog colony are overloaded, how the services are placed under such circumstances. After waiting for some time until the devices are overloaded in a particular fog colony, a service request is sent to a device in the colony. The resource provisioner checks the number of already deployed containers and device utilization. Once all the checks fail, it forwards the service request to the head fog node of the neighboring fog colony for service placement. If none of the fog nodes can be provisioned for a task request meant to be executed on fog devices, the task request is sent to the cloud.

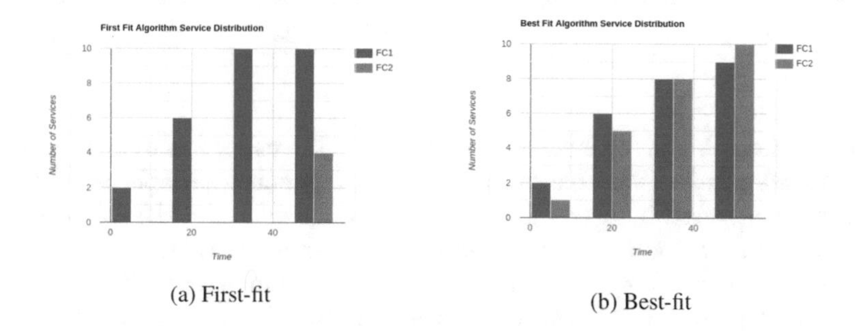

(a) First-fit

(b) Best-fit

Fig. 3. Service distribution between first-fit and best-fit

The Ethereum node is initialized using the genesis block with gas limit $(0xfffffff)$ and difficulty $(0x20000)$. Once the miner starts mining the blocks, a smart contract with getData() and setData() functions is added as a transaction block to the blockchain. The Ethereum blockchain is linked with the Oraclize API which is responsible for fetching external data into the blockchain network. Oraclize fetches the utilization data of the fog devices and adds it to the blockchain network. It maintains the integrity of the data inside the fog landscape. Using blockchain technology in fog landscape has a tradeoff in the form of performance. Incorporating the blockchain mechanism into the fog framework increases network traffic as well as the bandwidth of the network. Although it provides security from external intrusions, it also increases response times, bandwidth and network traffic.

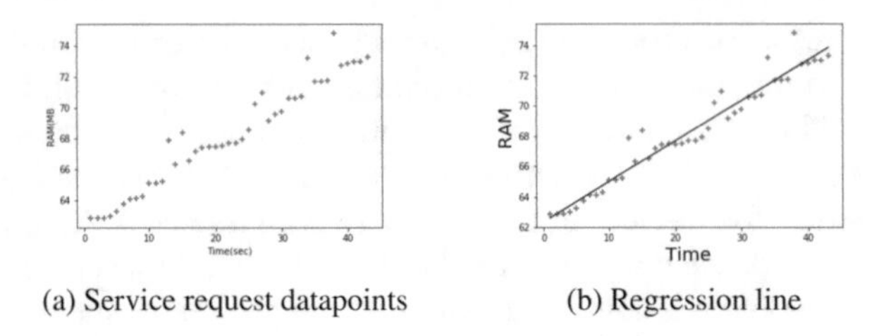

(a) Service request datapoints (b) Regression line

Fig. 4. Plot of service datapoints and linear regression model

Table 1. Evaluation results

Metrics	Algorithm	Random	Constant	Pyramid
Total deployment time per service (sec)	First-fit	6.41	2.80	2.56
	Best-fit	4.41	1.45	1.52
Total number of services deployed	First-fit	20	15	15
	Best-fit	22	15	15

The autoscaling in fog landscape is done with the use of machine learning. We take the independent variable as time and the dependent variable as RAM. The dataset is generated by extracting the RAM usage values of the fog cells under different service request arrival patterns. These values are then trained using the linear model of linear regression to get the slope and intercept. These values are then used to predict the RAM usage at any given time. In our experiments, we got a slope/coefficient equal to 0.266 and the intercept equal to 62.361. Figure 4a shows the data points plotted on the time vs RAM usage graph while Fig. 4b shows the regression line passing through those data points. The auto-scaling module runs alongside the fog landscape and monitors the resource utilization of each fog device. With the help of the trained dataset, the auto-scaler probes each fog device at regular intervals and checks the RAM utilization. If the RAM utilization is well within the defined limit (40–80), the auto-scaler takes no action. When the RAM usage of a fog device soars above 80%, the device over-utilization log is shown to the user and additional fog device is requested. If the RAM usage of a fog device declines below 40%, the device under-utilization log is shown to the user and the device is sorted to receive more service requests.

8 Conclusion

In this work, we propose a blockchain based fog framework with autoscaling capabilities. We enlighten the possibilities of using blockchain technology along with fog landscapes. Service placement in Fog is handled by resource provisioning algorithms first-fit and best-fit. Best fit algorithms proved to be better at balancing the service load and handling the requests. Blockchain technology provided data integrity by sending and receiving utilization details of the fog devices at the price of higher network traffic and bandwidth. Autoscaling is provided in the fog landscape with the help of machine learning using specific service arrival patterns. This work can be extended by using more sophisticated resource provisioning algorithms and using non-linear machine learning models for different test scenarios.

References

1. Hong, H., Tsai, P., Hsu, C.: Dynamic module deployment in a fog computing platform. In: 2016 18th Asia-Pacific Network Operations and Management Symposium (APNOMS), pp. 1–6, October 2016. https://doi.org/10.1109/APNOMS.2016.7737202
2. Ni, L., Zhang, J., Jiang, C., Yan, C., Yu, K.: Resource allocation strategy in fog computing based on priced timed Petri nets. IEEE Internet Things J. **4**(5), 1216–1228 (2017). https://doi.org/10.1109/JIOT.2017.2709814
3. Singh, V., Martin, J., Chandrasekaran, K.: Explicating fog computing: key research challenges and solutions. In: International Conference on Machine Learning, Image Processing, Network Security and Data, pp. 46–63 (2021)
4. Skarlat, O., Nardelli, M., Schulte, S., Borkowski, M., Leitner, P.: Optimized IoT service placement in the fog. Serv. Oriented Comput. Appl. **11**(4), 427–443 (2017). https://doi.org/10.1007/s11761-017-0219-8
5. Taneja, M., Davy, A.: Resource aware placement of IoT application modules in fog-cloud computing paradigm. In: 2017 IFIP/IEEE Symposium on Integrated Network and Service Management (IM), pp. 1222–1228, May 2017. https://doi.org/10.23919/INM.2017.7987464
6. Wen, Z., Yang, R., Garraghan, P., Lin, T., Xu, J., Rovatsos, M.: Fog orchestration for Internet of Things services. IEEE Internet Comput. **21**(2), 16–24 (2017). https://doi.org/10.1109/MIC.2017.36
7. Yousefpour, A., et al.: QoS-aware dynamic fog service provisioning. CoRR arXiv:1802.00800 (2018)

Cryptocurrencies Impact on Financial Markets: Some Insights on Its Regulation and Economic and Accounting Implications

Vanessa Jiménez-Serranía[1], Javier Parra-Domínguez[2](✉), Fernando De la Prieta[2], and Juan Manuel Corchado[2]

[1] Universitat Oberta de Catalunya, Barcelona, Spain
[2] BISITE Research Group, University of Salamanca, Edificio Multiusos I+D+i, 37007 Salamanca, Spain
javierparra@usal.es

Abstract. Cryptocurrencies have become a trend (and, sometimes, a nightmare) for the financial sector. There are a multitude of news, opinions and criticisms about cryptocurrencies that need to be contextualized and pondered. This paper aims to explain some of the key legal issues regarding cryptocurrencies in the financial market and to analyze the highlights of one of the most recent regulatory proposals. The article introduces the real legal and financial reality surrounding cryptocurrencies and without which they could not develop. In line with the development of technology, this paper clarifies the volatility and the negativity of volatility for the development of cryptocurrencies, which are favoured by the development of accounting systems. Through a concrete example, the need for consistent accounting legislation and cash principles and rules observed in the importance of the cryptocurrency wallet and its implications on the current and traditional cash and bank-based management is incorporated in the present study.

Keywords: Cryptocurrencies · Legal issues · Accounting and finance

1 Background

Cryptocurrencies have become the "trending topic" in financial matters of the moment. Although not long ago, they were considered marginal assets, nowadays cryptocurrencies are the object of desire of millions of investors (many of them without previous experience or knowledge of the financial system). This situation has been described in some forums as a democratization and decentralization of investment. However, recently (April–May 2021) we have witnessed the less "friendly" side of this new ecosystem: on the one hand, the incredible swing (drop) in cryptocurrency values following Elon Musk's statements involving Bitcoin [1] and the Chinese Government's announcement of a new ban related to cryptocurrencies [2]; on the other hand, the increase of judgments and complains on cryptocurrencies investments scams [3].

J. Prieto et al. (Eds.): BLOCKCHAIN 2021, LNNS 320, pp. 292–299, 2022.
https://doi.org/10.1007/978-3-030-86162-9_29

These issues, along with the fact that certain National Central Banks are considering the creation of national cryptocurrencies (crypto euro, digital yuan), lead us to wonder about the need for regulation of the impact of cryptocurrencies on the financial market.

In addition to regularisation, in this paper, we will elaborate on the implications of cryptocurrencies in the economic sphere from accounting and finance. Issues such as the type of transactions that can occur, the registration of cryptocurrency or the use of electronic money represent a real challenge for accounting [12, 13].

This paper aims to explain some of the key legal issues regarding cryptocurrencies in the financial market and to analyze the highlights of one of the most recent regulatory proposals (2). The article is accompanied by an interpretation of the economic implications of cryptocurrencies in the accounting and financial sphere.

The first part of the article includes the introduction; the second part deals with the financial market regulator's view of the reality of cryptocurrencies. The third section deals with the economic implications and ends with conclusions and future lines of research.

2 Cryptocurrencies from the Financial Markets regulator's Point of View

2.1 Legal Nature of Cryptocurrencies

One of the most controversial aspect related to cryptocurrencies has been its legal nature. This feature is fundamental due to its implications given its commercial, tax and regulatory implications.

Despite they are name as "(crypto)currencies", they are not equivalent to the currencies issued by a Central Bank (*fiat*). In fact, the European Central Bank (2019) defined cryptoassets as: "*any asset recorded in digital form that* ***is not and does not represent either a financial claim on, or a financial liability*** *of, any natural or legal person, and which does not embody a proprietary right against an entity. Yet, a crypto-asset* ***is considered valuable by its users (an asset) as an investment and/or means of exchange,*** *whereby controls to supply and the agreement over validity of transfers in crypto-assets are not enforced by an accountable party but are induced by the use of cryptographic tools*" (emphasis added) [4].

We find a similar description on article 3 of the MiCA Proposal for a Regulation: "*crypto-asset means a* ***digital representation of value or rights*** *which may be transferred and stored electronically, using distributed ledger technology or similar technology.*" (emphasis added) [5].

In both definitions we see that the description of cryptocurrencies as value is underlined, and they are only considered as a means of payment in the case that it is admitted as such by its users (for example, buyer and seller in a contract of sale of a good).

Following the reasoning provided by the Spanish Supreme Court ruling in June of 2019 [6], we can establish three characteristics that allow us to define the legal nature of cryptocurrencies:

- Cryptocurrencies are an intangible asset, in the form of a unit of account defined by means of computer and cryptographic technology called blockchain.

- Cryptocurrencies' value is the value that each unit of account or its portion reaches by the concert of supply and demand in the sale of these units is made through trading platforms.
- These intangible assets may have been considered as means of payment in any bilateral transaction in which the contracting parties accept it. Nevertheless, they can never be considered legal tender.

2.2 To Regulate or not to Regulate: This is the Question

Cryptocurrencies are considered by financial and securities market authorities as highly speculative tradable assets. For instance, in February 2018, the Spanish National Securities Market Commission pointed out that "although they (cryptocurrencies) are sometimes presented as an alternative to legal tender, they have very different characteristics, it is not obligatory to accept them as a means of payment of debts or other obligations". This authority also stated that the value of cryptocurrencies fluctuates strongly, so they cannot be considered as a good store of value or as a stable unit of account [7].

Nevertheless, cryptocurrencies have undoubtedly generated enormous attraction, especially in the case of "neophyte" investors, due to their apparent high profitability.

On the other hand, their role in obtaining financing is significant, especially in the case of ICOs. In addition, the development of stable coins as well as initiatives by certain countries to issue national cryptocurrencies have begun to moderate the perception of cryptocurrencies as mere speculative assets.

Therefore, we can easily deduce that the impact of cryptocurrencies on traditional financial markets has been a source of both concern and opportunity. As it has been pointed out: "(e)*ffective regulation of financial services promotes long-term economic stability and minimizes the social costs and negative externalities from financial instability*" [8]. Following this reasoning, certain regulators have begun to act and establish certain rules, with the aim of protecting both the financial market and (especially) investors.

In recent years we have seen several initiatives by different countries (i.a., France [9], Spain [10]) but we can say that 2020 was the turnaround year since a very important regulatory push on crypto assets was made by the European Union.

2.3 The EU Proposal: Digital Finance Package (with Special Reference to the MiCA Regulation Proposal)

The European Commission adopted on 24 September 2020 a digital finance package, including a digital finance strategy and legislative proposals on crypto-assets and digital resilience. This initiative aims to provide access to innovative financial products, while ensuring consumer protection and financial stability.

Among the proposals made, it is worth mentioning the MiCA Regulation Proposal [10]. The scope of this Regulation proposal are all crypto-assets not covered elsewhere in financial services legislation and e-money tokens (i.a., MiFID II financial instruments, deposits, structured deposits, securitization) as well as crypto-asset service providers, referred to as CASPs.

If adopted, the regulation will directly apply to all Member States, however, implementation in national law will not be required.

This regulation establishes four general objectives:

1. Legal certainty: it seeks to regulate previously unregulated assets in a common manner in the EU.
2. Support for innovation and fair competition.
3. Protection of consumers and investors and market integrity, trying to avoid the risks inherent to this type of assets that can be highly speculative.
4. Financial stability, notably by including safeguards to address the risks that could arise from "stable cryptocurrencies" in terms of financial stability and orderly monetary policy.

Following these four objectives, this new rule regulates (i) "utility tokens", (ii) asset-backed tokens, (iii) relevant asset-backed tokens, (iv) e-money tokens, or (v) relevant e-money tokens, It also establishes the need for CASPs to be authorised by the competition authority, and they must comply with a series of requirements depending on the activities they carry out (crypto asset custody, crypto asset issuance, exchange platforms, etc.).

The extensive regulation on CASPs contained in this Regulation is very relevant. Thus, Chapter 1 of Title 5 focuses on the authorisation of CASPs, establishing that "*Crypto-asset services shall only be provided by legal persons that have a registered office in a Member State of the Union and that have been authorised as crypto-asset service providers*" and detailing the content of the corresponding application (article 54), the assessment of this application (article 55) and the rights granted to the competent authorities with respect to the withdrawal of authorizations (article 56).

Furthermore, article 57 contains the details of the registration of CASPS and article 58 details all the information that a CASP that intend to provide cross-border crypto-asset services (in other words, crypto-asset services in more than one Member State) shall submit to the competent authority.

Particularly important is Chapter 2, which establishes the obligations to be fulfilled by the CASPs, such as the obligation to act honestly, fairly, and professionally in the best interest of clients and information to clients (article 59). This provision establishes that CASPs shall provide their clients with fair, clear and not misleading information, in marketing communications, and they also shall warn clients of risks associated with purchasing crypto assets.

Moreover, CASPs shall, always, have in place the prudential safeguards determined by article 60 and they must guarantee that, amongst other organizational requirements, that the members of their management bodies have the necessary good repute and competence, in terms of qualifications, experience and skills to perform their duties (article 61). Extremely important is the obligation of safekeeping of clients' crypto-assets and funds, especially in the event of the CASP's insolvency (article 63). Finally, CASPs shall establish and maintain effective and transparent procedures for the prompt, fair and consistent handling of complaints received from clients (article 64); they shall prevent, identify, manage and disclose conflicts of interest between the CASPs and their shareholders, their managers, their employees or their clients (65); and they will remain fully responsible in case of outsourcing.

Furthermore, it is also very important to point out the regulation on market abuse in relation to crypto assets propose by MiCA. The regulation of this figure in the MiCA is “inspired” by the current financial market regulations. Thus, insider dealing (article 78), unlawful disclosure of inside information (article 79) and market manipulation (article 80) are prohibited. In the other hand, issuers of crypto assets has the obligation to inform the public as soon as possible of inside information which concerns them, in a manner that enables the public to access that information in an easy manner and to assess that information in a **complete, correct and timely manner** (emphasis added, article 77).

The MiCA Regulation Proposal is the first rule at the international level that regulates the crypto sector and that clearly shows the change of stance towards the development of the crypto assets… as long as they comply with the parameters established by the regulator. Certainly, MiCA is a very ambitious regulation that seeks to put an end to regulatory arbitration and forum shopping in the crypto sector [11]. It remains to be seen whether it will become an international benchmark or whether crypto traders will prefer to look for alternatives … in any case, it is clear that a new global scenario for cryptocurrencies is emerging.

3 Economic Implications

One of the most important perspectives within the different scenarios that arise, in addition to the regulatory one, are the economic implications and the technological [14] or innovation policies [15].

Undoubtedly, after their legal coupling following the indications set out in this paper, the development of cryptocurrencies will have a major implication. The implication will be in the economic, financial and accounting sense, and it is in this last area where we will focus on the following.

In the economic sense, the impact is direct and arises from incorporating the digital economy [16].

Robust digital systems will only support the new economic reality of cryptocurrencies. All this opens up a new social implication understood as the cheapening of the maintenance of a currency in its digital version, thanks to the new technology, which would give rise to the birth of currencies with more community and business meanings [17].

From an economic point of view, the implication must be to maintain welfare in the face of possible volatility, a factor that could clearly affect the economic evolution of cryptocurrencies [18]. At this point, cryptocurrencies could be understood as digital assets [19], a term that would encompass a more global and affordable approach to cryptocurrencies.

3.1 Accounting Implications

Apart from the more general economic implications outlined in the previous section, a logical evolution is that of accounting in line with the development of cryptocurrencies.

In fact, the advance of cryptocurrencies will be accompanied not only by legal robustness but also by accounting systems being able to assume their role in managing cryptocurrencies [20].

The global monetary system is based on fluctuating exchange rates, depending on the supply and demand for money, but how economic data is recorded, sorted, manipulated and stored remains antiquated despite the technological evolution we are experiencing. One of the technologies coming into play in the logical evolution of new technological systems is the blockchain [21], which was created to improve the way we store and record data, especially in the financial world.

As currencies, they have a creator, maintenance and rules, the latter concerning their use and even the possibility of withdrawing units from circulation [22]. Although it is not the purpose of this section to develop the rules, as this has already been done in the previous section, it does allow us to consider, in addition to those already described above as implications, an implication whose concern is increasingly evident: the cryptocurrency wallet.

The case that we are going to develop for the practical purposes of accounting implications is the implication of the new concept of the wallet since it is, among other things, the axis of accounting being able to absorb the entire crypto world [23].

This point, which is exposed as a wallet, is vital to understand the scope and implication of one of the verticals of incorporating the cryptocurrency world into accounting. The characteristics of the cryptocurrency wallet are to establish control over the coins in question, provide a secure environment, control potential fees, and perform full validation [24].

Concerning accounting, a change begins to take shape here, since accounting, understood as a science, incorporates the accounting accounts "Cash" and "Banks" [25] and has to approach the wallet where, in fact, there are several types of wallets, such as desktop wallet, web wallet, app or mobile wallet and paper wallet [26]. With this example and implication, we can clearly see the differences between traditional accounting and digital accounting, which faces challenges beyond the one presented here.

4 Conclusions

Cryptocurrencies, likewise, the crypto sector, are in a clear expansion. Despite having been considered as highly volatile assets and, even, potential scams schemes, cryptocurrencies also offer significant investments and business opportunities, and they are starting to be really appreciated by the traditional financial sector.

Therefore, this is a key moment for this sector. Thus, regulators and market operators must carefully consider both the objectives to be achieved and the means to achieve them.

It is almost self-evident that some regulations is necessary, especially to prevent certain negative effects on markets and investors. Nevertheless, this regulation shall be designed in an internationally coordinated manner and, in our opinion, shall respect, as far as possible, the fundamentals of the crypto economy (decentralization and absence of intermediaries).

In addition to the regulatory environment, the development of cryptocurrencies has economic, financial and accounting implications. It is the new digital paradigm and the technologies incorporated into it that represent the real economic challenge. Still, we must also consider the development of accounting systems to accompany the better development of cryptocurrencies. Proof of this is the need for accounting support in cryptocurrency wallets, the development of which requires their incorporation into the current accounting system.

References

1. On May 13: Elon Musk announced that Tesla had suspended vehicles purchases using Bitcoin for environmental reasons. As a consequence, the value of Bitcoin and other cryptocurrencies drastically dropped. https://www.bbc.com/news/business-57096305. Accessed 27 May 2021
2. On May 18: China banned financial institutions and payment companies from providing services related to cryptocurrency transactions, and warned investors against speculative crypto trading. https://www.reuters.com/technology/chinese-financial-payment-bodies-barred-cryptocurrency-business-2021-05-18/. Accessed 27 May 2021
3. For instance, in Spain, there are already important condemnatory judgments (i.e. Supreme Court Judgment, 2109/2019, June 20th, 2019) and several on-going investigations are currently conducted by the National High Court (Audiencia Nacional) regarding pyramid investment schemes and fraud
4. ECB: Crypto-Assets: Implications for financial stability, monetary policy, and payments and market infrastructures, Occasional Paper Series, Nº223/May 2019, p. 7 (2019). https://www.ecb.europa.eu/pub/pdf/scpops/ecb.op223~3ce14e986c.en.pdf. Accessed 27 May 2021
5. Proposal for a Regulation of the European Parliament and of the Council on Markets in Crypto-assets and amending Directive (EU) 2019/1937 (COM/2020/593 final), MiCa
6. Supreme Court Judgment, 2109/2019, 20 June 2019 (2019)
7. See https://www.cnmv.es/loultimo/NOTACONJUNTAriptoES%20final.pdf. Accessed 27 May 2021. In the same vein, see the joint statement by the National Securities Market Commission and the Bank of Spain on the risk of cryptocurrencies as an investment, 9 February 2021 (2021). https://www.cnmv.es/Portal/verDoc.axd?t=%7Be14ce903-5161-4316-a480-eb1916b85084%7D. Accessed 27 May 2021
8. Cuervo, C., Morozova, A., Sugimoto, N.: Regulation of crypto assets. FinTech notes. International Monetary Fund, 1 (2019)
9. Act of May 22th, 2019 on the growth and transformation of companies (Loi relative à la croissance et la transformation des entreprises, know as Loi PACTE)
10. Royal Decree-Law 5/2021, March 12th, 2021 on extraordinary measures to support business solvency in response to the COVID-19 pandemic (Real Decreto-ley 5/2021, de 12 de marzo, de medidas extraordinarias de apoyo a la solvencia empresarial en respuesta a la pandemia de la COVID-19) that regulate advertising of cryto-assets
11. Marti Miravalls, J.: Aproximación a la propuesta de reglamento UE relativo a los mercados de criptoactiovos: MiCA. In: Belando Garín, B., Marimón Durá, R., Thomson-Reuters Aranzadi (dir) Retos del Mercado Financiero, p. 376 (2020)
12. Procházka, D.: Accounting for bitcoin and other cryptocurrencies under IFRS: a comparison and assessment of competing models. Int. J. Digit. Account. Res. **18**(24), 161–188 (2018)
13. Subačienè, R., Kurauskineè, N.: Evaluation of alternatives of cryptocurrency accounting. Buhalterinés apskaitos teroija ir praktika **22**, 4 (2020)
14. Filippova, E.: Empirical evidence and economic implications of blockchain as a general purpose technology. In: 2019 IEEE Technology & Engineering Management Conference (TEMSCON), pp. 1–8. IEEE (2019)
15. Allen, D.W., Berg, C., Markey-Towler, B., Novak, M., Potts, J.: Blockchain and the evolution of institutional technologies: implications for innovation policy. Res. Policy **49**(1), 103865 (2020)

16. Babkin Alexander, V., Burkaltseva Diana, D., Pshenichnikov Wladislav, W., Tyulin Andrei, S.: Cryptocurrency and blockchain-technology in digital economy: development genesis. St. Petersburg State Polytechn. Univ. J. Econ. **67**(5), 9–22 (2017)
17. Birch, D.G.: What does cryptocurrency mean for the new economy?. In: Handbook of Digital Currency, pp. 505–517. Academic Press (2015)
18. Saleh, F.: Volatility and welfare in a crypto economy. SSRN 3235467 (2019)
19. Pypenko, I.S., Kud, A.A.: Genesis of it economy: from cryptocurrency to digital asset (2019)
20. Chohan, U.W.: International law enforcement responses to cryptocurrency accountability: interpol working group. SSRN 3156531 (2018)
21. Hassani, H., Huang, X., Silva, E.: Big-crypto: big data, blockchain and cryptocurrency. Big Data Cogn. Comput. **2**(4), 34 (2018)
22. Miraz, M.H., Ali, M.: Applications of blockchain technology beyond cryptocurrency. arXiv preprint arXiv:1801.03528 (2018)
23. Suratkar, S., Shirole, M., Bhirud, S.: Cryptocurrency wallet: a review. In: 2020 4th International Conference on Computer, Communication and Signal Processing (ICCCSP), pp. 1–7. IEEE (2020)
24. Christiansen, N.B., Jarrett, J.E.: Forfeiting cryptocurrency: decrypting the challenges of a modern asset. US Att'ys Bull. **67**, 155 (2019)
25. Canning, J.B.: The economics of accountancy: a critical analysis of accounting theory. Doctoral dissertation, The University of Chicago (1929)
26. Salodkar, A., Morey, K., Shirbhate, M.: Electronic wallet. Int. Res. J. Eng. Technol. (IRJET) **2**(09), 975–977 (2015)

Usage of Multiple Independent Blockchains for Enhancing Privacy Using the Example of the Bill of Lading

Hauke Precht(✉) and Jorge Marx Gómez

Carl von Ossietzky University of Oldenburg,
Ammerländer Heerstraße 114-118, 26129 Oldenburg, Germany
{hauke.precht,jorge.marx.gomez}@uol.de

Abstract. Digitization of trade documents such as the Bill of Lading (B/L) is advancing. Several approaches have discussed blockchain as a suitable technology in general but state that they lack privacy aspects. Motivated by the antitrust-by-design approach from antitrust law, we propose a novel approach called *one B/L, one blockchain* in which we aim to create a dedicated blockchain for each B/L process. With this approach, we aim to enhance privacy and confidentiality. We gathered first insights by implementing the approach using Hyperledger Fabric. The first results show that the approach is technically feasible but requires manual steps and must be further refined.

1 Introduction

In today's international trade, ocean freight is a crucial backbone. Currently, 80% of the European Union imports and exports are done via ocean freight [18]. Recently, the G7 paved the way towards digitizing necessary paper-based documents and processes within international trade [33]. Considering ocean freight, the *Bill of Lading (B/L)* is one of the most important documents [6] as it groups multiple functions into a single paper: (1) It proves that the carrier took over the goods described on the B/L in the described form. (2) The B/L secures the obligation of the carrier to deliver the given goods to the consignee, to the stated place of destination. (3) It is a negotiable document of title, i.e., by transferring the B/L, the ownership of the goods is transferred as well. Already in 2016, the possibility to use blockchain technology to create an electronic B/L was discussed in [31] and later by Wunderlich and Saive in [34]. A conducted prototype based on Ethereum was presented in 2017 by [25], but the authors state that the prototype lacks speed and privacy of transactions. In general, private blockchains provide better performance and privacy than public blockchains [7,21,35]. However, as part of the privacy aspect, the specific field of antitrust law is not yet considered in current studies. Louven and Saive [20] point out that information exchange via respective systems can be subject to antitrust law, especially if data is shared within such systems, which would have been

J. Prieto et al. (Eds.): BLOCKCHAIN 2021, LNNS 320, pp. 300–309, 2022.
https://doi.org/10.1007/978-3-030-86162-9_30

otherwise undisclosed [20]. Currently, there exist different issues regarding privacy in blockchain systems. Especially deanonymization techniques and ways to prevent such attacks are researched within the scientific community. The main goal is to preserve the sender's and receiver's privacy of. Especially for the bitcoin protocol, several analysis have been carried out, e.g., [2,22,26,30]. Zhang et al. grouped the different approaches against deanonymization methods into three groups [36]: (1) Network: Users can use tools such as Tor[1] to connect. (2) Transaction: Users can leverage mixing to break the relationship between sending and receiving address (3) Application: Key material can be stored offline as cold storage to protect it against attacks. Note that [22] were still able to identify major institutions and the interactions, which could be subject to antitrust law as already stated. A different approach towards a privacy-enhancing technique (PET) is proposed by Parizi et al., as they propose to introduce Garlic routing and Onion routing in combination with the concept of a sidechain [5] which acts "[...] as a shield on stored data in the core blockchain" [27]. Based on the given introduction, we aim to develop a novel approach for further privacy enhancements in private blockchain systems using the example of the B/L, while addressing the antitrust by design introduced by the antitrust law.

This paper is structured as follows: First, in Sect. 2 we present an overview of the domain-specific background of the B/L process and its actors. Next, we present our conceptual approach which we called *one B/L, one blockchain* in Sect. 3. To gain first insights into the technical feasibility of the approach, we choose a suitable technology in Sect. 4 and describe a prototype implementation in Sect. 5. In Sect. 6 we discuss our first finding followed by a summary and outlook.

2 Domain-Specific Background

In order to understand the process and especially the involved actors, we provide a short overview of the most important actors of a B/L process. As there are multiple parties involved in each B/L process, there is a risk that competitors can extract data and information that would have otherwise been undisclosed from a blockchain system. The parties in a B/L process are shortly explained in the following:

- *Shipper and carrier* are the parties of the contract of carriage or the B/L-contract [1]. The shipper assigns the carrier to transport the goods. The carrier then executes the contract and transports the goods overseas. Shipper and carrier are not necessarily identical with the consignor and consignee of the goods.
- *Consignor and consignee* The consignor is entitled to possession of goods at the time they are loaded onto a ship, while the consignee is named in the B/L as to whom the goods shall be delivered [1]. If the goods afloat are sold, the buyer must be named as consignee in the B/L. Otherwise, he cannot receive

[1] https://www.torproject.org/.

the goods at the port of discharge. Thus, the current consignee must transfer the B/L to the buyer and appoints him as the new consignee.

- *The Port Agent* coordinates all communication concerning the goods and ship, and they take care of the process [23]. They prepare the B/L, review it for correctness and pass it on to the captain, who is then able to sign.
- *The Captain* or master is responsible for signing the B/L by which the captain states that the goods are loaded on board. Further, the captain may add remarks about the quantity and condition of the goods loaded.
- *The customs authorities* decide whether the goods are legally allowed to be exported or imported. They verify that the information given in the B/L complies with local regulations [32].
- *The Port authorities* ensure that all safety measures have been met [4]. Therefore, they need to know which ship is entering or leaving the port and what they have loaded. Port authorities review the B/L for information about the loaded goods.
- *Banks* serve as a third party handling payment with a Letter of Credit (L/C) [19]. Within the overseas purchase, the L/C serves as a form of guarantee for payment under the conditions of the presentation of the B/L.

3 Conceptual Approach for Privacy Enhancement

As stated in Sect. 1, private blockchain systems are considered to enhance privacy, as only before defined and identifiable parties are allowed to join the blockchain network and access the respective data. In our example of the B/L, especially the shippers, carriers and banks are considered competitors. In terms of privacy, our goal is to ensure that each of these parties are only allowed to access B/L data subject to their interest while denying information access to other processes. For our conceptual approach, we use the sidechain approach by Parizi et al. [27] as a starting point. The idea to split relevant information into a sidechain seems feasible. However, as shown in Sect. 2, the B/L process, as part of the international trade process, consists of a variety of actors and companies such as shippers, carriers or banks. Leveraging sidechains does not keep the system from getting large and consistently growing, meaning competitors from different domains, e.g., transport (shipper/carrier)or finance sector (banks), would partake in the same system and chain. With analytical tools as shortly described in Sect. 1, it could be possible for these participants to acquire knowledge of transactional behavior within the market represented by the chain, which would have been otherwise undisclosed. As shown in the introduction, this can be subject to the field of antitrust law. We raise the question of using multiple, independent private permissioned blockchains to be a suitable solution to decouple potential competitors within the same market further to ensure privacy and compliance to antitrust law. Based on discussions with legal experts and practitioners from within the ocean freight and logistics domain, we identified requirements, which such a system must meet. The three key requirements based on these discussions are shown in the following:

R1, multiple blockchain support The system must support multiple blockchains and must be able to connect on the fly to the respective one.
R2, programmatically creation of new blockchains As multiple B/L processes must be created on the fly, the creation of the respective blockchain should be done programmatically.
R3, confidentiallity The participants of a process should only have knowledge and access to the blockchain representing that specific process.

We call this abstract approach *one process, one blockchain* and in this example *one B/L, one blockchain* In Fig. 1, the this concept is presented. Note that even though we aim to provide for each process a dedicated blockchain, we consider the participants described in Sect. 2 as an overall part of the system. This is similar to the current market situation, where participants are known but not the business relationships. For example, the *Commerzbank* generally knows that the *Landesbank Baden-Württemberg LBBW* partakes in international trade finance but does not know their specific customers or business relations. Our approach aims to keep these relationships hidden while still leveraging blockchain properties such as immutability and integrity [35] as well as transparency among the parties subject to i.e., a specific blockchain. For example, in Fig. 1, Shipper 1 knows his relationship with consignee 1 in process 1 and N where shipper 1 is involved, but Bank 3 does not know B/L Process 1. The business relationships can remain hidden this way, following the antitrust-by-design concept. Due to the nature of private blockchains, properties such as transparency and equal rights [35] are adjusted to serve transaction speed and privacy. Note that, as shown in Sect. 2, the port agents have a prominent role in the current process serving as a middleman in the document part of the B/L process. This middleman role can be shrunk or even eliminated by introducing a decentralized blockchain system, leaving the port agents with only the coordination of goods at the port.

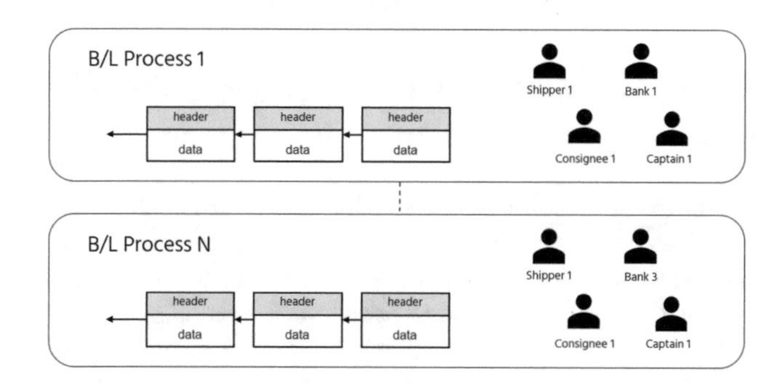

Fig. 1. The conceptual approach of *one B/L, one blockchain*

4 Blockchain Technology Selection

To evaluate the general approach described in Sect. 3, we aim to develop a prototype. In the first step, our goal is an analysis if the designed approach is functional in general. The interest in private blockchains grew for enterprises [28], leading to technologies emerging at a quick pace. A broad range of different systems is present in the market, which we need to evaluate for fitting the before described approach. Especially the identified requirements *R1, R2 and R3*, which are crucial for the concept *one B/L, one blockchain*, must be considered and checked. Polge et al. identified Hyperledger Fabric, Enterprise Ethereum, Quorum, MultiChain and R3 Corda as the five major private/permissioned blockchain frameworks [29]. Considering the concept of *one B/L, one blockchain*, only two of them mentioned blockchain framework could support this approach: MultiChain and Hyperledger Fabric. Polge et al. state that although MultiChain offers configurations regarding permissions and access control to the network, it will not introduce a smart contract-like concept until version 2.0.55 [29]. Further, privacy-enhancing features in particular, as well as scalability features, are only available in a paid, proprietary enterprise version of MultiChain [24]. This leads to our decision not to consider MultiChain as feasible for this research. Hyperledger Fabric is a permissioned blockchain for enterprises [10] and is one of the most used blockchain systems [8]. Within Hyperledger Fabric, the channel concept was introduced, which enables the creation of sub-nets that can only be accessed by a selected and configured set of network members [12]. Each channel represents an independent blockchain, supporting our designed approach. Note that Hyperledger Fabric provides two types of nodes. A peer node holds the actual blockchain and executes Smart Contracts [15] and can be part of multiple channels. An orderer node serves the sole purpose of ordering and creating blocks [14]. Hyperledger Fabric supports the usage of Smart Contracts, so-called chaincodes, in several programming languages [16].

5 Evaluating the Conceptual Approach Through Prototyping

In order to evaluate the technical feasibility of our described conceptual approach in Sect. 3, we aim to create a prototype implementation of said conceptual approach. Based on Sect. 4, Hyperledger Fabric, with its channel concept, seems like a perfect fit to support our approach of *one B/L, one blockchain.* We developed a three-tier architecture, shown in Fig. 2a consisting of a front-end that enables users to interact with the software and create B/Ls, a back-end that stores application information such as user credentials and draft B/Ls and finally, the blockchain component with a chaincode managing the B/L as an asset. The chaincode is written in Java, and its main functions are shown in Fig. 2b Our `BillOfLadingContract` implements the `ContractInterface` provided by Hyperledger Fabric. This `ContractInterface` provides utility functions for our contract allowing us to interact with the ledger and the system

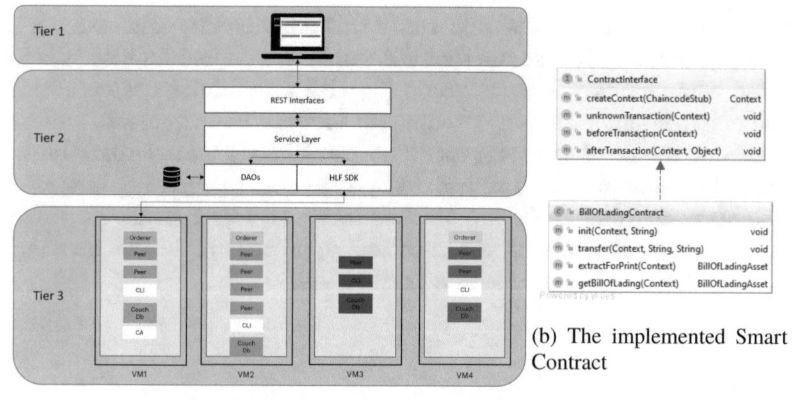

(a) The three-tier architecture

(b) The implemented Smart Contract

Fig. 2. The three-tier architecture and the developed Smart Contract

itself. Note that the `transfer` method can only be invoked by the current consignee as this is the only person legally allowed to transfer the B/L further. The B/L asset will be created via calling of the `init` method. Note that the B/L asset itself is created in the Java Application (back-end component/tier two) and is initially represented as Plain Old Java Object (POJO). Upon sending the B/L to the blockchain where it will be managed, it is serialized to JSON. Our chaincode takes this JSON and deserializes it again, and performs general checks, for example, the B/L indeed contains a delivery address. Via the `getBillOfLading` method, the respective B/L of this respective chain will be returned. However, our main object in this paper is to create further insights considering *R2*, the ability to programmatically create new blockchains as well as *R3*, fulfilling necessary confidentiality to push the privacy aspect further. First of all, we focus on *R2*. To do so, we started to design a test network. We set up four independent virtual machines (VM) for this test network where each represents an organization. In Fig. 2a, our tier-three architecture is shown. We varied in the number of orderer and peer nodes for each organization for a more realistic scenario. The network was built from scratch and is configured to only communicate via Transport Layer Security (TLS). In order to connect from our Java-based back-end to the Hyperledger Fabric network, we use their provided Software Development Kit (SDK). We created the interface `FabricChannelServices` along with an implementation called `FabricChannel-ServiceImpl` as well as the class `FabricWrappingClient`. The interface provides a `createChannel` method, which checks if the given user is valid when first called. Next, a new instance of the Hyperledger Fabric SDK provided class `HFClient` is created via the `FabricWrappingClient`. With this client object, the `create` method is called via the SDK provided class `ChannelConfiguration` and the `newChannel` method is called. Once the channel is created, the before-defined peers and orderers are joined, and finally, the channel is initialized. We wrote integration tests to further test and verify this functionality and start up multiple disjunct channels. This shows that *R2* can be considered fulfilled. We further conducted a preliminary a focus group [3] with practitioners representing carrier, shipper, consignor as well as banks to identify improvements and general acceptance of the created prototype. The participants stated that the private blockchain-based approach is considered positive and generated interest along with the approach *one B/L, one blockchain* which was considered positive as well. They further stated that a centralized approach would not be supported as they are unwilling to commit to a trusted third party.

6 Discussion

While implementing the approach with Hyperledger Fabric as the underlying technology, we found the implementation and setup of the system challenging. To create a new channel, i.e. a new blockchain within Hyperledger Fabric, the usage of the tool `configtxgen` along with a configuration file called `configtx.yaml` is required [13]. Within the configuration file, the initial members of the channel and participants who partake in the ordering are defined along with further system-based permissioned. Based on this configuration file, the `configtxgen` tool generates the respective genesis block for the new channel in which the members of the channel were initially defined. Therefore, we had to manually define this configuration file beforehand and call the `configtxgen` from within our Java Application. If a new channel with a different set of members should be created, the `configtx.yaml` must be adjusted. At the moment, there is no built-in programmatically way to do this from within the SDK. This limits our approach of *one B/L, one blockchain* as we need to be able to create multiple channels with different members on the fly. Further, the adjustment of the initial set of channel participants requires manually updating the channel definition by the administrator, limiting the programmatic extendability of once-create channels. Based on this experience, we think that the definitions of participants for each new channel can be challenging and creates additional overhead. Hyperledger Fabric leverages X.509 certificates for performing access control (part of the Membership Service Provider (MSP)) [17]. This further enables us to leverage an attribute-based access control (ABAC) on chaincode level [9]. In our current implementation, we leveraged self-signed X.509 certificates for a default MSP as suggest by Hyperledger. For a productive system, an organization should be formed that creates/signs such certificates, which can then be used to join respective channels. This organization needs to be independent and neutral with no access to actual data of the respective process to maintain neutrality. Via this organization, participants of B/L process could receive digital certificates (e.g. x.509 certificates), which could be used to identify the parties within the respective blockchain. In the current process, typically, the shipper starts to create the B/L, meaning this would be the case in our system as well, in which this actor should then add the already known participants as part of the configuration of the respective private blockchain so that only these given, identifiable participants can join. Further, note that up until the most recent Version 2.2 of Hyperledger Fabric, a so-called system channel is in place, which tracks the participants of each channel [11], meaning participants of the ordering service can still evaluate business relations which they themselves are not part of. This can be subject to antitrust law and foils our goal to provide confidentiality as stated in *R3*.

7 Summary and Outlook

In this paper, we showed that approaches towards electronic B/Ls lack privacy and that antitrust law is not yet considered. To comply with the antitrust-by-design approach, we developed an approach called *one B/L, one blockchain* in which we investigate the possibility of the usage of multiple independent blockchain for each B/L process. We found that within existing blockchain frameworks, Hyperledger Fabric, with its channel concept, fits our approach. By creating a prototype implementation consisting of a front-end, a back-end and Hyperledger Fabric as blockchain-component, we gained first insights regarding the technical feasibility of our approach. While we were able to show that the channel system can support our approach in general, we found limitations: (1) the current channel definition relies on a manual definition. (2) Up until version 2.2 of Hyperledger Fabric, a so-called system channel is in place through which information can be gathered, which organization transacts with another in a specific channel, therefore foiling our initial goal of confidentiality.

For future work, we plan to conduct case studies with multiple practitioners (e.g., shippers, carriers, banks, etc.) to gain more insights into the feasibility from a business perspective. Further, the discussed need for an organization managing digital identities for the system along with further development and maintenance needs to be analyzed and discussed as part of our future work. Also, incorporating other existing approaches towards enhancing privacy and confidentiality will be tracked and integrated into this approach as we believe that the approach is not yet a fully-fledged solution. Also, further areas of laws need to be analyzed, e.g., once the B/L process is finished and the goods are delivered, questions arise if the underlying blockchain and data could be deleted. In Germany for example, the legislator states in section 257 of the German Commercial Code, along with section 147 of the Fiscal Code of Germany, that documents such as the B/L need to be archived for 10 years. It needs to be analyzed in future work if a transfer of the data to a different system would be possible to clean up in-active blockchains.

References

1. Baughen, S.: Shipping Law, 5th edn. Routledge, Abingdon (2012)
2. Biryukov, A., Khovratovich, D., Pustogarov, I.: Deanonymisation of clients in bitcoin P2P network. In: Ahn, G.J., Yung, M., Li, N. (eds.) Proceedings of the 2014 ACM SIGSAC Conference on Computer and Communications Security, pp. 15–29. ACM, New York (2014). https://doi.org/10.1145/2660267.2660379

3. Brandtner, P., Helfert, M., Auinger, A., Gaubinger, K.: Conducting focus group research in a design science project: application in developing a process model for the front end of innovation. Syst. Signs Actions **9**, 26–55 (2015)
4. Brooks, M.R.: The governance structure of ports. Rev. Netw. Econ. **3**(2) (2004). https://doi.org/10.2202/1446-9022.1049
5. Dilley, J., Poelstra, A., Wilkins, J., Piekarska, M., Gorlick, B., Friedenbach, M.: Strong federations: an interoperable blockchain solution to centralized third-party risks (2016)
6. Fridgen, G., Guggenberger, N., Hoeren, T., Prinz, W., Urbach, N.: Chancen und herausforderungen von dlt (blockchain) in mobilität und logistik. https://www.bmvi.de/SharedDocs/DE/Artikel/DG/blockchain-grundgutachten.html
7. Gamage, H.T.M., Weerasinghe, H.D., Dias, N.G.J.: A survey on blockchain technology concepts, applications, and issues. SN Comput. Sci. **1**(2) (2020). https://doi.org/10.1007/s42979-020-00123-0
8. Hileman, G., Rauchs, M.: Global blockchain benchmarking study. https://www.ey.com/Publication/vwLUAssets/ey-global-blockchain-benchmarking-study-2017/$File/ey-global-blockchain-benchmarking-study-2017.pdf
9. Hu, V.C., et al.: Guide to attribute based access control (ABAC) definition and considerations: NIST special publication 800–162. https://doi.org/10.6028/NIST.SP.800-162. https://nvlpubs.nist.gov/nistpubs/SpecialPublications/NIST.SP.800-162.pdf
10. Hyperledger Fabric: Hyperledger fabric: open, proven, enterprise-grade DLT. https://www.hyperledger.org/wp-content/uploads/2020/03/hyperledger_fabric_whitepaper.pdf
11. Hyperledger Fabric: Channel configuration (configtx): orderer system channel configuration (2019). https://hyperledger-fabric.readthedocs.io/en/release-1.4/configtx.html#orderer-system-channel-configuration
12. Hyperledger Fabric: Channels (2020). https://hyperledger-fabric.readthedocs.io/en/latest/channels.html
13. Hyperledger Fabric: Craete a channel (2020). https://hyperledger-fabric.readthedocs.io/en/latest/create_channel/create_channel_participation.html
14. Hyperledger Fabric: Orderer (2020). https://hyperledger-fabric.readthedocs.io/en/release-2.0/orderer/ordering_service.html
15. Hyperledger Fabric: Peers (2020). https://hyperledger-fabric.readthedocs.io/en/release-2.0/peers/peers.html
16. Hyperledger Fabric: Smart contracts and chaincode (2020). https://hyperledger-fabric.readthedocs.io/en/release-2.0/smartcontract/smartcontract.html
17. Hyperledger Fabric: Membership service providers (MSP) (2021). https://hyperledger-fabric.readthedocs.io/en/release-2.2/msp.html
18. International Chamber of Shipping: Shipping and world trade: driving propserity. https://www.ics-shipping.org/shipping-fact/shipping-and-world-trade-driving-prosperity/
19. Jones, S.A.: Trade and Receivables Finance. Springer, Cham (2018). https://doi.org/10.1007/978-3-319-95735-7
20. Louven, S., Saive, D.: Antitrust by design - the prohibition of anti-competitive coordination and the consensus mechanism of the blockchain. Gewerblicher Rechtsschutz und Urheberrecht Internationaler Teil **6**, 537–543 (2019)
21. Mattila, J.: The blockchain phenomenon: the disruptive potential of distributed consensus architectures. ETLA Working Papers (38) (2016). http://pub.etla.fi/ETLA-Working-Papers-38.pdf

22. Meiklejohn, S., et al.: A fistful of bitcoins. In: Papagiannaki, K.D., Gummadi, K., Partridge, C. (eds.) Proceedings of the 2013 Conference on Internet Measurement Conference, pp. 127–140. ACM, New York (2013). https://doi.org/10.1145/2504730.2504747
23. Mukherjee, P.K., Brownrigg, M.: Farthing on International Shipping. Springer, Heidelberg (2013). https://doi.org/10.1007/978-3-642-34598-2
24. MultiChain: Download and install multichain: Start building your first blockchain application in minutes (2021). https://www.multichain.com/download-install/
25. Nærland, K., Müller-Bloch, C., Beck, R., Palmund, C.: Blockchain to rule the waves - nascent design principles for reducing risk and uncertainty in decentralized environments. In: ICIS (2017)
26. Ober, M., Katzenbeisser, S., Hamacher, K.: Structure and anonymity of the bitcoin transaction graph. Future Internet **5**(2), 237–250 (2013). https://doi.org/10.3390/fi5020237
27. Parizi, R.M., Homayoun, S., Yazdinejad, A., Dehghantanha, A., Choo, K.K.R.: Integrating privacy enhancing techniques into blockchains using sidechains. In: 2019 IEEE Canadian Conference of Electrical and Computer Engineering (CCECE), pp. 1–4. IEEE, 05–08 May 2019. https://doi.org/10.1109/CCECE.2019.8861821
28. Pawczuk, L., Massey, R., Holdowsky, J.: Deloitte's 2019 global blockchain survey: blockchain gets down to business. https://www2.deloitte.com/content/dam/Deloitte/se/Documents/risk/DI_2019-global-blockchain-survey.pdf
29. Polge, J., Robert, J., Le Traon, Y.: Permissioned blockchain frameworks in the industry: a comparison. ICT Express **59**, 134 (2020). https://doi.org/10.1016/j.icte.2020.09.002
30. Ron, D., Shamir, A.: Quantitative analysis of the full bitcoin transaction graph. In: Hutchison, D., et al. (eds.) Financial Cryptography and Data Security. Lecture Notes in Computer Science, vol. 7859, pp. 6–24. Springer, Heidelberg (2013). https://doi.org/10.1007/978-3-642-39884-1_2
31. Takahashi, K.: Blockchain technology and electronic bills of lading. J. Int. Marit. Law **22**, 202–211 (2016)
32. Veenstra, A.: Ocean transport and the facilitation of trade. In: Lee, C.Y., Meng, Q. (eds.) Handbook of Ocean Container Transport Logistics. International Series in Operations Research and Management Science, vol. 220, pp. 429–450. Springer, Cham (2015). https://doi.org/10.1007/978-3-319-11891-8_14
33. Wragg, E.: Exporters have '12-18 months' to prepare as G7 paves way for trade digitisation (2021). https://www.gtreview.com/news/fintech/exporters-have-12-18-months-to-prepare-as-g7-paves-way-for-trade-digitisation/
34. Wunderlich, S., Saive, D.: The electronic bill of lading. In: Prieto, J., Das, A.K., Ferretti, S., Pinto, A., Corchado, J.M. (eds.) Blockchain and Applications. Advances in Intelligent Systems and Computing, vol. 1010, pp. 93–100. Springer, Cham (2020). https://doi.org/10.1007/978-3-030-23813-1_12
35. Xu, X., et al.: A taxonomy of blockchain-based systems for architecture design. In: ICSA 2017, Piscataway, NJ, pp. 243–252. IEEE (2017). https://doi.org/10.1109/ICSA.2017.33
36. Zhang, Z., Yin, J., Liu, Y., Liu, J.: Deanonymization of litecoin through transaction-linkage attacks. In: 2020 11th International Conference on Information and Communication Systems (ICICS), pp. 059–065. IEEE, 07–09 April 2020. https://doi.org/10.1109/ICICS49469.2020.239510

A Dag Based Decentralized Oracle Model: Implementation and Evaluation

Ramy Gouiaa, Farouk Hdhili, and Marc Jansen(✉)

Computer Science Institute, University of Applied Science Ruhr West, Bottrop, Germany
{ramy.gouiaa,farouk.hdhili,marc.jansen}@hs-ruhrwest.de

Abstract. Blockchain comes up to solve centralization problems and to be an alternative solution to traditional finance. This technology seems to be very promising and has proven a big efficiency in ensuring trust among the network. It innovates and proposes new forms of governance and capital allocation, it promises decentralized financial services of credit, exchange, insurance, etc. It is the interweaving of computer programs, called smart contracts in a decentralized network, which replace traditional intermediaries and automate business processes. To make smart contracts in the blockchain context consistent and worthy, they must rely on solid data from the outside world, such as market prices observed on all exchange places for example. Blockchains are autonomous systems that are not natively designed to integrate data from the "outside", thus, oracles are external systems that solve the problem of feeding an isolated network with data from the outside. On the other hand, these systems are centralized, consequently, they grant a high risk of data corruption and unreliability issues.

In this paper, we present our decentralized oracle approach, based on a Dag (directed acyclic graph) model, its implementation as well as an evaluation of its accuracy in the context of live dynamic financial market data.

Keywords: Decentralized oracle · Blockchain · Tangle · DAG · Peer-to-peer

1 Introduction

Blockchain [1] is a decentralized peer-to-peer network that mainly guarantees trust. Based on this principle, this technology offers efficient solutions in various sectors, where trust plays a role. However, the blockchain does not allow data collection from an external source. It is the design of the blockchain that stands in the way. Initially, this technology was intended to be a secure ledger, in which each node in the network must be able to obtain the same final output from the same input. Furthermore, blockchain cannot rely on off-chain data incoming from any entity, because the verification might not ensure unanimity.

This infrastructure is intentionally designed to be deterministic to ensure trustlessness among the network. Indeed, it possible to dedicate a service responsible for manually entering external data into the blockchain. For instance, data can be retrieved from a specific data source and inserted into the blockchain at the location designated for it.

J. Prieto et al. (Eds.): BLOCKCHAIN 2021, LNNS 320, pp. 310–318, 2022.
https://doi.org/10.1007/978-3-030-86162-9_31

When the smart contract that requires this data is executed it will look for the data in the blockchain, at the address provided, and is executed according to this data. This service is defined as oracle for the blockchain.

Although this oracle's intervention raises an important question in practice, whether can we trust it or not. We can distinguish two types of oracles, on the one hand, a single entity that supplies off-chain data to a smart contract. Since this type of oracle works in a centralized manner and has an elementary infrastructure, it suffers from a bottleneck problem and can be an easy target for malicious attacks. The security of these systems is insufficient for widespread adoption by major players in the economy. If a contract is worth a big amount of money and the oracle that provides the data is hacked, the contract could, for example, transfer money to the wrong person. On the other hand in a decentralized oracle, many entities are involved to provide the blockchain with external data, which makes the network faster in operations and more reliable. So far research has shown various approaches of decentralized oracles and most of them are based on a reputation system, that rewards honest miners [2, 3].

In this paper, we will introduce an alternative approach for designing a decentralized oracle based on a directed acyclic graph and utilize a verification system adopted by the IOTA network [4].

2 Background

The concept of blockchain was first introduced in 2008. The term blockchain remained for years the prerogative of computer scientists and mathematicians, whose objective was to develop tools to guarantee respect for privacy on the internet. After the implementation of the first blockchain-based network called Bitcoin, this technology started reaching the attention of companies and the general public. Several researchers started exploring the blockchain and trying to expand its benefits from a virtual currency to touch a wider range of sectors. So far, the blockchain becomes a decentralized finance infrastructure managed by decentralized consensus and smart contracts.

A smart contract is the IT equivalent of a traditional contract. However, unlike a traditional contract, whose execution is governed by a legal framework, the smart contract does not require the intervention of any trusted third party and is, as a computer protocol, governed by computer code. A smart contract is a software program that has no legal authority as such. The smart contract relies on blockchain technology to secure and make the terms and conditions of its execution unforgeable.

In practice, a smart contract automatically executes predefined conditions that are recorded in a blockchain. Otherwise, the computer code based on the specified conditions or the available data stored in the blockchain, decides whether a contract will be fully or partially executed. To implement smart contracts some blockchain utilized existing programming languages and others designed new programming languages dedicated to smart contracts. Regardless of the programming language used, smart contracts are limited by the architecture of the network that runs them, knowing that the latter is completely isolated from the rest of the networks that make up the worldwide web. Consequently, smart contracts cannot communicate with external services or collect data from outside the network.

Off-chain services are implemented to provide smart contracts with desired data from the outside world, called oracles. As we mentioned in the previous section, there exist two types of oracles. The centralized oracles are a major stumbling block because blockchain as a decentralized network aims essentially to eliminate centralization and provide a trustless infrastructure.

Chainlink [2] is one of the most well-known decentralized oracles in the decentralized finance world. It is an off-chain oracle with a network of independent node operators that collects and delivers data to its middleware application. The middleware application aggregates and combines them into a single data output and delivers them to the blockchain. This allows them to be read by smart contracts that rely on them. It is designed to be modified to feed any blockchain but its widely used for Ethereum-based dApps (decentralized applications).

Band Protocol[1] is an on-chain oracle that runs on an independent Cosmos-based blockchain called BandChain. It allows developers to write custom oracle scripts to fetch data for use in their dApps. BandChain uses a delegated proof of stake consensus mechanism, and its validators fetch the queried data, which is published first to BandChain and then to the requesting platform chain. Similar to Chainlink, validators are paid with the native BAND token for providing the data and must stake these tokens, so any misreporting can be penalized.

Gravity Protocol [3] is also a decentralized blockchain-agnostic oracle technology for cross-chain communication between blockchains proposed by the Waves platform. This Protocol aims to overcome the weaknesses of other oracle approaches such as interoperability, token incentivization, network governance, and administration.

3 Research Question and Analysis Strategy

In the previous section, we introduced the most prominent decentralized oracle designs, in which the verification process is done by beforehand known validators.

This research aims to propose a new decentralized oracle approach allowing everyone to participate in the validation process and evaluates the accuracy of the data that our network provides. For this purpose, we implement an agnostic oracle that provides any blockchain with stock prices of financial markets. For the evaluation, we utilized the bitcoin price and we agreed to use "MarketWatch" and "Yahoo Finance" and "Wall Street Journal" as data sources. We will later compare the price provided by our network against the prices of the other sources at a given timestamp to examine the deviation. More precisely, we wrote a script that extracts data from our system and the mentioned data sources every second for 5 min. We use the gathered data to perform our statistical analysis. Our evaluation will investigate the following research question: To what extent is a DAG-based oracle system accurate?

4 Implementation

In this part, we introduce our decentralized oracle approach and its technical aspect. Our implementation consists of three main parts the oracle network, client and the evaluation script. The following figure (Fig. 1) illustrates the architecture of our oracle:

[1] https://docs.bandchain.org/whitepaper/.

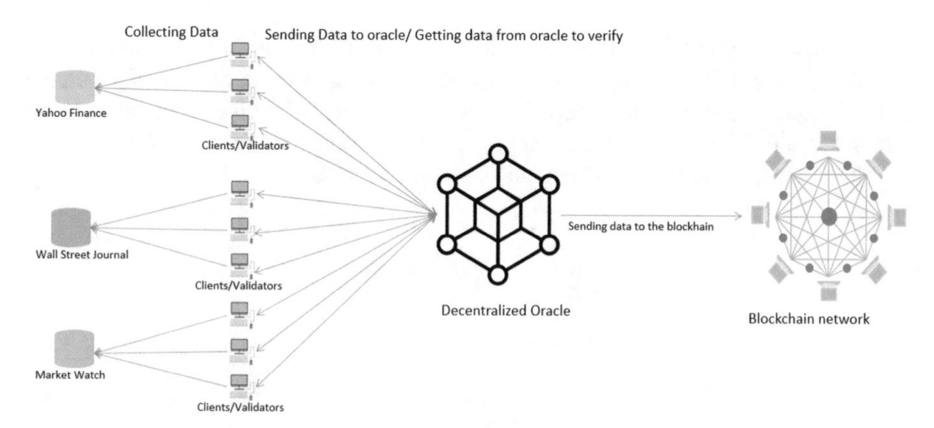

Fig. 1. Oracle architecture

As shown in Fig. 1, the first layer represents the sources from which the clients collect the data. In our case, we collect bitcoin prices from three different well-known financial platforms providing livestock and cryptocurrency rates.

In the second layer, we see the client-side, which is responsible for gathering data from our data sources and verifying two random unverified transactions. We consider that each data recorded in the oracle is a transaction. Each client/validator verifies the data collected from the source with which it is associated.

The next layer depicts our oracle system designed to be an interconnected peer-to-peer network, in which every node receives transactions from clients and store them in a Directed Acyclic Graph.

Every transaction will be diffused to the entire network, consequently, all peers have the same state of the Graph.

Finally, we could query any node from the decentralized oracle to get the last verified transaction of a given data source.

In the following subsections, we will describe main actors within our system.

4.1 Decentralized Oracle

Our oracle is a decentralized fully connected peer-to-peer network written in Javascript (Nodejs). Peers communicate with each other using HTTP requests. Moreover, our verification mechanism is inspired by the tangle architecture proposed by IOTA.

The IOTA network works using tangle technology which is similar to the blockchain as a fully decentralized network. The tangle is a technology that uses the concept of an acyclic directed graph, which eliminates the block validation delay. As a result transaction throughputs are super-fast and even accelerate as the number of network operations increases [5].

Validation of a transaction based on the tangle approach is the process by which other entities select that transaction using the "tips selection algorithm" [6], and then perform the proof of work necessary to approve that transaction. When several other transactions repeat this process one after the other, the validation of a transaction becomes increasingly important (A is directly validated by B, itself directly validated by C, so C indirectly validates A, and so on). Once the transaction exceeds a certain number of validations considered as "acceptable", it can be accepted by the recipient as being completely valid. The transaction then goes from "pending" to "confirmed".

The following figure (Fig. 2) depicts the tangle schema:

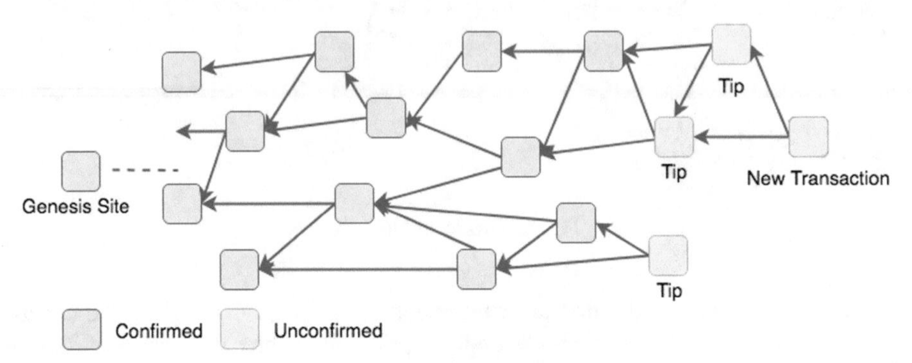

Fig. 2. Tangle schema (https://medium.com/@guofy15775100131/an-empirical-analysis-of-tip-selection-anomalies-in-the-iota-tangle-introduction-d9c3ac51494f)

We can summarize the workflow of each peer in the following points.

Each node of the network:

- Receives a transaction from a client/validator at a time T and registers it as a node in the Dag by assigning the value 'unverified' for this transaction.
- Broadcast the transaction to the whole network to ensure the synchronization of the Dag within the network.
- Sends two unverified recent transactions (Tips not older that one minute) randomly to the client/validator for verification.
- Adds an edge between the node and each Tip verified by the client.
- Broadcast the new state of the Dag to all other members of the network.
- Update the status of a transaction to 'verified' if a node in our graph reaches n In-degree (in our case n is equal to 3 for testing).
- Outputs the last verified transaction of a given data source.

To ensure data persistence, each peer stores dag changes in a graph database (neo4j) as transactions progress. While a graph based database seems to provide a natural data structure to store DAG based data, Neo4j also allows us to visualize the current state of the dag and to query the database for further analysis as shown in the following Figure (Fig. 3).

To prevent too much expansion of the tangle and an excessive size which would increase the execution time of finding tips, we considered only unverified transactions not older than 1 min as tips, this may of course differ from use-case to use-case and could flexibly be adapted. Old transactions will be ignored.

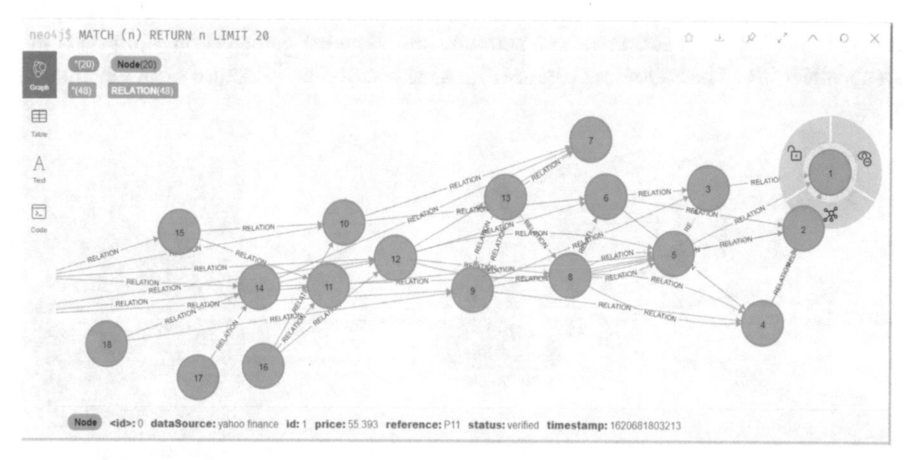

Fig. 3. Tangle illustration in Neo4j Graph database showing bitcoin price

4.2 Clients

A Client is a stand-alone intermediary program that works as follows:

- Permanently scraps bitcoin price of a specific data source (website)
- Collect actual bitcoin price from that data source
- Sends actual price to an oracle's Node
- Verify tips incoming from the oracle after each transaction.

A transaction is considered as verified only if the difference between the tips' price coming from the oracle and the actual price coming from the data source is lower than 0.05. Here, of course, the acceptance interval could also be flexibly adapted to the use-case.

In this research, we used a web scraping tool to get the bitcoin price from the mentioned external data source (see Fig. 1). We assume that APIs are faster and more efficient to obtain instantaneous data but they are generally associated with high costs. To use multiple data sources, we implemented a scraping tool that extracts the price. In this case, the querying time will increase, which can noteworthy influence the results in our evaluation.

5 Evaluation and Results

To evaluate our system, we installed three peers on different computers. On each computer, we executed one client for each peer that will collect the data from one of the presented data sources. Another program has been implemented to perform the assessment. This program will query every second for 5 min, one peer of our decentralized oracle including all mentioned data sources, and store all gathered bitcoin prices in a spreadsheet file. The following figure (Fig. 4) shows the architecture of our evaluation system:

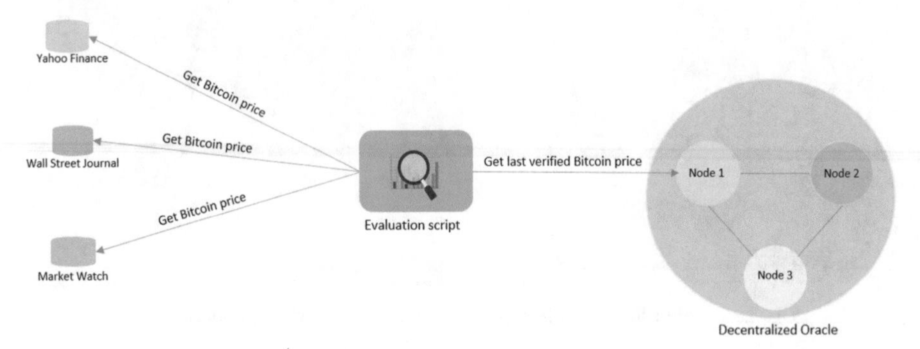

Fig. 4. Evaluation workflow

For the assessment, the following statistical analysis will be performed:

- How often the oracle's price matches at least one of the external data sources' price
- Average bitcoin price from the 3 external data sources for each record
- Deviation from the mean
- (Deviation from the mean)^2
- Standard deviation.

The first point will help us count approximately the number of times our oracle will provide an accurate bitcoin price against our data source.

We will compare the Average bitcoin price coming from our data sources with the value provided by our oracle. We determined the deviation about the mean in percent to make the values more interpretable. Finally, we will determine the standard deviation to get a general idea of the accuracy of our oracle.

To consolidate the evaluation we repeated the experiment three times in different periods.

5.1 Evaluation Results

In this subsection, we describe how we analyzed our spreadsheet file where all records are stored and finally we depict the evaluation results for each execution.

The following table (see Table 1) shows an example of a row from our spreadsheet file. We executed our script during 5 min, which at the end will create a file of 300 rows of data. In this example, we observe that our external data sources themselves do not provide the same bitcoin price at a given timestamp. Otherwise, when the oracle price matches at least with one of our external data sources, can be considered sufficient for our evaluation. To measure the deviation of the prices in the entire file, we used the average of all prices coming from our three external data sources.

Table 1. An example of a row in our dataset with at least one data external source matches the oracle price.

Timestamp	Oracle	Market watch	Yahoo finance	Wall street journal	Oracle Price = 1 × external source Price	AVG
1611238427503	31,829$	31,829$	32,032$	31,829$	1	31,896$

The table below (Table 2) shows an example of a data row where the oracle price do not match any external data source value. Though the deviation about the mean in this case do not exceed 0.22%.

Table 2. Example of a row in our dataset where the oracle's price does not match any external data's price

Timestamp	Oracle	Market watch	Yahoo finance	Wall street journal	Oracle Price = 1 × external source Price	AVG
1611238499868	31,899$	31,808$	31,867$	31,808$	0	31,827$

The following table (Table 3) shows the final results after executing the evaluation script 3 times.

Table 3. Execution results after three experiments

	Oracle's price = (1 × data Source's price)	Variance	Standard deviation
First execution	130/300	0,17%	0,42%
Second execution	143/300	0,15%	0,39%
Third execution	207/300	0,12%	0,28%

After the first execution, we notice that nearly half of node prices match at least one value of the used data sources. The second execution does not differ greatly from the first. We also witness a significant price fluctuation during the first and second execution.

In the third execution, we see that two-thirds of our oracle prices match with our data sources. In contrast to the two previous executions, during the last execution, we observed less variation in bitcoin price.

We could conclude that the accuracy of our system will slightly decrease about 0.4% within a period of 5 min when the data provided by the sources shifts considerably.

6 Conclusion

This work proposes a different approach to the existing decentralized oracle systems and tries to strengthen the trust among the network by increasing the number of validators and allowing everyone to contribute to the process.

The evaluation results have shown that we successfully implemented our oracle providing a minimal error rate. It is considered that the scenario chosen to evaluate this approach offers reliable results because the cryptocurrency market is by its nature very dynamic. A further experiment can be done with the proposed approach by using an appropriate API for cryptocurrencies offering real-time data.

Other less volatile data can be examined and for longer periods to study the efficiency of such a system, as oracles are designed to be permanently operational. This will provide us with a more accurate evaluation of the system.

References

1. Nakamoto, S.: Bitcoin: a peer-to-peer electronic cash system (2008). https://bitcoin.org/bitcoin.pdf
2. Ellis, S., Juels, A., Nazarov, S.: Chainlink a decentralized oracle network. Accessed 15 Mar 2021
3. Pupyshev, A., et al.: Gravity: a blockchain-agnostic cross-chain communication and data oracles protocol. arXiv preprint arXiv:2007.00966 (2021)
4. Silvano, W.F., Marcelino, R.: Iota Tangle: a cryptocurrency to communicate Internet-of-Things data. Future Gener. Comput. Syst. **112**, 307–319 (2020)
5. Popov, S.: The Tangle (2018). https://assets.ctfassets.net/r1dr6vzfxhev/2t4uxvsIqk0EUau6g2sw0g/45eae33637ca92f85dd9f4a3a218e1ec/iota1_4_3.pdf. Accessed 5 Mar 2021
6. Carlin, B.P., Chib, S.: Bayesian model choice via Markov chain Monte Carlo methods. J. R. Stat. Soc.: Ser. B (Methodol.) **57**(3), 473–484 (1995)

Blockchain-Enabled Next Generation Access Control

Yibin Dong, Seong K. Mun, and Yue Wang(✉)

Virginia Polytechnic Institute and State University, Arlington, VA 22203, USA
{yibin.dong,munsk,yuewang}@vt.edu

Abstract. In the past two decades, longitudinal personal health record (LPHR) adoption rate has been low in the United States. Patients' privacy and security concerns was the primary negative factor impacting LPHR adoption. Patients desire to control the privacy of their own LPHR in multiple information systems at various facilities. However, little is known how to model and construct a scalable and interoperable LPHR with patient-controlled privacy and confidentiality that preserves patients' health information integrity and availability. Understanding this problem and proposing a practical solution are considered important to increase LPHR adoption rate and improve the efficiency as well as the quality of care. Even though having the state-of-the-art encryption methodologies being applied to patients' data, without a set of secure access control policies being implemented, LPHR patient data privacy is not guaranteed due to insider threats. We proposed a definition of "secure LPHR" and argued LPHR is secure when the security and privacy requirements are fulfilled through adopting an access control security model. In searching for an access control model, we enhanced the National Institute of Standards and Technology (NIST) next generation access control (NGAC) model by replacing the centralized access control policy database with a permissioned blockchain peer-to-peer database, which not only eases the race condition in NGAC, but also provides patient-managed access control policy update capability. We proposed a novel blockchain-enabled next generation access control (BeNGAC) model to protect security and privacy of LPHR. We sketched BeNGAC and LPHR architectures and identified limitations of the design.

Keywords: Secure LPHR · Blockchain-enabled next generation access control · Privacy · Confidentiality · Security

1 Introduction

The longitudinal personal health record (LPHR) is "an electronic, lifelong resource of health information needed by individuals to make health decisions" [1] and to "improve the quality and efficiency" of their own health care [2]. There are three types of personal health record (PHR) implementations: standalone PHR, tethered PHR, and untethered [3] PHR.

J. Prieto et al. (Eds.): BLOCKCHAIN 2021, LNNS 320, pp. 319–328, 2022.
https://doi.org/10.1007/978-3-030-86162-9_32

- *Standalone PHR* is managed by patients without health care providers' inputs [4]. Patients have full control of data in a standalone PHR. However, standalone PHR is not connected to any of the electronic health record (EHR) systems [4].
- *Tethered PHR* is confined in a single EHR system [3]. In this setting, patients and providers form a partnership. Patients are allowed to directly access their clinical data with read-only permission. Additionally, patients can request to add supplementary information to their own patient records. The hosting health care provider of the tethered PHR has full control of the patient records [4].
- When patients receive care from multiple providers that use different EHRs, a cross-organizational PHR or integrated non-tethered PHR or *untethered PHR* is required to provide a complete patient PHR dataset [3]. There has been more interest to develop untethered PHR due to its capability of building longitudinal patient health records [3]. Furthermore, patients showed strong interest in untethered PHR, with expectations of easy access to a centralized PHR system and full control of their own PHR [4], where patients, providers, payers, and stakeholders are sharing the same platform [3].

Digitizing patients' medical records was started and mandated in "the Health Insurance Portability and Accountability Act of 1996 (HIPAA)" [5]. One of the objectives of HIPAA was "to simplify the administration of health insurance" [6] that led to use of EHR. The first HIPAA Privacy Rule was released [5] to "improve privacy standards and to restrict the disclosure of Protected Health Information (PHI) and personal identifiers to unauthorized individuals" [5]. There was no incentive to use untethered PHRs to build LPHR [7] at that time. To promote implementations of EHR and fix a loophole in the HIPAA privacy rule, "the Health Information Technology for Economic and Clinical Health Act ("the HITECH Act") was enacted" [8] with PHR adoption incentives. However, despite the U.S. federal government's laws and regulations, including incentives and penalties, LPHR adoption rate has been low (Fig. 1) in the past two decades [9, 10]. A.A. Abd-alrazaq et al. conducted a comprehensive PHR literature review of peer-reviewed publications between 2000 and 2018 and concluded patients' privacy and security concerns was the primary negative factor [9]. Moreover, health care providers are using some type of EHR systems that are managed by various EHR vendors. Hence patients' health care records can be scattered in different information systems at various facilities. Patients showed strong interest in untethered PHR with expectations of easy access to and full control of their own LPHR [4]. Patients want to be assured that nobody can access their LPHR without patients' authorization [9]. Furthermore, on March 9, 2020, "the ONC Cures Act Final Rule" [11] and "the CMS Interoperability and Patient Access Final Rule" [12] were released to truly give patients authority to control access to their health care data. However, little is known how to model and construct a scalable and interoperable LPHR with patient-controlled privacy and confidentiality that preserves patients' health information integrity and availability. This motivated us to research an effective novel design of secure LPHR that can ease the patients' security and privacy concerns in order to increase LPHR adoption rate in the Unites States.

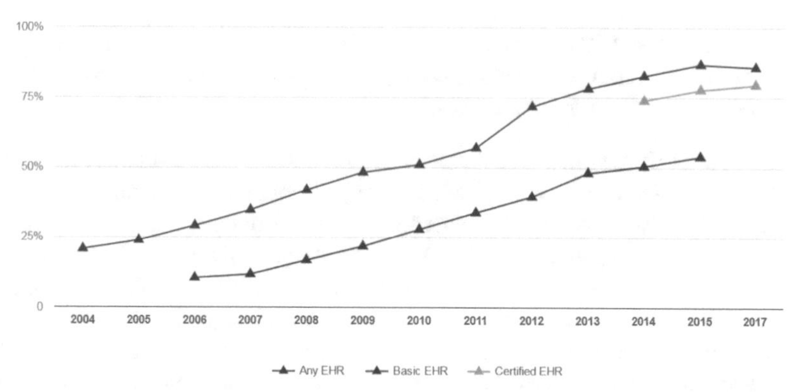

Fig. 1. "Office-based Physician Electronic Health Record Adoption (2004–2017)" [10].

During reviewing works accomplished by peers from 1970 to 2020, we found encryption, which was used as the primary methodology to protect privacy and confidentiality of patients' data, could only prevent external risks but not insider threat [13]. A secure access control model along with implementation mechanism are desired to remediate internal threats. After analyzing LPHR requirements and both traditional as well as next generation access control models, we explained our preference of the National Institute of Standards and Technology (NIST) next generation access control (NGAC) model [14]. However, NGAC inherited a drawback of "race condition" from distributed systems. We overcame this by integrating permission blockchain Hyperledger Fabric (HF) [15] with NGAC and reconstructed a novel blocked-enabled NGAC (BeNGAC) model. NGAC is also complementary to HF and boosts the HF confidentiality protection capability through next generation access control policies. The BeNGAC offers granular patient consent capability so it can improve the trustfulness between patients and providers who manage the untethered PHR. We sketched BeNGAC functional architecture as well as LPHR architecture. Key properties of BeNGAC were identified that can fulfill the requirements of a scalable and interoperable LPHR with patient-controlled privacy and confidentiality that preserves patients' health information integrity and availability. A high-level example of BeNGAC policy configuration was given to demonstrate the privacy protection capability of BeNGAC.

2 LPHR Privacy and Security Literature Review

This systematic review was conducted using instructions from the "Preferred Reporting Items for Systematic Reviews and Meta-Analyses (PRISMA) Statement" [16] to avoid publication selection bias [17]. Library worldwide union catalog WorldCat [18] and Google Scholar were employed as the main searching tools. The search started on May 10th, 2020 and finished on May 23rd, 2020. The search terms were based on keywords: longitudinal [personal | patient] [health | medical] record [security | privacy | confidentiality]. Publications were chosen from year 1970 to 2020. Studies were excluded if they were not peer-reviewed to avoid selection bias [17]. Only English language was included in the search. The publications worldwide were included. 1,232 results in

WorldCat libraries and 32 outcomes in Google Scholar were returned. Duplicated studies were excluded. After reviewing titles and abstracts, 23 peer-reviewed publications were chosen to conduct further analysis.

Review Results: Encryption was the primary approach to protect privacy and security from external threats. However, Miller & Tucker [13] conducted an empirical research of the relationship between patient data encryption and loss of data. The authors concluded applying encryption to patient data does not decrease proclaimed incidents of data loss. On the contrary, there were increased likelihood of public announcements of patient data loss because of human errors, from either computer equipment stolen or insider security threats. Miller & Tucker argued that vulnerability of misused privileges inside organizations are bigger than external threats. The authors recommended policymakers to expand the scope of security measure by including user access control which is more suitable than encryption to tackle the insider vulnerability or threat.

3 LPHR Requirements

A) **LPHR Security and Privacy:** LPHR security rules are developed to protect information system against intrusion, either intentional or by accident, while LPHR privacy rules are built to hide information from unauthorized people or processes. Patients' health information availability, integrity, and confidentiality are the core of LPHR security. Under the HIPAA Security Rule, LPHR availability means personal health records "are accessible and usable on demand by authorized persons" [19]. LPHR integrity means personal health records "are not altered or destroyed in an unauthorized manner" [19]. LPHR confidentiality, by its security definition, preventing unauthorized access and disclose of LPHR, serves as a means to protect patients' privacy so that patients' health information is not revealed to unauthorized subjects. Therefore, confidentiality is a property of both LPHR security and privacy. The HIPAA Security Rule's [19] confidentiality property of LPHR reinforces the privacy requirements in the HIPAA Privacy Rule, which disallow unauthorized use and disclosure of LPHR.

B) **LPHR Authorization:** LPHR consists of two independent components, standalone PHR and untethered PHR. A patient has control of both parts of the patient's health records stored in two partitions of the storage under different rules and regulations. The LPHR authorization requirements are summarized in Table 1.

C) **Other Requirements:** changes to LPHR are tamper resistant; D) access is fully auditable; E) LPHR is enterprise scalable; F) LPHR is distributed; G) LPHR is interoperable and integrable to other EHRs.

Table 1. LPHR authorization.

LPHR patients' data source	Patients' permission	LPHR providers' or administrators' permission	Third parties' permission
Data generated by patients	Full control	Manage on behalf of patient upon patients' authorization	Can request to access individuals' information with patients' authorization
Patients' EHR data in covered entity	Read	Disclose to patients' LPHR with patients' authorization	Can request to access individuals' information with patients' authorization
Untethered PHR providers (or covered entities)	Read access Can request to submit supplemental information to be added to PHR Can control access to the patients' PHR by third parties Can control access to the patients' PHR by the covered entities of the untethered PHR	Can use individuals' PHI permitted by HIPAA Privacy Rule Can disclose individuals' PHI to third parties permitted by HIPAA Privacy Rule	Can request to access individuals' information with patients' authorization

4 System Design and Methods

Access control model is a formal representation of access control policies and their implementation reinforcement mechanisms on the systems being designed [20]. "By proving the access control model is secure (w.r.t. meeting the policies compliance requirements), demonstrating the model is correctly implemented through access control mechanisms which fulfil the security requirements, we can convince users and vendors that the system adopting the access control model is secure" [21, 22]. Hence, the LPHR adopted the secure access control model is a *secure LPHR*.

There are two categories of access control models. Traditional access control model is established on a closed centrally controlled and server-oriented access control environment, in which users are well-known [23]. "Discretionary access control (DAC), mandatory access control (MAC), and role-based access control (RBAC)" [20] belong to traditional access control model. Next generation access control model is based on open access control surroundings, where users can be unknown or centrally known. "Attribute-based access control (ABAC)" [20] is next generation access control model. The NIST NGAC, a type of ABAC, was standardized by the "InterNational Committee for Information Technology Standards (INCITS)" [14]. NGAC was chosen to construct secure LPHR because firstly, users of LPHR can be unknown or centrally known, which fits the property of NGAC. Secondly, NGAC provides unifying access control policies while the resources reinforcing the access control polices locally are distributed [20], which aligns with data distributedness requirement of LPHR. Thirdly, NGAC is enterprise scalable [20] that fulfills the requirement of LPHR scalability. Lastly, NGAC model is "inherently policy neutral" [24] and applies to any users or objects that have common attributes described by a policymaker, which gives much flexibility when implementing in organizations.

However, in the NGAC model and the NIST Policy Machine [24] mechanism, since the decision making and policy expression are disjointed, it can cause a "race condition" in distributed systems [24]. We addressed this problem by replacing the NGAC policy database with a peer-to-peer decentralized blockchain database and reshaped the NGAC model to be BeNGAC. We chose "permissioned" blockchain HF [15] because it fits the property of LPHR that patients and providers forming a trusted network. The immutability of HF guarantees data integrity which is timestamped. Moreover, the "concurrency control" [25] in HF eases the race condition in NGAC model by utilizing the "HF consensus" [25]. The access control data is stored in off-chain private database, while the transactional audit logs are on-chain. On the other side, the data in HF is protected with little degree of confidentiality, and this can be perfected by NGAC access control policies. Commonly, both NGAC and HF are finite state machines [15, 26], which are naturally married together as a new model BeNGAC.

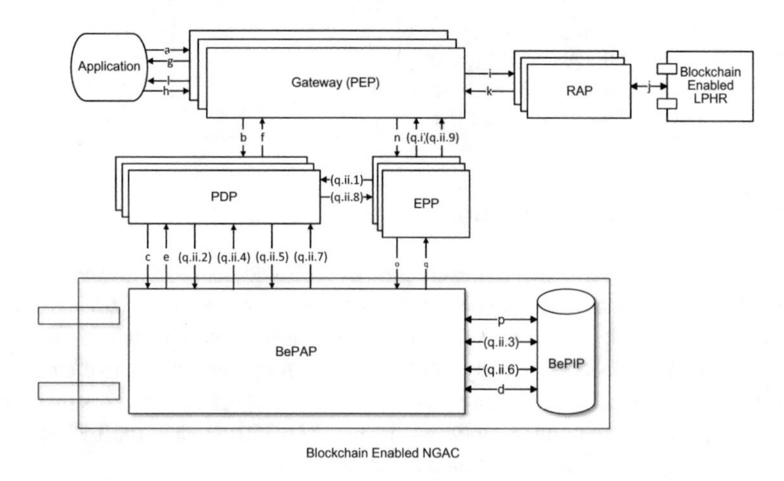

Fig. 2. Blockchain-enabled NGAC functional architecture [24].

BeNGAC Functional Architecture (Fig. 2): In this new model, resources ("policy enhancement point (PEP), policy decision point (PDP), event processing point (EPP), resource access point (RAP), policy administration point (PAP), and policy information point (PIP)" [24]) forcing the access control policies are locally distributed. "PEP" [24] checks the authorization permissions via "PDP" [24], which queries the blockchain-enabled PIP (BePIP) policy database via blockchain-enabled PAP (BePAP). BePAP acts as an endorsing or committing peer. BePAP and BePIP are decentralized and constitute a BeNGAC policy administration unit. The applications accessing the same policy database share the identical BeNGAC policies. During "event response (obligation)" or "prohibition" [24], "EPP" [24] queries the same BeNGAC policy database BePIP.

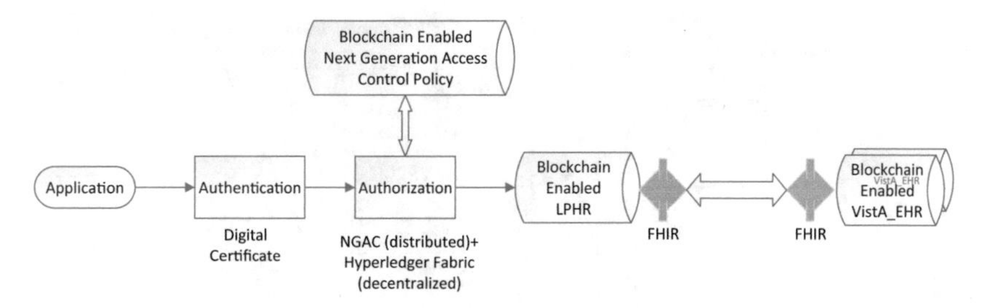

Fig. 3. LPHR architecture.

LPHR Architecture (Fig. 3): A user accesses his or her LPHR through an application. The user is assigned with a digital certificate which is pre-generated by the certificate administrators. Upon successful authentication, the user is presented with a view of his or her own LPHR. This list of authorized view of information is retrieved from BeNGAC policy database.

The authentication via digital certificate governs the LPHR login security with the user's identity. The BeNGAC policy protects the LPHR from being "altered or destroyed in an unauthorized manner" [19], i.e. ensuring *LPHR integrity*. The policy also guarantees the LPHR "is accessible and usable on demand by an authorized person" [19], which provides *LPHR availability*. The policy disallows unauthorized use and disclosure of LPHR to unauthorized parties, which safeguards *LPHR confidentiality*. Therefore, with digital certificate guarded secure authentication and BeNGAC policies, we can meet the LPHR security and privacy as well as access authorization requirements, which include U.S. privacy policies, laws and regulations.

Unauthorized changes are prohibited by BeNGAC policies. This meets the tamper resistance change requirement. BeNGAC is enterprise scalable because both NGAC and HF are enterprise scalable [15, 24].

Shared Access Control Policy Database: The patient has full control of the permissions of his or her own records in the LPHR. A patient is able to give very granular consent to use their LPHR by trusted providers, which will improve the trustfulness between patients and providers. The access control information is stored in a dedicated BeNGAC database, which is distributed yet decentralized among trusted parties in a secure blockchain network. Any member in the patient's BeNGAC network share the same access control information that is updated in the order of timestamp which is tamper resistant and immutable. The changes are recorded in blockchain audit logs. The changes are non-repudiable. The blockchain based peer-to-peer BeNGAC database avoids racing condition during policy reinforcement.

The patients and health care providers are granted permissions to use the applicant client to access the HF network. The patient and one or more designated health care providers in the HF network are assigned with endorsing peer role. The patient and all health care providers in the HF network are committing peers.

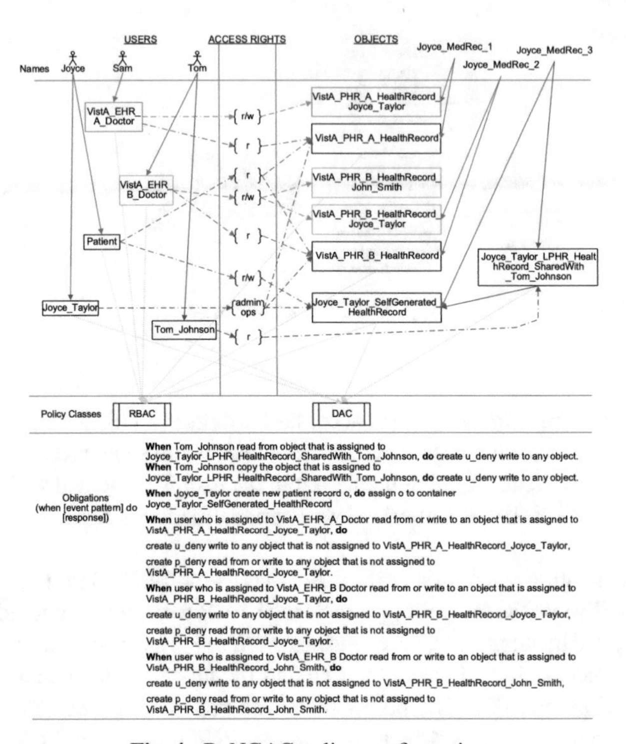

Fig. 4. BeNGAC policy configuration.

An example of BeNGAC policy configuration has been provided in Fig. 4 to illustrate the capability of privacy protection. In this implementation, from a patient point of view, the LPHR consists of three parts: patient health records stored in EHR VistA PHR A, patient health records contained in EHR VistA PHR B, and patient self-generated health records from personal devices or manual inputs. RBAC and DAC policy classes are governed by policy administrators. Two RBAC data leakage prevention obligation rules are configured:

Configuration Rule 1:
When process p performs (r, o) where $o \rightarrow med\_rec$ **do** create $u\_deny(p, \{w\}, \neg o)$.

Configuration Rule 2:
When process p performs $(copy, o)$ **do** create $u\_deny(p, \{w\}, \neg o)$.

The derived privileges of the configuration are the following:

(Joyce, r, Joyce_MedRec_1), (Joyce, r, Joyce_MedRec_2), (Joyce, r, Joyce_MedRec_3), (Joyce, w, Joyce_MedRec_3), (Sam, r, Joyce_MedRec_1), (Sam, w, Joyce_MedRec_1), (Tom, r, Joyce_MedRec_2), (Tom, w, Joyce_MedRec_2), (Tom, r, Joyce_MedRec_3).

5 Discussions

There are few limitations of the design. Firstly, being the kernel of the security features offered by the LPHR, owner's secret private key is essential and the passport to manage his or her LPHR securely. All transactions are dependent on the secret private key for authentication. This security feature is inherited from the blockchain technology itself, and naturally the problem when the owner losing the private secret key presented a road blocker if there is no solution to this challenge.

Secondly, in general, LPHR owner can grant permissions to a legitimate third party, for instance a specialist doctor he or she will visit during a referral encounter. There are situations such as emergency departments visit, where LPHR access is desired by the ER physicians to make better decision of a care plan by using the patient's LPHR information such as medications taken, allergy conditions, recent doctors' visits, chronic diseases, and recent laboratory test results. However, as a limitation that the patient needs to directly grant the permissions of LPHR to the doctors in ER, it is not uncommon that the patient is unconscious and cannot authorize the access of his or her LPHR to the doctors in ER facility.

These two problems can be solved by using proposed BeNGAC model. The former incident is rare, so the recovery of the owner's private secret key can be purposely designed with a sophisticated and secure execution plan by using BeNGAC and RBAC separation of duty (SoD) [20] capability. The latter can be worked out by using BeNGAC and DAC policy.

In this research, we proposed a solution to ease the "privacy and security concerns of patients" [9] when adopting LPHR in the United States. We provided a formal definition of "secure LPHR" and introduced a novel BeNGAC model. We designed a scalable and interoperable LPHR with patient-controlled privacy and security. While "blockchain-based self-sovereign identity" [27] focuses on identity management (decentralized authentication), the proposed solution contributes to the knowledge by emphasizing granular patient-controlled decentralized authorization, which has gone beyond the trust of identity, i.e. governance of trust around confidentiality, availability, integrity, and privacy of LPHR.

We are in the process of implementing the design and applying to a use case of patient data access control and platform sharing among providers. We are also developing a prototypical evaluation and plan to measure the properties of the design such as processing delay, security and privacy control, scalability, auditability, and tamper resistibility.

References

1. AHIMA. Defining the personal health record. J. AHIMA **76**(6), 24–25 (2005)
2. Personal health records and the HIPAA privacy rule. https://www.hhs.gov/sites/default/files/ocr/privacy/hipaa/understanding/special/healthit/phrs.pdf
3. Key considerations, Venesco and Personal Health Records Community of Practice, in Health Information Exchange (HIE) Shared Collaborative Learning Combine. Venesco LLC (2015)
4. Assadi, V.: Adoption of integrated personal health record systems: a self-determination theory perspective. M. University, Hamilton, pp. 1–195 (2013)

5. HIPAA history. https://www.hipaajournal.com/hipaa-history/
6. Pub. L. No. 104-191, 110 Stat. 1936 (1996)
7. Sterud, B.: Practitioner application: the challenges in personal health record adoption. J. Healthc. Manage. **64**(2), 109–110 (2019)
8. H.R.1 - American Recovery and Reinvestment Act of 2009 PLAW-111publ5 (2009)
9. Abd-Alrazaq, A.A., et al.: Factors that affect the use of electronic personal health records among patients: a systematic review. Int. J. Med. Inform. **126**, 164–175 (2019)
10. ONC. Office-based physician electronic health record adoption (2017). https://dashboard.healthit.gov/quickstats/pages/physician-ehr-adoption-trends.php
11. ONC cures act final rule, 9 March 2020. https://www.healthit.gov/curesrule/
12. CMS interoperability and patient access final rule, 9 March 2020. https://www.cms.gov/Regulations-and-Guidance/Guidance/Interoperability/index
13. Miller, A.R., Tucker, C.E.: Encryption and the loss of patient data. J. Policy Anal. Manage. **30**(3), 534–556 (2011)
14. INCITS-499-comments-due-2–28–2017 Next Generation Access Control - Functional Architecture (NGAC-FA) (2016)
15. Androulaki, E., et al.: Hyperledger fabric: a distributed operating system for permissioned blockchains. In: Proceedings of the Thirteenth EuroSys Conference, Association for Computing Machinery, Porto (2018)
16. Moher, D., et al.: Preferred reporting items for systematic reviews and meta-analyses: the PRISMA statement. Int. J. Surg. **8**(5), 336–341 (2010)
17. Knobloch, K., Yoon, U., Vogt, P.M.: Preferred reporting items for systematic reviews and meta-analyses (PRISMA) statement and publication bias. J. Cranio-Maxillofacial Surg. **39**(2), 91–92 (2011)
18. OCLC. WorldCat. https://www.worldcat.org/
19. HIPAA Security Rule (2013). https://www.hhs.gov/hipaa/for-professionals/security/laws-regulations/index.html
20. Hu, V.C., Ferraiolo, D.F., Kuhn, D.R.: Assessment of Access Control Systems, pp. 20899–8930. NIST, Gaithersburg (2006)
21. Samarati, P., de Vimercati, S.C.: Access control: policies, models, and mechanisms. In: Focardi, R., Gorrieri, R. (eds.) FOSAD 2000. LNCS, vol. 2171, pp. 137–196. Springer, Heidelberg (2001). https://doi.org/10.1007/3-540-45608-2_3
22. Landwehr, C.E.: Formal models for computer security. ACM Comput. Surv. (CSUR) **13**(3), 247–278 (1981)
23. Jaehong, P., et al.: Towards usage control models: beyond traditional access control. In: Symposium on Access Control Models and Technologies, pp. 10121–0701. ACM, New York (2002)
24. Ferraiolo, D.F., et al.: Policy machine: features, architecture, and specification. NIST (2015)
25. IBM. Hyperledger Fabric a Blockchain platform for the enterprise. https://hyperledger-fabric.readthedocs.io/en/latest/
26. Ferraiolo, D., Atluri, V., Gavrila, S.: The policy machine: a novel architecture and framework for access control policy specification and enforcement. J. Syst. Architect. **57**(4), 412–424 (2011)
27. Houtan, B., Hafid, A.S., Makrakis, D.: A survey on blockchain-based self-sovereign patient identity in healthcare. IEEE Access **8**, 90478–90494 (2020)

Attacks on Blockchain Based Digital Identity

Akshay Pillai, Vishal Saraswat(✉),
and Arunkumar Vasanthakumary Ramachandran

Robert Bosch Engineering and Business Solutions Pvt. Ltd., Bangalore, India
{Akshay.Pillai,Vishal.Saraswat,Arunkumar.VR}@bosch.com

1 Introduction

An identity is a representation, in a given context, of a user, device, process or service which makes it unique or distinctive. It consists of attributes, traits, or preferences unique to an individual and may contain their date of birth, bank account number(s), social security number, passport information, family members' information, (cryptographic) credentials, biometrics, or other personally identifying information (PII) necessary to uniquely identifying the individual. An attacker in possession of such information can damage the individual in various ways such as financially by accessing their bank accounts, or infiltrate their organizations, or putting blame on the individual for some wrongdoing.

Blockchain based digital identity management systems provide a peer-to-peer (P2P) and decentralized platform and allows a user to completely control their identity information. Identity management is generally done using *DApps* (decentralized applications) which store an individual's complete identity and for each addition, updation and deletion of information, these DApps use crypto tokens for completing the transaction in blockchain. While in traditional digital identity management, user identity is proven by associating the user-id with a password, in blockchain based identity management systems, the proofs are based on user-id and password as well as public and private keys. Each user gets a public address which is unique for each user and a private key, which is stored securely in either with user's own system or the organization server.

1.1 Our Methodology

In this paper, we survey the popular attacks on traditional and blockchain based identity management systems and describe the mechanism of these attacks. We study the attack goals, attackers' incentives in carrying out the attack, and their requirements (cost or otherwise) in successfully carrying out the attack. We then review the possible damage scenarios and damage consequences of these attacks. Finally, we do an attack cost vs benefit analysis based on which an organization may do risk analysis.

J. Prieto et al. (Eds.): BLOCKCHAIN 2021, LNNS 320, pp. 329–338, 2022.
https://doi.org/10.1007/978-3-030-86162-9_33

2 Attacking Digital Identity via Client User Interface

2.1 Credential Stuffing

Credential stuffing is a type of brute force attack in which an attacker takes advantage of the struggle of a user to select different passwords for different accounts. If account credentials are compromised in a data breach, attackers try these compromised credentials on other websites or applications to test if the credentials are re-used. These attacks are usually done by bots [9].

Mechanism: The attacker gains/collects credentials from data breach and password dump site and uses automated tools to test credentials across different websites to steal digital identity. When a successful login occurs, the attacker accesses website/application and steals digital identity from target website to misuse/sell digital identity information to interested organizations.

Usecase: Attackers can gain different password sets from password dump site and can use those credentials to gain access to any private repository from compromised user-ids and passwords.

Requirements: The attacker needs a used password from password dump and a team to test all credentials in different accounts for a successful attack.

Goal: To use the credentials from data breach to gain sensitive information from a random organization or user.

Benefits: The attacker can sell the sensitive information of the target organization to a rival organization. The attacker can steal financial information (bank account details) of the target user.

Benefits vs cost: An attacker needs to get used passwords either from password dump or data breaches. In this attack, attacker does not require any special skilled hacking technique to execute the attack. There is 40–50% chance of a successful attack.

Damage consequence: Sensitive information (bank details, organization details, etc.) theft either from an organization or a user. Organization will suffer from reputation and money loss and user may suffer from loss of sensitive financial information.

2.2 Password Spraying

Password spraying [2] is a form of brute force attack which takes advantage of a user's tendency to rely on common passwords such as '`password123`'.

Mechanism: The attacker gathers a list of commonly used passwords and tries to use these passwords across different accounts. Once a login is successful, the attacker extracts the digital identity from target source.

Usecase: Many organizations are vulnerable to password spraying because employees keep weak passwords or a common password which can be easily guessed by attackers. If the first attempt fails, then the attacker can try again with different passwords by waiting around a minute or hour.

Requirements: The attacker needs to enter common or weak passwords in different websites and accounts and needs a big team to increase the attack success rate or attacker can design botnets and broadcast them in target network.

Goal: To use weak passwords to steal user's sensitive or financial information.

Benefits: An attacker can either blackmail user about compromised sensitive data, or they can steal financial information like social security number and identity card details.

Benefits vs cost: An attacker needs a huge team for increasing the success rate of this attack and damage caused is little as compared to phishing attack.

Damage consequence: Leak of user personal or sensitive information or financial data theft. User might suffer from mental harassment or money loss due to financial data theft.

2.3 Malware and Trojan Attacks

Malware and trojan are the software which is designed to damage or steal information from the system, client, server, or any network. In a blockchain, if a user is using public-private key based login system, then malware and trojan will play a major role for the successful attack (stealing or destroying private key) [11].

Mechanism: The attacker detects flaws on user's system (for example, a system without anti-virus system or with a weak anti-virus) for installing malware software on that system. After installation of malware, the attacker hijacks the system to find the location of the private key or try to find other information which will be useful for stealing digital identity. If the private key is encrypted, then the attacker tries to find the seed [11] in same system or they destroy the private key in that system. On the other hand, if the private key is not encrypted then the attacker uses the key for accessing user account and steal digital identity.

Usecase: For public-private key based login system, user enters some numbers to access the account using private key. If user has stored his/her financial information in blockchain then private key is stored either in organization server or in user's system. If user does not have a good anti-virus, then the attacker uses malware and trojan horse to steal or destroy the private key. In this system once private key is compromised then the attacker gains access on victim's account and all its information.

Requirements: For a successful attack, an attacker should be a full-fledged application developer to develop a trojan or malware software. To build/develop malwares he needs skills in identification of threats and its mitigation, how to play with operating systems (Windows, Linux, etc.), intrusion detection and penetration testing tools and knowledge of common viruses with code analysis. Further, the attacker must have good skills on social engineering to send the malicious software to the remote host (victim).

Goal: To steal the private keys from the victim's system and execute next stage of attack.

Benefits: Once the private key of the DApp is compromised, then the attacker has complete access on victim's account. Then the attacker can either blackmail victim (if private key is encrypted and victim did not take a backup for private key) for the money or steal financial information from victim's account.

Benefits vs cost: An attacker may or may not require a team for successful execution of attack and benefits will depend on the victim. That is, an attacker will either get a large sum of money or sometimes they get nothing from the victim's account.

Damage consequence: Financial and sensitive information theft or blackmailing victim for money in return of private key. Victim may suffer from mental harassment because of sensitive information leak, organization or other users related to victim may also suffer money loss.

3 Attacking Digital Identity via APIs or DApps

3.1 Phishing Campaigns

In phishing campaigns, the attackers recognize that they need to acquire only a few normal users or one admin account to compromise an entire organization. An attacker needs to send phishing mails broadly so attack can successfully compromise 5 out of 100 users from a well-secured organization [10].

In a blockchain, an attacker can target public addresses of user who are doing more transactions (the attacker can guess by number of transactions that the user is high-graded or not) and acquire information from public address and broadly send phishing mails to all users.

Mechanism: The attacker gains email addresses from the target organization from which it wants to gain digital identity and designs the phishing message or a *link* relevant to that application. The attacker distributes the link or malicious message broadly and waits for user to enter the credentials. After getting credentials, the attacker accesses sensitive data or use that data (identity) to target desired identity.

Usecase: Phishing attacks target individuals with specific mission. An attacker can craft email that looks extremely legitimate and compelling. Once the target clicks on a link or opens an attachment, the attacker establishes a foothold in the network, enabling them to complete their illicit mission [10].

Requirements: The attacker requires proficiency in HTML, CSS, JQUERY, PHP, ASP.NET, etc. and front-end web development and/or design to create a phishing website or online applications. The attacker must use messages like account deactivation for creating sense of urgency. The phishing mail should mimic actual mails from a spoofed organization by using same phrases, logos and signature that makes the message more legitimate.

Goal: To send the phishing mails broadly to an organization so that 1 out of 20 users get scammed by this attack so that the attacker can access the sensitive information about that user, or he can use that information to target higher authorities of that organization.

Benefits: The attacker may blackmail user to provide some sensitive information of the target organization. He can also sell the information to other organization in good amounts.

Benefits vs cost: The attacker needs some professional web designers and developers for spreading phishing mails broadly in a legitimate way. If the attack gets successful, the attacker can gain a large sum of money from the target organization.

Damage consequence: Sensitive information theft from the target organization. Organization will suffer financial loss as well as reputation loss and it might result in shutting down of organization.

3.2 Cryptojacking

Cryptojacking is similar to web phishing in which an attacker hijacks a target user's browser to mine cryptocurrencies in blockchain. In this attack, an attacker uses JavaScript code and malicious link to hijack a user's browser [6].

Mechanism: The attacker need to detect a flaw in the website to execute attack against victim. The attacker uses their own scripts also known as cross-site scripting [4] to make the phishing website. To execute the attack without sending any mails or links to victim, an attacker can do network attacks (BGP hijacking [3]) for better results.

Usecase: An attacker can send mails or links related to the respective professions of the victim about expiration of account or deactivation of cards.

Requirements: The attacker needs a team to find the vulnerabilities in different websites. The attacker must have a very knowledge of designing websites and JavaScript.

Goal: To steal the login credentials or to steal the private key from the decentralized application.

Benefits: With the compromised credentials and private keys, an attacker can steal sensitive data from the target organization. Profit is based on organization.

Benefits vs cost: For a successful attack, an attacker needs a team who have good knowledge on websites and networking whereas benefits depend on the organization whose credentials got compromised. That is, *bigger the organization, bigger the benefits.*

Damage consequence: Sensitive information theft from the target organization. Organization will suffer financial loss as well as reputation loss and it might result in shutting down of organization.

3.3 Stealing Clicks

Stealing clicks are generally seen in advertisement click frauds but an attacker can use this attack when there is multi factor authentication feature is active in account. In public-private key based login system, user requires public address and private key passcode to access account. In multi-factor authentication after entering passcode, one more authentication process is there in which organization may use OTP (One-time password), 4–6 digit PIN in virtual keyboard or selection of images by user at time of registration. Therefore, entering PIN and selecting images on that system screen user use clicks so with stealing clicks attack, attacker can determine the multi factor authentication PIN or image selection [6].

Note: This attack is commonly applicable on android keyboard applications.

Clickjacking and Cursorjacking [1] are also click based attacks in which attacker places some hidden buttons in website so when victim clicks the button, it gets redirected to the malicious websites.

Mechanism: The attacker searches for the flaw in the organization server to gain access to the organization server. After gaining access to the server, attacker extracts information from virtual keyboard which may contains sensitive information like bank account details, email address, mobile phone numbers, profile photos, etc.

Requirements: An attacker needs a good knowledge on networking and FTP server and little knowledge on HTML and PHP (Hypertext preprocessor).

Goal: To extract information from virtual keyboard or android keyboard application to steal sensitive information of victim.

Benefits: Benefits for attacker will depend on victim, but approximately benefits will be very low as compared to insider and phishing attacks.

Benefits vs cost: For successful attack attacker needs good knowledge on networking which is actually a difficult task. On the other hand, if an attacker does not take proper precautions, then they may get identified or getting caught. In terms of benefits, it will totally depend on the victim (for example, the server user is using and the keyboard application they are using) and approximate damage rate is also very low for this attack as compared to other attacks.

Damage consequence: Leak of sensitive information like bank account number, ATM pin, email address, IMEI number, personal sensitive information. Victim may suffer from mental harassment and money loss and may also affect the other users related to victim.

4 Attacking Digital Identity via Network

4.1 Network Based Attacks

In network based attacks, an attacker identifies and exploits vulnerabilities which results in DNS spoofing and DDoS attack then combined with phishing attack to steal user credentials. This results in compromise of application and leads to privacy loss (Credentials may contain user's identity/personally identifiable information).

Mechanism: The attacker needs to detect the flaw in network using networking tools like Wireshark to execute the network attack against victim. In the next step, attacker can exploit the flaw/vulnerability and executes the attack.

Usecase: In terms of network attacks in online blockchain wallet organization, such an attack happened recently. Attackers used BGP hijacking to steal more than $150k worth of Ether from MyEtherWallet using DNS spoofing and phishing attack.

Requirements: The attacker needs a team who has expertise on networking tools, protocols, and programming to identify faults in different networks. The attacker also requires knowledge on proxies, routers, switches and hubs and network ports for exploiting faults on the target network.

Goal: To compromise server of the target organization so the attacker can execute the next step of the main attack, for example, phishing attack.

Benefits: With the compromised server, an attacker can execute phishing attack in that target server, so whenever user uses the target server, it will redirect to the malicious site or phishing site.

Benefits vs cost: For a successful attack, an attacker needs a full-fledged network as well as web and application engineer whereas benefits will depend on the user or organization. The chances of attack success are very high as compared to any other phishing attacks previously discussed.

Damage consequence: Depends on the network attack; if the attack is DDoS, then organization service to customers will get delayed and organization will lose its reputation; if the attack is like DNS spoofing and BGP hijacking then the consequences will be same as phishing attacks.

4.2 Man-in-the-Middle Attack (MitM)

A MitM attack executed on an organization or a user can result in credentials compromise [7]. An attacker can intercept the network connection via session hijacking which compromises web sessions by stealing session tokens. In terms of blockchain, the attacker can steal session tokens from decentralized applications (DApps).

Mechanism: The attacker intercepts an insecure network connection that a user's device unknowingly connects, for example, via evil twin attack. If data is encrypted, then the attacker installs malicious certificate by tricking the user for data decryption. Then the attacker steals the session token to authenticate user's account and steal digital identity.

Usecase: An adversary can either inject an invalid list of seeder nodes in the open source Blockchain software, or poison DNS cache at the resolver. By default, the Blockchain software client has a list of seeders that allow the network discovery. If the attacker injects a fake list of seeders, the user will be compromised. As a result, the adversary can potentially isolate Blockchain peers and lead them to a counterfeit network.

Requirements: The attacker needs the strong knowledge on networking and cryptography to intercept or eavesdrop the keys between users. The attacker needs a team, knowledgeable in Wireshark for detecting weak or vulnerable networks.

Goal: To intercept or eavesdrop keys in weak network to steal sensitive data from decentralized applications.

Benefits: An attacker can isolate Blockchain peers and lead them to a counterfeit network or use the sensitive information from DApp data breach to blackmail normal user or employee from organization for money.

Benefits vs cost: An attacker needs a well experienced networking engineer and good knowledgeable cryptographer for executing attack whereas benefits depend on the DApp and web security of that organization. If popular organization has low network properties in DApp then benefits are huge otherwise they are low.

Damage consequence: Leakage of sensitive communication between high profiles of that organization or financial information from DApp. Organization will suffer financial loss and reputation loss which may result in shut down of the organization.

5 Attacking Digital Identity by Exploiting the Blockchain

5.1 Insider Attacks

An insider may be an employee or people within the organization who have inside information of organization's security and sensitive information. An insider is a malicious threat to a company or organization because insider can either sell sensitive information to other organization or steals financial information of customers for his/her own benefits [8]. In permissioned blockchain, miners/validators of transaction can play as insider role. The miners/validators can leak sensitive information from blockchain which results in loss of trade information and user's identity loss which make huge negative impact on customer's trust. In permissioned blockchain miners/validators will have access on node setup operations which includes addition or removal of members, identity and access control and consensus (depends on limitations on consensus algorithm).

Mechanism: Some of the blockchain organization stores customers private key in their server. An insider can steal customers transaction information to select a customer which give more benefit to insider. An insider steals their private keys from organization server and use for own benefits.

Usecase: Attacks on private keys will occur on DApp/APIs layer and network layer but in blockchain layer validators can leak Personally Identifiable Information (PII) of other organization members who are collaborating the business in blockchain network which result in compromise identity. Validators may not require gain majority on other nodes to leak sensitive information.

Requirement: An attacker needs to be an employee or associate in an organization to execute attack from inside. In blockchain layer, attacker should play the role of validator to gain access on blockchain nodes.

Goal: To steal an organization's or customers sensitive information to use for its own benefits.

Benefits: An insider can sell sensitive inside information to a rival company. Also, an insider can misuse customers' information to gain money.

Benefits vs cost: This attack contains big risk of getting caught as the attacker is an associate and there is more chance of getting caught due to a smaller number of suspects. So, an attacker tries to extract most information which can make a large gain for the attacker.

Damage consequence: Highly sensitive information theft from organization and stealing of financial data of high-level clients. An organization and its customers related to that organization can suffer money loss and reputation loss, or even lead to shutting down of organization.

5.2 51% Attack

51% or majority of malicious nodes is the attack on blockchain which can be done by the group of miners/validators if they gain 51% of the total network hashrate (This will work in Proof-of-work consensus algorithm). In terms of Byzantine Fault Tolerance (BFT), if more than one-third (33%) miners agree to make changes in blockchain, then they can get power to append, update or delete the data in the blockchain [5]. In permissioned blockchain also, dishonest majority of validator nodes in consortium can gain complete access over blockchain network which covers every layer, that is, Blockchain layer, DApps layer, Network layer, client-user interface layer and smart contracts.

Mechanism: First, the attacker gains control of a majority of blockchain network. Then the attacker mines blocks to make another chain. The other chain runs parallel with the chain on which rest of the networks are mining. So, when an attacker mines a block, they do not announce the other 49% of the network. Now the malicious miners/validators can make changes in headcount of node operators since addition and removal of node operators is based on majority in consensus. Changing node operators or miners frequently will result in huge leakage of digital identity. Now the false chain becomes the longest chain, so the attacker announces the chain to the rest of the nodes on network.

Attack Scenarios:

1. If a DApp is based on public blockchain then in terms of identity management, users can face privacy issues because in public blockchain, anyone can become a miner and the data is shared among all the miners. On the other hand, possibility of 51% attack on public blockchain is more as compared to permissioned or private blockchain. In PoW (Proof-of-Work) consensus algorithm, an attacker needs 51% of hashrate of entire blockchain network to execute this attack.
2. On the other hand, if DApps are based on a permissioned or private blockchain, the privacy issue gets mitigated but if the blockchain is BFT consensus based, then a malicious miner needs support of other miners

to execute this attack. In permissioned blockchain, suppose there are 8 miners, so an attacker would need 4 more miners on his/her side (other miners should agree decision of attacker miner) then according to BFT rule they can add, modify, or delete the data from blockchain which also results in forking.

Requirements: An attacker only needs the support of high power (hashrate) miners for successful attack. In permissioned blockchain consortium, an attacker playing the role of validator/miner node from an organization to gain access to PII of other organization to compromise members identity.

Goal: To compromise 51% hashrate of blockchain network to modify or forging identity.

Benefits: An attacker can use fraudulent activities with victim's identity.

References

1. Chancel, J.: Mozilla foundation security advisory 2015–35: cursor clickjacking with flash and images, March 2015. https://www.mozilla.org/en-US/security/advisories/mfsa2015-35/
2. FireEye: Overruled: containing a potentially destructive adversary, July 2019. https://www.fireeye.com/blog/threat-research/2018/12/overruled-containing-a-potentially-destructive-adversary.html
3. Gavrichenkov, A.: Breaking HTTPS with BGP hijacking. In: Black Hat USA Briefings (2015)
4. Grossman, J.: XSS Attacks: Cross-site Scripting Exploits and Defense. Syngress Media (2007)
5. Hern, A.: Bitcoin currency could have been destroyed by '51%' attack, June 2014. https://www.theguardian.com/technology/2014/jun/16/bitcoin-currency-destroyed-51-attack-ghash-io
6. McAfee: Cryptojacking. In: Blockchain Threat Report, August 2018. https://www.mcafee.com/enterprise/en-us/assets/reports/rp-blockchain-security-risks.pdf
7. Pandya, G.: Nokia's MITM on HTTPS traffic from their phone, January 2013. https://gaurangkp.wordpress.com/2013/01/09/nokia-https-mitm/
8. Rusinek, D.: Blockchain: new types of insider threat, January 2020. https://www.securing.pl/en/blockchain-new-types-of-insider-threat/
9. UK National Cyber Security Centre: Advisory: Use of credential stuffing tools, November 2018. https://www.ncsc.gov.uk/news/use-credential-stuffing-tools
10. Verizon: Data breach investigations report, April 2015. https://doi.org/10.13140/RG.2.1.4205.5768
11. Wuille, P.: Hierarchical deterministic wallets seed phrase security (2012). https://github.com/bitcoin/bips/blob/master/bip-0032.mediawiki#Security

Cryptocurrencies and Price Prediction: A Survey

Yeray Mezquita(✉), Ana Belén Gil-González, Javier Prieto, and Juan Manuel Corchado

BISITE Research Group, University of Salamanca, Edificio Multiusos I+D+i, 37007 Salamanca, Spain
yeraymm@usal.es

Abstract. Cryptocurrencies are fungible digital assets whose market capitalization has not stopped growing since the appearance of their first use case in 2009, Bitcoin. However, one of the biggest problems facing cryptocurrencies is the enormous fluctuation of their value in the market. To help understand different patterns in cryptocurrency ecosystems, several machine learning-based solutions have been proposed in the literature. This paper aims to study in detail the solutions proposed in the literature for the detection of patterns and anomalies in cryptocurrency ecosystems. The aim is to bring together different proposals and studies to help users of this market to understand how it works.

Keywords: Cryptocurrencies · Price · Prediction · Survey

1 Introduction

Cryptocurrencies are fungible digital assets whose market capitalization has not stopped growing since the appearance of their first use case in 2009, Bitcoin [17]. Many investors are looking to put their money in cryptocurrencies, considering them an asset that will gain value over time because of the possibilities they offer, either by the possibility of payment automation in distributed platforms with the use of smart contracts, or by the creation of a more democratic money market thanks to being considered distributed assets [13,14,14,15].

However, one of the biggest problems facing cryptocurrencies is the enormous fluctuation of their value in the market. This prevents them from being used in people's daily lives as a means of transaction. Instead, they are used as means of investment, being exposed to speculation by investors, sometimes even expossed to pump and dump schemes [8,18].

To help understand different patterns in cryptocurrency ecosystems, several machine learning-based solutions have been proposed in the literature [10,19,21]. Identifying these patterns can help users not to enter the game of pump and dump schemes, eliminating some risk and attracting more people to the market, which would help stabilize crypto prices [22].

J. Prieto et al. (Eds.): BLOCKCHAIN 2021, LNNS 320, pp. 339–346, 2022.
https://doi.org/10.1007/978-3-030-86162-9_34

This paper aims to study in detail the solutions proposed in the literature for the detection of patterns and anomalies in cryptocurrency ecosystems. The aim is to bring together different proposals and studies to help users of this market to understand how it works. To this end, the Sect. 2 shows a theoretical context of cryptocurrencies and the technology that underlies them: the blockchain. In Sect. 3 the research questions and search criteria are defined. In addition, the papers found are detailed and the section is developed by grouping them according to what they propose. Finally, in the Sect. 4, the findings of the work are discussed and future work is concluded.

2 Theoretical Context

Cryptocurrencies are fungible digital assets whose main purpose is to being used in transactions as fiat money. Blockchain technology is used to store cryptocurrency transactions and maintain user balances. This technology is a type of distributed ledger, the ledger where the transactions are stored is supported and maintained by a network of nodes that communicate with each other through a series of protocols. Thanks to this, it is possible to have a tamper-proof ledger of transactions and balances, earning the trust of their users.

Blockchain technology makes use of a public key signature mechanism, thanks to which it is possible to easily verify the source of the data generated, guaranteeing the integrity of the data and easily verifying its origin. Due to the open, decentralized, and cryptographic nature of a blockchain, it is possible to i) avoid intermediaries in transactions between non trusted parties; and ii) generate an immutable ledger kept by the network of nodes, by using consensus protocols, nodes of the network make the ledger tamper-proof [15].

There are being designed more types of consensus algorithms and their variations, each depending on the type of the blockchain network. If a blockchain is public, everyone can access it and connect to it, or if it is permissioned, only identified players can connect to it and maintain the ledger. The most widespread algorithms are [16]: i) the **Proof-of-Work** (PoW) algorithm, a network node must solve a cryptographic problem to add a new block to the blockchain. The computational cost and difficulty of solving the problem, the energy expended in finding its solution (work), and the simplicity of its verification, are sufficient reasons to deter nodes that add new blocks (miners) from engaging in illegal transactions; ii) the **Proof-of-Stake** (PoS) is a consensus algorithm, in which miners take turns adding new blocks to the blockchain. The probability of a miner getting his turn to add a block depends on the number of coins deposited for the miner as escrow (Stake). This algorithm assumes that a node is going to be honest in creating a block to avoid losing escrow; the **Practical Byzantine Fault Tolerance** (PBFT). Round is the process of adding a new block to the blockchain. In each round, a node is selected to propose a new block, and then the block is broadcasted to the network, to let the network validate it. The block is validated by each node in the network, getting one vote for each node that successfully validated it. When a block receives 2/3 of the votes from all nodes in the network, it is considered valid and is added to the blockchain.

Another feature of blockchain technology that some cryptocurrencies like Ethereum are taking most profit from, is the possibility of deploy smart contracts. Smart contracts are a sequence of code stored within a blockchain that can be executed in a distributed manner alongside the network, reaching a consensus among the nodes on the result obtained from its execution. These programs facilitate, verify, and enforce an agreement on a set of predefined conditions [7]. Smart contracts are self-enforcing and self-verifying contractual agreements that automate the lifecycle of a contract to improve compliance, mitigate risk and increase efficiency on any platform where entities with different interests have to interact with each other [16].

Although blockchain technology offers a great variety of possibilities, there also poses some challenges. E.g. each consensus algorithm has its risks and vulnerabilities, like PoW that spends a huge amount of energy to solve cryptographic problems to produce new blocks, and it is very limited in terms of scalability [1]. In the case of the PoS algorithm, its nothing at stake theory causes forks to occur on the blockchain more frequently than with other consensus algorithms [11]. Meanwhile, the main risk of PBFT is that it is a permissioned protocol and not a truly decentralized one, making it not suitable for public networks [24]. Finally, the use of smart contracts is not covered by the law, making it difficult to resolve disputes that are not well covered in the contract.

To address the scalability problems that PoW based blockchain platforms, are arising PoS based blockchain platforms, and some of its variations like the delegated Proof-of-Stake (dPoS) algorithm, in which participants vote, based on the stake they hold, which nodes can add new data to the blockchain. Some cryptocurrencies, like Tron, Cardano, EOS, and possibly Ethereum, adopt this approach to provide solutions to the scalability problems of the PoW based solutions and make it possible for Decentralized Applications (DApps) to become mainstream.

3 Literature Review

The main objective of this paper is to study the possibility of predicting cryptocurrency market fluctuations and how they can be done. To achieve the objectives, a study has been carried out in the Science direct database to find papers that help to find answers to the following hypotheses:

- **RQ1**: What has been used in the literature to forecast prices in the crypto market?
- **RQ2**: Are the methods used reliable?

The keywords used were: i) *("distributed ledger technology" OR blockchain* OR cryptocurrenc*)*, identified as Cryptocurrency; ii) *price*; iii) and *prediction*. Using the keywords, the following search string has been formulated: *Cryptocurrency AND Price AND Prediction*. Using this search string, 1004 results were obtained, of which the most relevant ones were used in this work, obtaining 13 papers in which prediction in cryptocurrency markets is studied.

In previous studies such as the one conducted in [20], the influence that news and social networks have on emerging technology, such as the case of distributed Ledger Technologies (DLT), more specifically blockchain technologies, and somewhat unknown to the general public, has been tested. In that study, only English-language news referring to such technologies from the period from 2010 to 2018 have been used. It is revealed a pattern of DLT diffusion in transition from its infancy, through a growing awareness of its potential, and favor and opposition to its applications. Additional sentiment analysis reveals that the likelihood of unfavorable sentiment toward Bitcoin remained higher than favorable sentiment.

Following the steps of the previous study, in [10] we seek to use news, more specifically published Tweets, and use them as a means to predict cryptocurrency prices. Authors obtained daily and hourly Twitter sentiment polarity for 9 major cryptocurrencies: Bitcoin (BTC), Ethereum (ETH), Bitcoin Cash (BCH), Litecoin (LTC), Cardano (ADA), Stellar (XLM), EOS, TRON, Ripple (XRP). Causality tests indicate Twitter sentiment help predict the returns of BTC, BCH, LTC. Using a bullishness ratio, predictive power is found for the returns of EOS TRON. Tweet volume only has predictive power for the price returns of LTC and XRP. Finally, it is estimated that 14% of Tweets were identified as posted by Twitter bot accounts.

In [23], it has been tried to assess the properties of the cryptocurrency market and the associated phenomena, by studying the characteristics of the complexity of exchange rates on the cryptocurrency market and comparing it to traditional and mature markets, such as stocks, bonds, commodities or currencies. With the help of statistical physics methods like the multifractal cross-correlation analysis and the dependent detrended cross-correlation coefficient, the non-linear correlations and multiscale characteristics of the top 100 cryptocurrencies of the cryptocurrency market are analyzed. Thanks to the study carried out, it has stated that due to the Covid-19 pandemic it has been witnessed a "phase transition" of the cryptocurrencies from being a hedge opportunity for the investors fleeing the traditional markets to become a part of the global market that is substantially coupled to the traditional financial instruments like the currencies, stocks, and commodities.

Cryptocurrency price prediction has become a trending research topic globally. Many machine learning and deep learning algorithms such as Gated Recurrent Unit (GRU), Neural Networks (NN), and Long short-term memory (LSTM) have been used by the researchers to predict and analyze the factors affecting the cryptocurrency prices [19]. In the referenced paper, a LSTM and GRU-based hybrid cryptocurrency prediction scheme is proposed, which focuses on only two cryptocurrencies, namely Litecoin and Monero. The results depict that the proposed scheme accurately predicts the prices with high accuracy, revealing that the scheme can be applied in various cryptocurrencies price predictions.

Regarding the ETH cryptocurrency, [21] has proposed the use of two machine learning methods, namely linear regression (LR) and support vector machine (SVM), by using a time series consisting of daily ETH closing prices. When using the proposed model of this work, the SVM method has higher accuracy (96.06%) than the LR method (85.46%).

Meanwhile, in [3], it is made an attempt to apply machine learning techniques on the index and constituents of cryptocurrency with a goal to predict and forecast prices thereof. In particular, the purpose of the referenced work is to predict and forecast the closing price of the cryptocurrency index 30 and nine constituents of cryptocurrencies using machine learning algorithms and models so that it becomes easier for people to trade these currencies. Results show a 92.4% accuracy using an ensemble learning method, which is considered as the best among all the models used in the referenced paper. Using such prediction and forecasting methods, people can easily understand the trend and it would be even easier for them to trade in a difficult and challenging financial instrument like cryptocurrency.

The study of [4], employed a variety of mathematical tools for the analysis of six significant cryptocurrencies: BTC, ETH, XMR, LTC, XRP, ETC. It used fractional and fractal mathematical tools as an instrument that can help market agents and investors to more clearly assess the cryptocurrencies price dynamics and, guiding investment decisions more assertively while mitigating risks. The BTC was the only cryptocurrency that presented more consistent long-memory behavior. The LTC exhibited the lowest predictable horizon compared to the other cryptocurrencies, pointing to chaotic behavior. The ETH and the XMR behaved mainly as an anti-persistent process showing a short memory effect. In this study, it can be concluded that, with exception of BTC, the other five cryptocurrencies analyzed are mean reverting, showing lower predictability.

When it comes to making predictions, the literature gives us a warning on the subject of regulations in the different countries [2]. In the referenced paper, it is examined whether cryptocurrency traders perceive the market regulation beneficially. Using an event study methodology for daily data covering the period 2015-2019, it is assessed how regulatory news and events have affected returns in the cryptocurrency market. The results suggest that events that increase the probability of a regulation adoption are associated with a negative abnormal return for concerning cryptocurrencies. It is also stated that investors reacted less negatively for most illiquid cryptocurrencies, as well as for cryptocurrencies that had more information asymmetry. From this work, it can be concluded that the performance in the pre-event period is positive and significant; however, it appears to be not significantly different from zero in the post-event period.

In [6], it has been examined the returns and volatility spillover with the cryptocurrency market over the period 1 September 2017 to 7 June 2018, on cryptocurrency-linked stocks (CLS) engagement with blockchain technology in Australia. In the referenced work, it is used the spillover methodology of [5], finding significant unidirectional return spillover and weak volatility spillover from the cryptocurrency market to CLS, after controlling return dynamics of the Australian dollar, gold, and commodity. The results indicate that investors incorporate the price dynamics of cryptocurrencies into their trading decisions for CLS.

From [9], can be found that several corporate blockchain developments have generated significant spillovers. So, after analyzing whether corporate blockchain patent developments influence BTC volatility, it can be concluded that its usage by corporations reaffirms investor confidence in BTC. Those results also provide evidence of the maturity of the cryptocurrency markets.

In [8] it is gathered the most comprehensive dataset on cryptocurrency pump and dump schemes. From that study it is found that i) pumped coins experience a modest price increase, but it is higher for less popular coins; ii) brazen pumps are more "successful" than those obscuring their corrupt intent; iii) and high concentration among pump channels creates an opportunity for regulators. In the work [18], it has been found that i) fraud groups use Telegram and other social media to organize and conduct pumps; ii) market signals are better than administrative social signals as predictive factors; iii) and changes in market movements could influence robustness in performance over time.

To conclude our study, in [12], it is stated that Directed Acyclic Graphs (DAGs) are an increasingly valuable substitution for blockchain technology, possessing several key strategic advantages over blockchains, namely their lack of transaction fees and increased speeds. Three cryptocurrencies already employ DAGs: IOTA, NANO, and Obyte. All three assets are built around a different data structure (The Tangle, a Block Lattice, and a Main-Chain) and are resistant to quantum computing, meaning they already possess an important long-term advantage over currencies such as BTC, ETH, and LTC. In the referenced paper, it is examined the volatility reaction of these groups of currencies to market shock events (in the form of regulatory announcements), stating that DAG-based assets have become increasingly responsive to market shocks as they mature, evidencing signals of substantial market maturity in recent years.

4 Conclusion

This paper aims to provide, in the most detailed way possible, information on the current state of the cryptocurrency market and the patterns that can be extracted from its activity. For this purpose, a study has been made of the literature that focuses on the possible prediction of prices in this market. A total of 13 papers have been studied that focus their work in this area, in one way or another: some seek to find patterns in the activity of social networks and news; others seek the best machine-learning algorithm to predict its price; or simply analyze the behavior of the market in the face of external factors such as its regulation by different governments.

In this study, responding to RQ1, works have been found that make use of different methods of machine learning or statistical analysis to find the market closing prices. We have also found works that analyze different factors such as i) the sentiment of Tweets, ii) of news appearing in the media, iii) how investors react to announcements of regulatory measures affecting the market, iiii) and the technology behind each crypto.

Answering RQ2, it can be said that methods have been seen with a very good percentage of predicting the market closing price, although other works have also been studied that deny that one can ever predict the price of all crypto. Moreover, if external and social factors, such as regulatory measures and community sentiment in social networks, are not taken into account, it is not possible to accurately identify prediction patterns. In addition, there are pump and dump schemes that make prediction tasks difficult. Therefore, it can be concluded that, separately, the works are not effective, but as a whole, important conclusions can be drawn for users who intend to operate in this market.

This work is far from perfect, in the future we intend to extend the study to more databases and articles. We will also emphasize the possibilities offered by each of the algorithms used and what attributes they use to create their models.

Acknowledgement. The research of Yeray Mezquita is supported by the pre-doctoral fellowship from University of Salamanca and co-funded by Banco Santander. This research was also partially Supported by the project "Computación cuántica, virtualización de red, edge computing y registro distribuido para la inteligencia artificial del futuro", Reference: CCTT3/20/SA/0001, financed by Institute for Business Competitiveness of Castilla y León, and the European Regional Development Fund (FEDER).

References

1. Beikverdi, A., Song, J.: Trend of centralization in bitcoin's distributed network. In: 2015 IEEE/ACIS 16th International Conference on Software Engineering, Artificial Intelligence, Networking and Parallel/Distributed Computing (SNPD), pp. 1–6. IEEE (2015)
2. Chokor, A., Alfieri, E.: Long and short-term impacts of regulation in the cryptocurrency market. The Quarterly Review of Economics and Finance (2021)
3. Chowdhury, R., Rahman, M.A., Rahman, M.S., Mahdy, M.: An approach to predict and forecast the price of constituents and index of cryptocurrency using machine learning. Phy. A Stat. Mech. Appl. **551**, 124569 (2020)
4. David, S., Inacio Jr., C., Nunes, R., Machado, J.: Fractional and fractal processes applied to cryptocurrencies price series. J. Adv. Res. (2021)
5. Diebold, F.X., Yilmaz, K.: Better to give than to receive: predictive directional measurement of volatility spillovers. Int. J. Forecast. **28**(1), 57–66 (2012)
6. Frankovic, J., Liu, B., Suardi, S.: On spillover effects between cryptocurrency-linked stocks and the cryptocurrency market: evidence from Australia. Glob. Financ. J. 100642 (2021)
7. Gazafroudi, A.S., Mezquita, Y., Shafie-khah, M., Prieto, J., Corchado, J.M.: Islanded microgrid management based on blockchain communication. In: Blockchain-based Smart Grids, pp. 181–193. Elsevier (2020)
8. Hamrick, J., et al.: An examination of the cryptocurrency pump-and-dump ecosystem. Inf. Process. Manag. **58**(4), 102506 (2021)
9. Hu, Y., Hou, Y.G., Oxley, L., Corbet, S.: Does blockchain patent-development influence bitcoin risk? J. Int. Financ. Mark. Inst. Money **70**, 101263 (2021)
10. Kraaijeveld, O., De Smedt, J.: The predictive power of public twitter sentiment for forecasting cryptocurrency prices. J. Int. Financ. Mark. Inst. Money **65**, 101188 (2020)
11. Martinez, J.: Understanding proof of stake: The nothing at stake theory (June 2018). https://medium.com/coinmonks/understanding-proof-of-stake-the-nothing-at-stake-theory-1f0d71bc027. Accessed 9 October 2019

12. Meegan, A., Corbet, S., Larkin, C., Lucey, B.: Does cryptocurrency pricing response to regulatory intervention depend on underlying blockchain architecture? J. Int. Financ. Mark. Inst. Money **70**, 101280 (2021)
13. Mezquita, Y., Casado, R., Gonzalez-Briones, A., Prieto, J., Corchado, J.M., AETiC, A.: Blockchain technology in iot systems: review of the challenges. Annals of Emerging Technologies in Computing (AETiC), pp. 2516–0281 (2019). Print ISSN
14. Mezquita, Y., Casado-Vara, R., GonzÁlez Briones, A., Prieto, J., Corchado, J.M.: Blockchain-based architecture for the control of logistics activities: pharmaceutical utilities case study. Log. J. IGPL (2020)
15. Mezquita, Y., González-Briones, A., Casado-Vara, R., Chamoso, P., Prieto, J., Corchado, J.M.: Blockchain-based architecture: a mas proposal for efficient agri-food supply chains. In: Novais, P., Lloret, J., Chamoso, P., Carneiro, D., Navarro, E., Omatu, S. (eds.) Ambient Intelligence–Software and Applications–,10th International Symposium on Ambient Intelligence. ISAmI 2019. Advances in Intelligent Systems and Computing, vol. 1006, pp. 89–96. Springer, Cham (2019). https://doi.org/10.1007/978-3-030-24097-4_11
16. Mezquita, Y., Valdeolmillos, D., González-Briones, A., Prieto, J., Corchado, J.M.: Legal aspects and emerging risks in the use of smart contracts based on blockchain. In: Uden, L., Ting, I.H., Corchado, J. (eds.) Knowledge Management in Organizations. KMO 2019. Communications in Computer and Information Science, vol. 1027, pp. 525–535. Springer, Cham (2019). https://doi.org/10.1007/978-3-030-21451-7_45
17. Nakamoto, S., et al.: Bitcoin: a peer-to-peer electronic cash system (2008)
18. Nghiem, H., Muric, G., Morstatter, F., Ferrara, E.: Detecting cryptocurrency pump-and-dump frauds using market and social signals. Expert Syst. Appl. 115284 (2021)
19. Patel, M.M., Tanwar, S., Gupta, R., Kumar, N.: A deep learning-based cryptocurrency price prediction scheme for financial institutions. J. Inf. Secur. Appl. **55**, 102583 (2020)
20. Perdana, A., Robb, A., Balachandran, V., Rohde, F.: Distributed ledger technology: its evolutionary path and the road ahead. Inf. Manag. **58**(3), 103316 (2021)
21. Poongodi, M., et al.: Prediction of the price of ethereum blockchain cryptocurrency in an industrial finance system. Comput. Electr. Eng. **81**, 106527 (2020)
22. Valdeolmillos, D., Mezquita, Y., González-Briones, A., Prieto, J., Corchado, J.M.: Blockchain technology: a review of the current challenges of cryptocurrency. In: Prieto, J., Das, A., Ferretti, S., Pinto, A., Corchado, J. (eds.) Blockchain and Applications. BLOCKCHAIN 2019. Advances in Intelligent Systems and Computing, vol. 1010, pp. 153–160. Springer, Cham (2020). https://doi.org/10.1007/978-3-030-23813-1_19
23. Wątorek, M., Drożdż, S., Kwapień, J., Minati, L., Oświęcimka, P., Stanuszek, M.: Multiscale characteristics of the emerging global cryptocurrency market. Physics Reports (2020)
24. Witherspoon, Z.: A hitchhiker's guide to consensus algorithms (November 2017). https://hackernoon.com/a-hitchhikers-guide-to-consensus-algorithms-d81aae3eb0e3. Accessed 9 October 2019

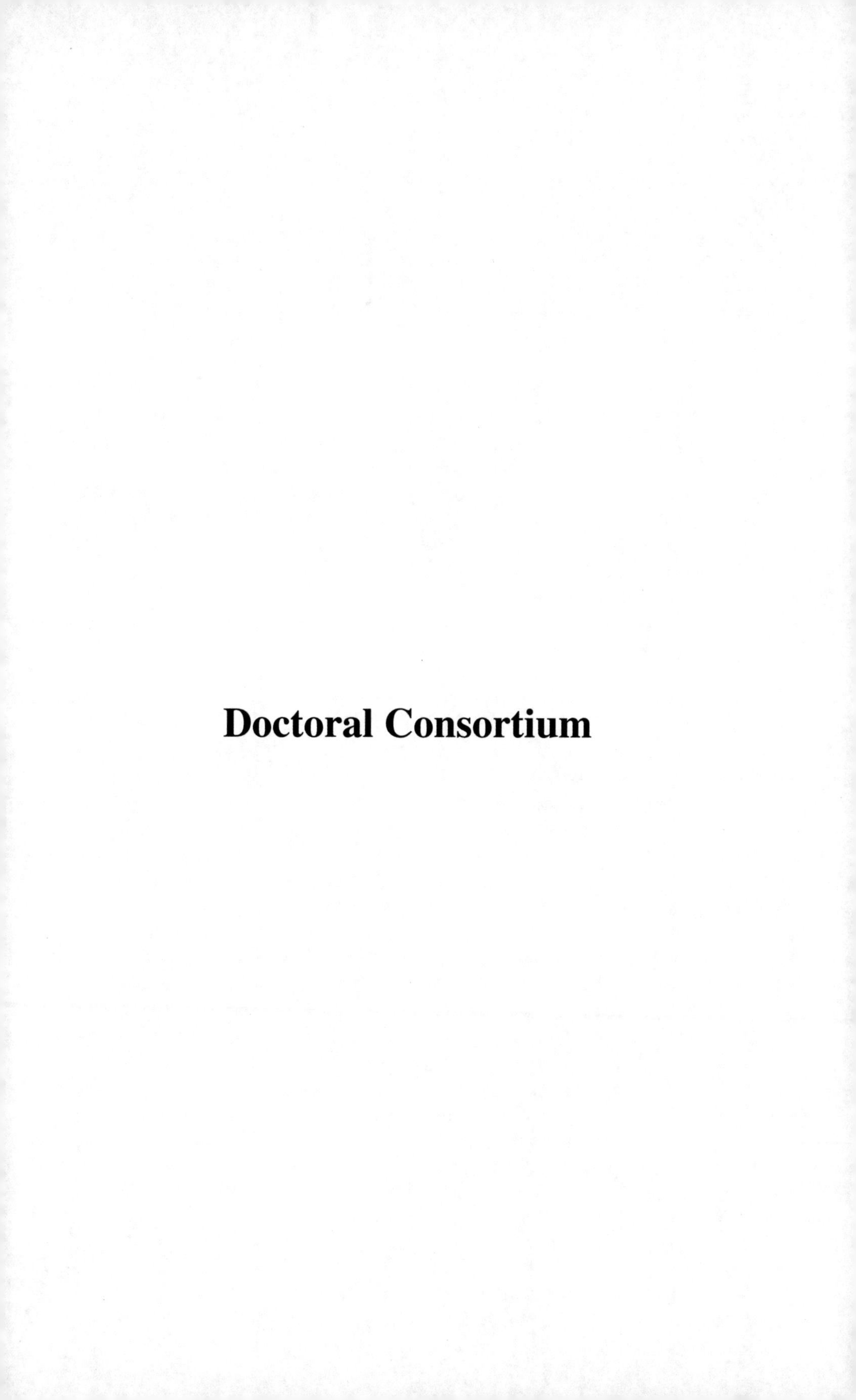

Doctoral Consortium

A Fully Anonymous e-Voting Protocol Employing Universal Zk-SNARKs and Smart Contracts

Aritra Banerjee(✉)

ADAPT Centre, School of Computer Science and Statistics,
Trinity College Dublin, Dublin, Ireland
abanerje@tcd.ie

Abstract. The idea of smart contracts has been around for a long time. The introduction of Ethereum has taken the concept of smart contracts to new heights because of its integration with Blockchain technology. As a result, the applications of smart contracts have also surged in areas such as e-Voting, Insurance, Crowdfunding, etc. In this paper, we aim to present the construction of a "Fully Anonymous e-Voting" protocol using the concepts of zkHawk and Zcash. zkHawk is a novel smart contract protocol designed during this Ph.D. that improves upon the Hawk protocol by solving the underlying anonymity problem of a trusted manager. We will leverage the concept of zk-SNARKs in Zcash to carry out the voting phase of the election and the zkHawk smart contract protocol to tally the results of the election. The voting phase employing Zcash will be initially designed with Non-Universal zk-SNARKs and improved upon with Universal zk-SNARKs.

Keywords: Zcash · e-Voting · Blockchain · Hawk · Smart contracts · zk-SNARKs

1 Introduction

e-Voting is a classic example of a smart contract evaluation [1]. Advancements in Blockchain have made e-Voting more private and decentralized [2,3]. There have been some recent developments using the Ethereum smart contracts [4] for private boardroom voting. Related works have shown leveraging a multi-party computation (MPC) program for the tallying phase in the e-Voting procedure [5–7]. Smart contracts are generally used during the "tallying" procedure in e-Voting. The design of zkHawk [8] facilitates off-chain Smart contract execution that ensures privacy. Therefore, zkHawk can be used in e-Voting for such cryptocurrency blockchains which do not support Smart contracts like Zcash [9] or Monero [10,11] instead of Ethereum [12]. The construction of zkHawk was inspired by Hawk [13] but it added in a feature whereby there was no need to

J. Prieto et al. (Eds.): BLOCKCHAIN 2021, LNNS 320, pp. 349–354, 2022.
https://doi.org/10.1007/978-3-030-86162-9_35

have the trust assumption of a manager. Instead, an MPC program would execute the smart contract while maintaining the zero-sum constraint[1]. The smart contract execution is done off-chain between the participants by running the MPC and the blockchain is just used to validate the transactions and execute contract closure[2].

1.1 Problem Statement

The research goal of this Ph.D. is to improve upon the underlying concerns of Privacy and Scalability [2,14] in e-Voting using zkHawk [8] and Zcash [9] protocols. Zcash e-Voting was first suggested in 2017 by Pavel et al. [14] where each voter is given a Zcash wallet from which they can send one coin to a candidate that is counted as one vote for the candidate, and then use the JoinSplit transaction [15] of Zcash to tally the votes and declare the results. The problem with this approach is the massive number of zk-SNARK computations that will occur if there are a large number of voters. To solve this problem, we suggest leveraging the zkHawk protocol in the "tallying" phase such that there is a significant number of zk-SNARK computations while maintaining the privacy and anonymity we get from Zcash.

Assumption: The voters are registered, and they have been invited to vote.

Goal: To develop a fully anonymized e-Voting protocol that uses Zcash tokens for e-Voting and leverages zkHawk for tallying the votes and declaring the results.

2 Proposal

We now present an overview of the novel Universal zkHawk e-Voting protocol (Figs.1, 2 and 3).

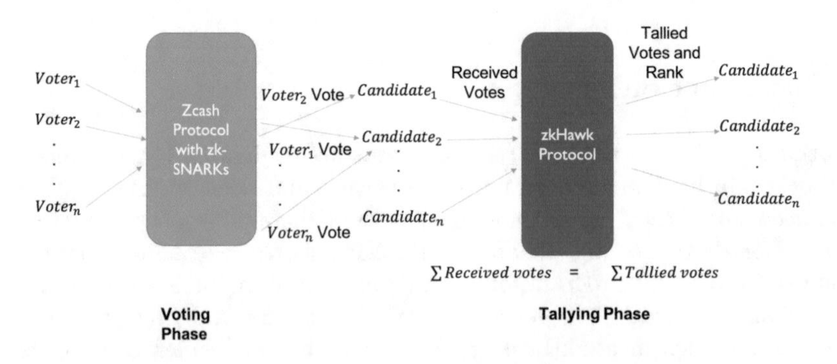

Fig. 1. An overview of the suggested zkHawk e-Voting protocol

[1] The sum of the input balance of a smart contract is equal to the sum of the output balance.

[2] Contract closure signifies that a smart contract has closed and all the payouts have been completed.

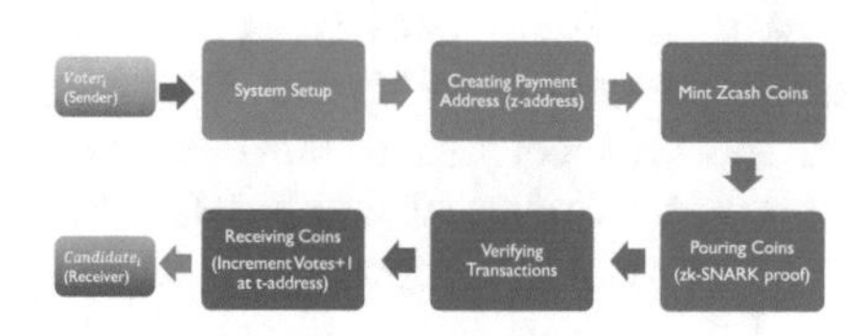

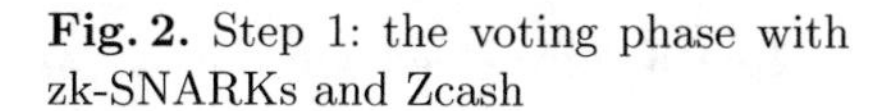

Fig. 2. Step 1: the voting phase with zk-SNARKs and Zcash

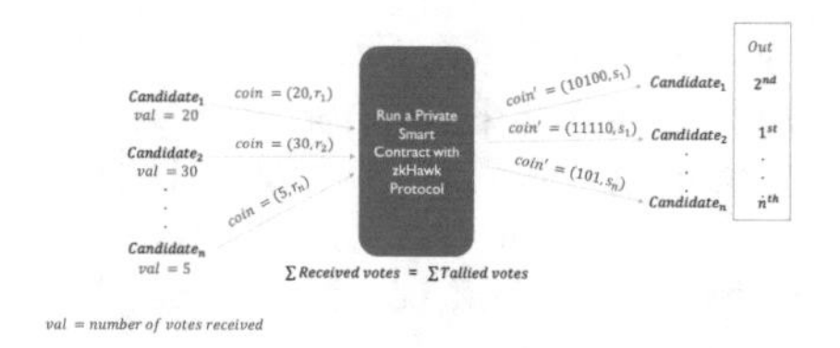

Fig. 3. Step 2: the tallying phase with zkHawk

2.1 Computational Advantage

This section discusses exactly how much we can improve computationally when using zkHawk e-Voting instead of Zcash JoinSplit e-Voting. The privacy remains the same but the number of zk-SNARK [16–18] computations drastically reduces when using zkHawk e-Voting. Let the number of zk-SNARK computations in vanilla Zcash e-Voting be E and the number of computations in zkHawk e-Voting be F, then mathematically speaking: if,

$$\begin{aligned} Voters &= x \\ Candidates &= y \end{aligned} \tag{1}$$

then,

$$\begin{aligned} E &= x * (y + 1) \\ F &= x \end{aligned} \tag{2}$$

Suppose, there are 500 Voters and 10 Candidates, then using Eq. (1) and (2) instead of 5500 zk-SNARK computations (E) only 500 zk-SNARK computations (F) have to be performed, i.e. this protocol reduces the number of zk-SNARK computations by $1/10^{th}$ as compared to vanilla Zcash e-Voting.

3 Evaluation Plan

Currently, the fastest and the smallest known zk-SNARK is the Groth16 [17] algorithm which is used in Zcash. But this algorithm is Non-Universal and hence the trusted setup is always tied to one specific circuit e.g. one-to-one money transfer. Zcash recently performed a zk-SNARK trusted setup ceremony in 2018 for a one-to-one direct money transfer relation i.e., you can transfer money from A to B, B to C, etc. But, for Zcash to support smart contracts or other relations each time a zk-SNARK trusted setup ceremony has to be performed which is computationally very expensive. Hence, the concept of Universal zk-SNARKs [19–24] is introduced. Universal zk-SNARKs imply that all circuits (i.e., relations) can be validated using just one trusted setup. The CRS (Common Reference String) [25] generated in Universal zk-SNARKs is "updatable" and "unending" (also called SRS) [26]. The SRS can support one-to-one token transfer, smart contracts, and other relations without needing to create a separate trusted setup for each relation. And since the SRS is Updatable, anytime any information is leaked any honest party can go to the SRS and update the parameters. We will first design the protocol using Non-Universal zk-SNARKs for the Voting phase, then move on to Universal zk-SNARKs for its feasibility and computational inexpensive nature. Contemporary Universal zk-SNARKs that will be explored during this Ph.D.:

- **Sonic** (2019) [19] is the first Universal zk-SNARK introduced which uses the universal and updatable SRS [26].
- **Plonk** (2019) [20] is an improvement on Sonic which has a significantly lower prover time than Sonic (around 5 times better)
- **Marlin** (2020) [21] is another improvement on Sonic with 10 times better prover time and 4 times better verification time.
- **Mirage** (2020) [22] was suggested by the authors of Hawk and this protocol has linear Universal circuit operations instead of quasi-linear universal circuits. It is built on the Groth16 [17] protocol instead of Sonic.
- **Lunar** (2020) [23] used polynomial holographic IOPs[3] (a new variant of IOPs) to give a very small proof size and prover time.

4 Conclusion

The first step of this Ph.D. was to design a fully anonymous off-chain Smart Contract protocol inspired by Hawk [13] but would omit the requirement to trust a manager. This construction was achieved as designed in [8] and is necessary for the next step of the Ph.D. After successfully designing zkHawk, our next step is to leverage this concept and design an e-Voting protocol that is fast, private, and scalable. This novel protocol also uses the concepts of zk-SNARKs (Non-Universal and Universal) and Zcash. Furthermore, we will provide an elaborate construction of the designed protocol in our future work and implement the algorithm for bench-marking against other e-Voting protocols in terms of privacy and computational complexity.

Acknowledgement. This publication has emanated from research conducted with the financial support of Science Foundation Ireland grants 13/RC/2106 (ADAPT) and 17/SP/5447 (FinTech Fusion). This work was also supported in part by Science Foundation Ireland grant 13/RC/2094 (Lero). I would also like to take the opportunity to thank my supervisor Dr. Hitesh Tewari for his flawless guidance in this PhD. A special acknowledgement is due for Dr. Michael Clear, a postdoc in Trinity working closely with Hitesh and me for his guidance.

References

1. Tso, R., Liu, Z.Y., Hsiao, J.H.: Distributed e-voting and e-bidding systems based on smart contract. Electronics **8**(4), 422 (2019)

[3] Interactive Oracle Proofs [27].

2. Hjálmarsson, F., Hreiarsson, G.K., Hamdaqa, M., Hjálmtýsson, G.: Blockchain-based e-voting system. In: 2018 IEEE 11th International Conference on Cloud Computing (CLOUD), pp. 983–986. IEEE (2018)
3. Kshetri, N., Voas, J.: Blockchain-enabled e-voting. IEEE Softw. **35**(4), 95–99 (2018)
4. McCorry, P., Shahandashti, S.F., Hao, F.: A smart contract for boardroom voting with maximum voter privacy. In: International Conference on Financial Cryptography and Data Security, pp. 357–375. Springer (2017)
5. Khader, D., Smyth, B., Ryan, P., Hao, F.: A fair and robust voting system by broadcast. In: Proceedings-Series of the Gesellschaft fur Informatik (GI). Lecture Notes in Informatics (LNI), pp. 285–299 (2012)
6. Küsters, R., Liedtke, J., Müller, J., Rausch, D., Vogt, A.: Ordinos: a verifiable tally-hiding e-voting system. In: 2020 IEEE European Symposium on Security and Privacy (EuroS&P), pp. 216–235. IEEE (2020)
7. Cortier, V., Gaudry, P., Yang, Q.: A toolbox for verifiable tally-hiding e-voting systems. Cryptology ePrint Archive, Report 2021/491 (2021). https://eprint.iacr.org/2021/491
8. Banerjee, A., Clear, M., Tewari, H.: zkhawk: Practical private smart contracts from MPC-based hawk. arXiv preprint arXiv:2104.09180 (2021)
9. Sasson, E.B., et al.: Zerocash: Decentralized anonymous payments from bitcoin. In: 2014 IEEE Symposium on Security and Privacy, pp. 459–474. IEEE (2014)
10. Van Saberhagen, N.: Cryptonote v 2.0 (2013)
11. Noether, S.: Ring signature confidential transactions for monero. IACR Cryptol. ePrint Arch. 2015, 1098 (2015)
12. Buterin, V., et al.: A next-generation smart contract and decentralized application platform. White paper 3(37) (2014)
13. Kosba, A., Miller, A., Shi, E., Wen, Z., Papamanthou, C.: Hawk: the blockchain model of cryptography and privacy-preserving smart contracts. In: 2016 IEEE symposium on security and privacy (SP), pp. 839–858. IEEE (2016)
14. Tarasov, P., Tewari, H.: The future of e-voting. IADIS Int. J. Comput. Sci. Inf. Syst. **12**(2) (2017)
15. Hopwood, D., Bowe, S., Hornby, T., Wilcox, N.: Zcash protocol specification. GitHub, San Francisco (2016)
16. Banerjee, A., Clear, M., Tewari, H.: Demystifying the role of zk-SNARKs in zcash. In: 2020 IEEE Conference on Application, Information and Network Security (AINS), pp. 12–19. IEEE (2020)
17. Groth, J.: On the size of pairing-based non-interactive arguments. In: Annual International Conference on the Theory and Applications of Cryptographic Techniques, pp. 305–326. Springer (2016)
18. Groth, J.: Short pairing-based non-interactive zero-knowledge arguments. In: International Conference on the Theory and Application of Cryptology and Information Security, pp. 321–340. Springer (2010)
19. Maller, M., Bowe, S., Kohlweiss, M., Meiklejohn, S.: Sonic: zero-knowledge snarks from linear-size universal and updatable structured reference strings. In: Proceedings of the 2019 ACM SIGSAC Conference on Computer and Communications Security, pp. 2111–2128 (2019)
20. Gabizon, A., Williamson, Z.J., Ciobotaru, O.: Plonk: Permutations over lagrange-bases for oecumenical noninteractive arguments of knowledge. IACR Cryptol. ePrint Arch. 2019, 953 (2019)

21. Chiesa, A., Hu, Y., Maller, M., Mishra, P., Vesely, N., Ward, N.: Marlin: preprocessing zkSNARKs with universal and updatable SRS. In: Annual International Conference on the Theory and Applications of Cryptographic Techniques, pp. 738–768. Springer (2020)
22. Kosba, A., Papadopoulos, D., Papamanthou, C., Song, D.: {MIRAGE}: succinct arguments for randomized algorithms with applications to universal zk-SNARKs. In: 29th {USENIX} Security Symposium ({USENIX} Security 20), pp. 2129–2146 (2020)
23. Campanelli, M., Faonio, A., Fiore, D., Querol, A., Rodrıguez, H.: Lunar: a toolbox for more efficient universal and updatable zkSNARKs and commit-and-prove extensions. ERINT IACR, Report 1069(2020), 101 (2020)
24. Ráfols, C., Zapico, A.: An algebraic framework for universal and updatable SNARKs. Cryptology ePrint Archive, Report 2021/590 (2021). https://eprint.iacr.org/2021/590
25. Blum, M., Feldman, P., Micali, S.: Non-interactive zero-knowledge and its applications. In: Providing Sound Foundations for Cryptography: On the Work of Shafi Goldwasser and Silvio Micali, pp. 329–349. ACM (2019)
26. Groth, J., Kohlweiss, M., Maller, M., Meiklejohn, S., Miers, I.: Updatable and universal common reference strings with applications to zk-SNARKs. In: Annual International Cryptology Conference, pp. 698–728. Springer (2018)
27. Ben-Sasson, E., Chiesa, A., Spooner, N.: Interactive oracle proofs. In: Theory of Cryptography Conference, pp. 31–60. Springer (2016)

A Secure Email Solution Based on Blockchain

Diego Piedrahita(✉), Javier Bermejo, and Francisco Machío

Universidad Internacional de la Rioja, Logroño, Spain
diego.piedrahita@comunidadunir.net, {javier.bermejo, francisco.machio}@unir.net

Abstract. Email is one of the most important online communication services between individuals and businesses. The large amount of information that passes through this medium is the object of desire for many attackers who use a whole series of malicious maneuvers to get hold of it. The number of emails circulating in the world per day is growing, going from 319.6 billion in 2021 to more than 361 billion by the end of 2024. Attackers use all kinds of attacks to participate in these communications, perpetuating attacks that range from spam to very sophisticated attacks such as phishing, scams, among others. There are many solutions available in the market and the email security problem continues to grow and there seems to be no solution in sight. We are working on a different approach. First, we have deeply analyzed the state of the art of email in terms of its design, the security problems identified and the proposed solutions. Subsequently, we have identified the security requirements necessary to address the problem and based on them we have designed an architecture based on a Blockchain platform, whose components interact through different protocols to achieve the objectives proposed in our research.

Keywords: Email · Blockchain · Security

1 Introduction and Problem Statement

The integrity, confidentiality and availability of information resources are the primary objectives of information security. Electronic mail (email), defined as a digital message that is transmitted between a sender and a receiver, contains information and any vulnerability in these three key areas will expose the system to any exploitation [1]. Email is still the most widely used electronic communication method on the internet by many users [2]. In 2004 SANS Institute conducted a study that reflected what were the email security threats at that time. Viruses, unwanted mail (spam) and phishing were the main problems and most of the viruses were hidden in the attachments of the mail messages [1]. After many years of the publication of that report, the same problems persist [3]. These attacks generate large economic losses to organizations and individuals [4].

2 Related Work

We have studied the countless applications of Blockchain in different domains [5], and we have identified that many companies do not take risks when approaching a project with

J. Prieto et al. (Eds.): BLOCKCHAIN 2021, LNNS 320, pp. 355–358, 2022.
https://doi.org/10.1007/978-3-030-86162-9_36

a technology where security aspects are under study. We see in the scientific literature different approaches to address this problem using Blockchain [6–8].

There are several commercial proposals available in the market offering secure mail on Blockchain, but there is little technical documentation on their construction, such as Mailchain [9]. Blockchain has been used as an extension to email to perform activities such as validation of unknown senders [10], validation of emails to verify integrity and to know if the email has not been forged [11], the use of payment with cryptocurrencies to be able to send emails [12], the use of decentralized reputation systems on Blockchain to avoid spam [13] or the use of Blockchain for sending certified mail without a trusted third party [14]. This type of work focuses on the use of Blockchain as an additional module to the traditional email system that provides a service to mitigate some of the problems that we have previously discussed.

Another type of proposal focuses on providing a comprehensive solution using Blockchain as the central email service. One of the papers proposes the use of the same identifiers used in Blockchain transactions as user identifiers [15]. Another proposal makes use of Blockchain to store messages using a proprietary server [16]. Finally, one of the analyzed works is the one presented by González in which he proposes to replace email protocols with smart contracts [17]. We must consider that some proposals are in development and apparently the architecture has not been designed in its entirety, but aspects are defined as progress is made.

3 Hypothesis

The research work hypothesis is stated as follows: Email today presents innumerable security vulnerabilities that can be resolved using a distributed and secure digital alternative mail system based on a new protocol, according to the technological advances of recent times in computer security and that complies with the basic principles of information security and the requirements of the different participants who make use of the solution.

4 Proposal

Our goal is to design a secure email solution based on a flexible and integrable architecture that is usable and integrable by users and that solves the problem of email security by providing services of confidentiality, integrity, availability, non-repudiation, authentication, control, and audit. The architecture will use Blockchain as one of its core components and the interaction of its components will be regulated by a series of distributed protocols that will be responsible for providing security to all processes (Fig. 1).

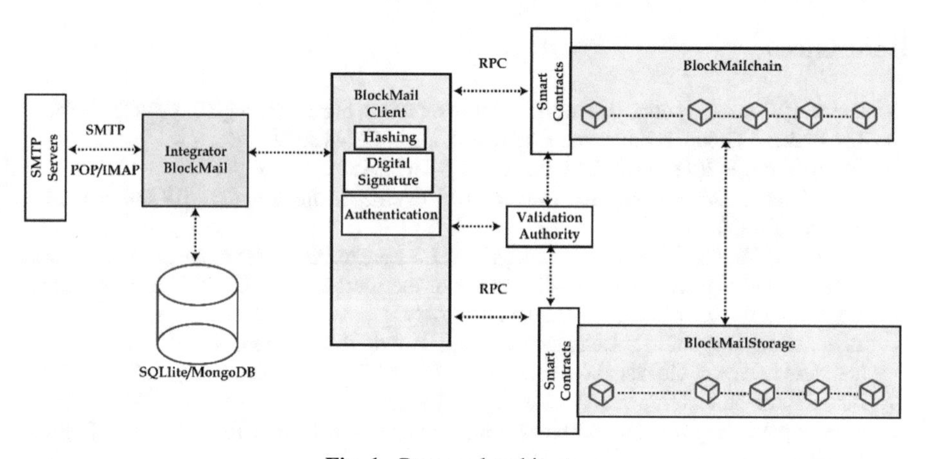

Fig. 1. Proposed architecture

5 Preliminary Results

We have developed the state of the art of email and the different proposals developed in the last 5 years. We are developing the proof of concept with different use cases that we have been defining. In our study we have identified eight aspects that must be considered when addressing security in the context of email and they are confidentiality, integrity, non-repudiation, authentication, availability, control, auditing, usability, and integration. Based on these eight aspects and the literature reviewed on email security, we have identified 43 security requirements that an email solution must meet to be considered secure. We have designed an architecture using a public blockchain that controls security aspects and another private blockchain that manages email information. The protocols are smart contracts that provide solutions to the identified security requirements. The architecture also provides integration with SMTP servers through a gateway called the integrator, but it is a partial solution while the new model is adopted.

6 Reflections

Blockchain technology can be used to provide security through smart contracts. The use of smart contracts seen as a protocol facilitates the standardization of many processes. However, Blockchain is not ideal for intensive user-centric systems or for data models whose rules do not require strict and detailed historical state management, as in the case of email. The decentralization feature offered by this technology is ideal for minimizing availability and integrity issues, but not for solving confidentiality issues or for storing large volumes of data such as those handled in email. To solve this problem, we had to develop a hybrid solution between private and public Blockchains, but that meets the identified requirements. We hope that our work will be a starting point for many other works and together we can offer an ideal solution to this problem that affects us all.

References

1. Cocca, P.: E-mail secutiry threats. Sans Intitute (2004). https://doi.org/10.9780/22307850
2. The Radicati Group, I. Email statistics report, 2020–2024 (2020)
3. Proofpoint: Q1–2018 quarterly threat report 1–16 (2018)
4. FBI's Internet Crime Complaint Center: 2019 Internet Crime Report. 2019 Internet Crime Report 1–28 (2019)
5. Pournader, M., Shi, Y., Seuring, S., Koh, S.C.L.: Blockchain applications in supply chains, transport and logistics: a systematic review of the literature. Int. J. Prod. Res. **58** (2020). https://doi.org/10.1080/00207543.2019.1650976
6. Taylor, P.J., Dargahi, T., Dehghantanha, A., Parizi, R.M., Choo, K.K.R.: A systematic literature review of blockchain cyber security. Digital Commun. Netw. **6**(2), 147–156 (2020)
7. Bernal Bernabe, J., Canovas, J.L., Hernandez-Ramos, J.L., Torres Moreno, R., Skarmeta, A.: Privacy-preserving solutions for blockchain: review and challenges. IEEE Access **7**, 164908–164940 (2019)
8. Vassilakis, C.: Blockchain technologies for leveraging security and privacy. Homo Virtualis. **2** (2019). https://doi.org/10.12681/homvir.20188
9. Mailchain https://mailchain.xyz/
10. Dennis, S.M.: Blockchain based email procedures (US Patent No. US010305833B1) (2019)
11. Varghese, A.: Email verification service using blockchain 0–7 (2019)
12. Sheikh, S.A., Banday, M.T.: A cryptocurrency-based E-mail system for SPAM control. Int. J. Adv. Comput. Sci. Appl. **12** (2021). https://doi.org/10.14569/IJACSA.2021.0120139
13. Kannan, L., Jebakumar, R.: Public Sender Score System (S3) by ESPs for Email spam mitigation with score management in mobile application. Int. J. Interact. Mob. Technol. **14** (2020). https://doi.org/10.3991/ijim.v14i17.16609
14. Hinarejos, M.F., Ferrer-Gomila, J.: A solution for secure multi-party certified electronic mail using blockchain. IEEE Access **8**, 102997–103006 (2020). https://doi.org/10.1109/ACCESS.2020.2998679
15. Khacef, K., Pujolle, G.: Secure peer-to-peer communication based on blockchain. In: Barolli, L., Takizawa, M., Xhafa, F., Enokido, T. (eds.) WAINA 2019. AISC, vol. 927, pp. 662–672. Springer, Cham (2019). https://doi.org/10.1007/978-3-030-15035-8_64
16. Menegay, P., Salyers, J., College, G.: Secure communications using blockchain technology. In: Proceedings of the IEEE Military Communications Conference MILCOM (2019)
17. González, J.C., García-Díaz, V., Núñez-Valdez, E.R., Gómez, A.G., Crespo, R.G.: Replacing email protocols with blockchain-based smart contracts. Clust. Comput. **23**(3), 1795–1801 (2020). https://doi.org/10.1007/s10586-020-03128-9

Artificial Intelligence Decision and Validation Powered Smart Contract for Open Learning Content Creation

Frederick Ako-Nai[1(✉)], Enrique de la Cal Marin[2], and Qing Tan[1,2]

[1] University of Oviedo, Oviedo, Spain
qingt@athabascau.ca
[2] Athabasca University, Athabasca, Canada
delacal@uniovi.es

Abstract. Educational learning contents, usually developed by instructional designers to provide learning specification and impart knowledge and skills to learners, are often only updated or changed in the yearly course revisions. This can be presented with some challenges, especially to courses on emerging subjects and catering to diversified learners, which includes the ability to provide adaptive and updated learning contents to the learners, and the opportunity to continually incorporate feedback from all stakeholders. Characteristic of Blockchain, such as secure, privacy, and consensus makes the technology adaptable for open and collaborative applications. In this paper, we propose a framework that uses artificial intelligence with blockchain smart contracts to support an open platform that gives stakeholders flexible access to content creation where individual learners can obtain extra contents as needed and updated knowledge can be shared and built upon easily to enrich the learning contents.

Keywords: Blockchain · Smart contracts · Learning content creation · Artificial intelligence · Prescriptive learning analytics

1 Introduction

Traditional course development is usually handled by instructors/professors and learning designers, and only updated or changed in the yearly course revisions. This sometimes makes the contents outdated as information and technology change rapidly. This research proposes an artificial intelligence (AI) enhanced smart contract for open learning content creation platform to handle authentication, content recommendation, content validation, change tracking and authorization to speed up the consensus process, thus allowing learning contents to be updated more frequently as well as meet the individual learner's needs.

The use of blockchain and smart contracts will open the creation of learning contents to all interested stakeholders and provide fairness, openness, security and efficiency for smart education outcomes [1]. AI based decisions can help incorporate learner feedback [2], and other artifacts proposed by learners to satisfy the lack of exact artifact needed

J. Prieto et al. (Eds.): BLOCKCHAIN 2021, LNNS 320, pp. 359–362, 2022.
https://doi.org/10.1007/978-3-030-86162-9_37

to complete learning activities [3, 4]. Blockchain and smart contracts will be used to verify these proposals and speed up their inclusion for use in the course. AI and Smart Contracts can help with validating such proposed contents so to provide a rich learning experience for learners.

The structure of the paper is as follows. Section 2 lists some literature and related topics in this research area. Section 3 presents the proposed architecture and discussion of a prototype design. Section 4 concludes the paper.

2 Related Work

The main challenges proposed in this paper are: blockchain and smart contracts in education, and data validation using artificial intelligent decision-making algorithms. A literature review conducted shows some ways in which these topics can be used in learning content creation.

Most uses of blockchain in education have been for certificate, credential, and records management [5–9]. In emerging cases, [10] proposes a content creation platform for e-books, CHiLO, that uses blockchain technology to help solve the challenges of copyright and virtual currency (reward) for acquisition and dissemination of knowledge. Zhao *et al.* [11], propose using smart contracts to authenticate student identities in a blockchain based e-portfolio system for student growth information. Using AI with Blockchain, though there exist some challenges like, privacy, scalability, security, smart contract vulnerabilities, and governance etc., it can help remediate these challenges [12]. AI on the other hand can provide reliable and intelligent decision making in blockchain systems by acting as a recommendation system, analyzing past data to recommend which smart contract actions should be taken [13].

3 The Open Learning Content Creation Platform

An artificial intelligence (AI) enhanced smart contract for open learning content creation platform is presented in this paper. Its architecture is shown in Fig. 1. In the proposed platform, there are three original modules to ensure the learning content creation is open, effective, accurate, and creditable. The prescriptive analytics (PA) module automatically suggests extra learning content to meet learners' personalized needs by analyzing their learning data. The AI validation (AIV) module validates any suggested learning content and decides if the content could be included into the course or not. The smart contract (SC) module checks, and controls validated content according to the terms of the contract eventually allowing or refusing a suggested learning content being added to the blockchain. The smart contract is an intelligent way to arrive at a consensus and keep track of contributions of stakeholders involved in the learning environment to collaboratively contribute to the creating of contents.

The main components of this architecture include input from stakeholder which will include learners, Subject Matter Experts (SME), instructors, and content designers; presentation (user interface) layer; application layer including the PA module; AI Decision/Validation layer consisting of AIV module; data layer consisting of a staging database (MySQL) and Interplanetary File System (IPFS). As IPFS, like Blockchain is immutable, collaboration and consolidation of ideas on the initial content proposals will be stored in the staging database. The final consolidated content will then be stored in IPFS); and Blockchain layer residing SC module.

To experiment the research idea, we developed a prototype platform using a simple web form for submitting and approving content proposals. In the prototype the smart contract is implemented by stakeholders' voting on the proposed content with a "yes" or "no" choice. After the voting time is reached, the block is added to the chain if the yes votes are greater than 51% of all votes cast. In this implementation, a proposal once submitted is considered as a validated content and creates a block. Once the block complies with the terms of the contract, it is registered by the smart contract then it is allowed to be added into the blockchain. The smart contract address is then added to the corresponding entry for the proposed content in our smart content database.

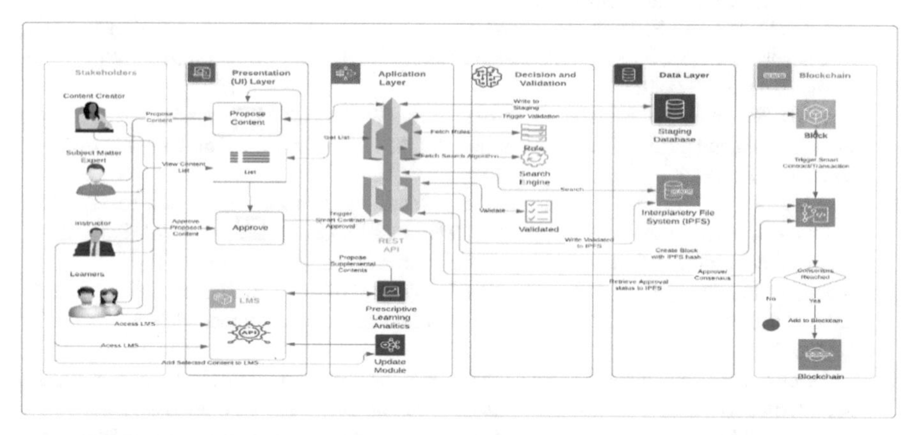

Fig. 1. Conceptual architecture diagram of the learning content creation platform

4 Conclusion

The proposed framework represented by the architecture provides an interface that gives stakeholders an opportunity to contribute to the content creation process where updated knowledge can be shared and built upon easily during the course. With the help of AI decision and validation algorithms, and smart contracts, proposed learning content can be validated and added to the course and accessed through the LMS in a timely fashion.

The prototype shows how stakeholders can propose and validate contents using a web form and blockchain. In the next steps, we will study AI algorithms for decision and validation, and learning analytics algorithms for prescriptive analytics to help with more frequents course updates. Based on the research results, we will create the detail terms for the smart contract for optimizing the proposed open learning content platform.

Acknowledgement. This research has been funded partially by Spanish Ministry of Economy, Industry and Competitiveness (MINECO) under grant TIN2017-84804-R/PID2020-112726RB-I00.

References

1. Gong, X., Liu, X., Jing, S., Xiong, G., Zhou, J.: Parallel-education-blockchain driven smart education: challenges and issues. In: Chinese Automation Congress (CAC) 2018, pp. 2390 2395 (2018)
2. Ako-Nai, F., Tan, Q., de la Cal Marin, E.A.: Employing blockchain technology in instructional design and learning content creation. In: Rønningsbakk, L., Wu, T.-T., Sandnes, F.E., Huang, Y.-M. (eds.) ICITL 2019. LNCS, vol. 11937, pp. 581–588. Springer, Cham (2019). https://doi.org/10.1007/978-3-030-35343-8_61
3. Ako-nai, F., Tan, Q.: Location-based learning management system for adaptive mobile learning. Int. J. Inf. Educ. Technol. **3**(5), 529–535 (2013)
4. Ako-Nai, F., Tan, Q., Pivot, F.C., Kinshuk: The 5R adaptive learning content generation platform for mobile learning. In: Proceedings of the 2012 IEEE 4th International Conference on Technology for Education, T4E 2012 (2012)
5. G. Zhao, H. He, and B. Di, "Design and implementation of the digital education resources authentication system based on blockchain," in *ACM International Conference Proceeding Series*, 2020, pp. 100–104.
6. Bore, N., Karumba, S., Mutahi, J., Darnell, S.S., Wayua, C., Weldemariam, K.: Towards Blockchain-enabled school information hub. In: ACM International Conference Proceeding Series, vol. Part F132087, pp. 1–4 (2017)
7. Han, M., Wu, D., Li, Z., Xie, Y., He, J.S., Baba, A.: A novel blockchain-based education records verification solution. In: SIGITE 2018 Proceedings of the 19th Annual SIG Conference on Information Technology Education, vol. 18, pp. 178–183 (2018)
8. Ocheja, P., Flanagan, B., Ogata, H.: Connecting decentralized learning records: a blockchain based learning analytics platform. In: ACM International Conference Proceeding Series, pp. 265–269 (2018)
9. Jirgensons, M., Kapenieks, J.: Blockchain and the future of digital learning credential assessment and management. J. Teach. Educ. Sustain. **20**(1), 145–156 (2018)
10. Hori, et al. M.: Learning system based on decentralized learning model using blockchain and SNS. In: CSEDU 2018 Proceedings of the 10th International Conference on Computer Supported Education, vol. 1, pp. 183–190 (2018)
11. Zhao, G., He, H., Di, B., Xia, Q., Fu, Z.: A blockchain-based system for student E-portfolio assessment using smart contract. In: 2020 4th International Conference on Computer Science and Artificial Intelligence (2021)
12. Salah, K., Rehman, M.H.U., Nizamuddin, N., Al-Fuqaha, A.: Blockchain for AI: review and open research challenges. IEEE Access **7**, 10127–10149 (2019)
13. Almasoud, A.S., Eljazzar, M.M., Hussain, F.: Toward a self-learned smart contracts. In: Proceedings of the 2018 IEEE 15th International Conference on e-Business Engineering, ICEBE 2018, pp. 269–273 (2018)

Towards a Holistic DLT Architecture for IIoT

Denis Stefanescu(✉)

Ikerlan Technology Research Centre, 20500 Arrasate-Mondragon, Spain
distefanescu@ikerlan.es

Abstract. The problem I intend to address in my PhD research is related to the application of DLTs in resource-constrained industrial environments. Could it be feasible and effective to design a holistic multi-DLT architecture in order to tackle most of the challenges that are encountered when applying DLTs in IIoT? Here I present a multi-DLT architecture for an Industry 4.0 case scenario where I try to tackle the limitations of DLTs for every level of the aforementioned scenario.

1 Problem Statement and Hypothesis

The Industrial Internet of Things (IIoT), which can be defined as the use of the Internet of Things (IoT) in industrial sectors and applications, is one of the most important parts of Industry 4.0 [1]. IIoT integrates many different technologies such as smart sensors, robots, machine-to-machine (M2M) communication, big data and artificial intelligence into traditional industrial procedures. Thus, IIoT aims to optimize the existing production procedure, enhancing customer experiences, reducing costs and increasing efficiency.

However, the IIoT has several challenges that need to be addressed [2]. First, security is a major problem, as IIoT devices can be easily attacked, data can be tampered by malicious third parties, and are more vulnerable to denial of service attacks. Second, data privacy is also a major issue, since the data can be easily exposed. Finally, centralization is also a problem, since traditional industrial systems rely on centralized servers, which not only are commonly threatened by weak connections and security vulnerabilities, but also become a significant bottleneck when the number of devices and data flow increases (limiting the performance of the network).

To solve aforementioned challenges, many researchers have proposed the use of blockchain in IIoT. Blockchain is a type of Distributed Ledger Technology (DLT) in which all the transactions are stored in a chain of blocks that are cryptographically linked. A DLT consists of replicated, shared, and synchronized digital data spread across multiple sites, with no central authority. The participating nodes reach an agreement over the state of the ledger by following a consensus algorithm. This technology has many benefits such as decentralization, security, privacy, treaceability and interoperability, which makes it ideal for the deployment of the Industry 4.0 [3].

J. Prieto et al. (Eds.): BLOCKCHAIN 2021, LNNS 320, pp. 363–366, 2022.
https://doi.org/10.1007/978-3-030-86162-9_38

Nonetheless, currently, blockchain is not well suited for resource-constrained environments such as the IIoT, since most of the existing consensus algorithms offer low throughput, require high resource consumption, have a lack of energetic efficiency, low scalability and a considerable delay in storing transactions [4]. Furthermore, blockchain also requires huge storage capacity and network bandwidth, as the information has to be replicated in each one of the participating nodes. Indeed, in a recent systematic literature review (currently under review for publication) I have conducted, I acknowledge that applying blockchain to IIoT is still a major challenge, and there is no DLT architecture that completely fulfills the requirements of IIoT scenarios in terms of computational power, storage, throughput and energetic efficiency.

Taking the main problems that current DLT solutions present when applied to IIoT contexts, this PhD thesis aims to investigate the following hypothesis:

Combining an improved version of various types of DLTs in one architecture in order to tackle the challenges of an industrial scenario would result in an appropriate solution to securely cover all the process from when the data is generated in IIoT up until the data is processed at a business level.

2 Related Work

In this Sect. 1 summarize the main characteristics and contributions of other three relevant DLT-based architectures for IIoT.

Seok *et al.* [4] propose a blockchain architecture for the IIoT that improves the performance of the cryptographic hash algorithm that is typically used in the mining process. This proposal reduces the computational burden and latency of blockchain, while reducing the enormous energy requirements of the mining process. Liu *et al.* [3] introduce a lightweight, IIoT oriented blockchain system that improves many problematic aspects of blockchain such as the energy, storage and bandwidth consumption. First, they propose a more efficient PoW based consensus which employs a reputation system in order to reduce its energy consumption. Then, they introduce a more lightweight data structure for the blocks to streamline broadcast content, reducing the network bandwidth. Finally, they design a novel "unrelated Block offloading filter" to reduce the size of the ledger. Yu *et al.* [5] use the Edge and Cloud technologies to create a layered architecture for the IIoT. The authors of this paper propose three layers: the IIoT devices layer, the Edge layer and the Cloud blockchain layer. The IIoT layer comprises heterogeneous IIoT devices that send their data to the Edge layer. The Edge layer processes and temporarily stores the data that comes from the IIoT layer. Finally, the Cloud blockchain layer stores the offloaded data that is sent from the Edge layer. This proposal claims to improve the scalability and efficiency of blockchain by distributing the load among various layers that are meant to perform a specific task.

However, none of the existing works propose a holistic approach that covers the different layers of a smart factory scenario, nor they consider a specifically lightweight and trustworthy DLT-based approach suited for an environment from when the data is generated in IIoT up until the data is processed at a business level. Furthermore, no architecture takes advantage of multiple DLT technologies in order to create a complete lightweight ecosystem for the Industry 4.0.

3 Proposal

I propose a lightweight multi-DLT architecture that covers all the process from when the data is generated in IIoT up until the data is processed at an Industry 4.0 business level. The proposed architecture consists of three layers and it is deployed on top of an Industry 4.0 case scenario that is defined based on practical experience. The scenario includes various levels within an industrial environment: machine level, production line level, plant level and industrial consortium level.

The first layer of the architecture is at the machine level where IIoT and control devices are located, and the production line level, which is typically formed by various machines. This layer is in charge of processing the data from the IIoT devices. The middle layer of the architecture includes the plant level of the defined case scenario. This layer connects in a DLT network all the production lines of one or several plants. Furthermore, this intermediate network also acts like a data bridge between the other layers of the architecture. Finally, the consortium DLT network includes the top level of the case scenario, which is the consortium-business level. This network aims at establishing a lightweight secure DLT network between various smart factories, which would form a connected smart factory ecosystem in the context of Industry 4.0. All the layers of the architecture have different challenges and needs, thus this justifies the need to employ more than one type of DLT. Also, improving some aspects of the DLTs that I employ in the proposed architecture is necessary in order to deliver a truly lightweight solution for IIoT.

4 Evaluation Plan

In order to evaluate the architecture I will implement a proof of concept of each one of the three defined layers. The implementation will be carried out using two possible approaches: (1) DLT platforms (e.g., Ethereum), or (2) simulators (e.g., NS3).

The first layer of the architecture, which is the closest to IIoT, will implement several enhancements aimed at reducing the burden of DLT technology on lightweight IIoT devices. Therefore, in this case, I will compare my enhanced DLT with a baseline. Thus, in this layer, I will evaluate the metrics that are related to the performance of the DLT for resource-constrained environments:

- (1) Computational burden, (2) Throughput, (3) Energy consumption, (4) Storage usage and (5) Cryptographic algorithms performance.

In the uppermost layers, there is no need for the DLTs to be as lightweight or as efficient as in the IIoT level. However, speed and efficiency still have to be relatively high. However, other metrics such as these related to the interoperability between DLTs are also relevant here. Therefore, I will evaluate the following metrics:

- (1) The interconnection between DLTs, (2) Gateways efficiency, (3) Throughput, (4) Data encryption performance and (5) Storage usage.

5 Reflections

The feasibility of DLTs in Industry 4.0 is still a major challenge that needs further research. However, a lot of progress has been made in this field, especially due to the creation of alternative DLTs to blockchain (e.g., Directed Acyclic Graph (DAG)). Even though DAG DLTs are not yet ideal for IIoT, further improvements on this technology could finally create a truly lightweight and efficient DLT. Nonetheless, DAG DLTs are not suitable for an entire Industry 4.0 ecosystem. Therefore, it is highly probable that multiple DLTs would have to coexist in an Industry 4.0 ecosystem. Thus, interoperability between multiple DLTs also has to be taken into account.

References

1. Xu, L.D., Xu, E.L., Li, L.: Industry 4.0: state of the art and future trends. Int. J. Prod. Res. **56**(8), 2941–2962 (2018)
2. Sisinni, E., Saifullah, A., Han, S., Jennehag, U., Gidlund, M.: Industrial Internet of Things: challenges, opportunities, and directions. IEEE Trans. Ind. Inf. **14**(11), 4724–4734 (2018)
3. Liu, Y., Wang, K., Lin, Y., Xu, W.: Lightchain: a lightweight blockchain system for industrial Internet of Things. IEEE Trans. Ind. Inf. **15**(6), 3571–3581 (2019)
4. Seok, B., Park, J., Park, J.H.: A lightweight hash-based blockchain architecture for industrial IoT. Appl. Sci. (Switzerland) **9**(18) (2019)
5. Yu, Y., Liu, S., Yeoh, P., Vucetic, B., Li, Y.: LayerChain: a hierarchical edge-cloud blockchain for large-scale low-delay IIoT applications. IEEE Trans. Ind. Inform. **3203**(C), 1 (2020)

Author Index

J. Prieto et al. (Eds.): BLOCKCHAIN 2021, LNNS 320, pp. 367–368, 2022.
https://doi.org/10.1007/978-3-030-86162-9

Printed in the United States
by Baker & Taylor Publisher Services